RISK MODELING FOR HAZARDS AND DISASTERS

RISK MODELING FOR HAZARDS AND DISASTERS

Edited by

GERO MICHEL
Chaucer Syndicates, Denmark

Elsevier
Radarweg 29, PO Box 211, 1000 AE Amsterdam, Netherlands
The Boulevard, Langford Lane, Kidlington, Oxford OX5 1GB, United Kingdom
50 Hampshire Street, 5th Floor, Cambridge, MA 02139, United States

Library of Congress Cataloging-in-Publication Data
A catalog record for this book is available from the Library of Congress

British Library Cataloguing-in-Publication Data
A catalogue record for this book is available from the British Library

ISBN: 978-0-12-804071-3

For information on all Elsevier publications visit our website at https://www.elsevier.com/books-and-journals

ELSEVIER
Book Aid
International
Working together
to grow libraries in
developing countries
www.elsevier.com • www.bookaid.org

Publisher: Candice Janco
Acquisition Editor: Candice Janco
Editorial Project Manager: Hilary Carr
Production Project Manager: Paul Prasad Chandramohan
Designer: Greg Harris

Typeset by TNQ Books and Journals

Contents

II

MODEL CREATION, SPECIFIC PERILS AND DATA

5. The Use of Historic Loss Data for Insurance and Total Loss Modeling

JAMES E. DANIELL, FRIEDEMANN WENZEL, ANDREAS M. SCHAEFER

6. Indirect Loss Potential Index for Natural Disasters for National and Subnational Analysis

JAMES E. DANIELL, BIJAN KHAZAI, FRIEDEMANN WENZEL

7. Probability Gain From Seismicity-Based Earthquake Models

KRISTY F. TIAMPO, ROBERT SHCHERBAKOV, PAUL KOVACS

8. Big Data Challenges and Hazards Modeling

KRISTY F. TIAMPO, SETH MCGINNIS, YELENA KROPIVNITSKAYA, JINHUI QIN, MICHAEL A. BAUER

9. Progress Toward Hyperresolution Models of Global Flood Hazard

PAUL D. BATES, JEFF NEAL, CHRIS SAMPSON, ANDY SMITH, MARK TRIGG

III

MODEL INSURANCE USE-CASES

10. Insurance Pricing and Portfolio Management Using Catastrophe Models

HELENA BICKLEY, GERO MICHEL

List of Contributors

Paul D. Bates School of Geographical Sciences, University of Bristol, Bristol, UK; SSBN Flood Risk Solutions, Cardiff, UK

Michael A. Bauer Department of Earth Sciences, Department of Computer Sciences, Western University, London, ON, Canada

Helena Bickley Sompo International

Laure Cabantous Faculty of Management, Cass Business School, City, University of London, London, UK

Karen Clark Karen Clark & Company, Boston, MA, USA

Ian Cook Willis Towers Watson, Willis Research Network, London, UK

James E. Daniell Geophysical Institute & Center for Disaster Management and Risk Reduction Technology (CEDIM), Karlsruhe Institute of Technology (KIT), Karlsruhe, Germany

Jayanta Guin AIR Worldwide, Boston, MA, USA

Alasdair Hunter Barcelona Supercomputing Centre, Barcelona, Spain

Ioanna Ioannou Department of Civil, Environmental & Geomatic Engineering, University College London, London, UK

Steve Jewson Risk Management Solutions, Newark, CA, USA

Bijan Khazai Geophysical Institute & Center for Disaster Management and Risk Reduction Technology (CEDIM), Karlsruhe Institute of Technology (KIT), Karlsruhe, Germany

Paul Kovacs Institute for Catastrophic Loss Reduction, Toronto, ON, Canada

Yelena Kropivnitskaya Department of Earth Sciences, Department of Computer Sciences, Western University, London, ON, Canada

Seth McGinnis Institute for Mathematics Applied to Geosciences, National Center for Atmospheric Research, Boulder, CO, USA

Patrick McSharry ICT Center of Excellence, Carnegie Mellon University Africa, Kigali, Rwanda; University of Rwanda, Kigali, Rwanda; Smith School of Enterprise and the Environment, University of Oxford, Oxford, UK

Gero W. Michel Chaucer Copenhagen, Copenhagen, Denmark

Gero Michel Western University, London, Ontario, Canada; Chaucer Underwriting A/S, Copenhagen, Denmark

Jeff Neal School of Geographical Sciences, University of Bristol, Bristol, UK; SSBN Flood Risk Solutions, Cardiff, UK

Jinhui Qin Department of Earth Sciences, Department of Computer Sciences, Western University, London, ON, Canada

Mohsen Rahnama Risk Management Solutions, Newark, CA, USA

Tiziana Rossetto Department of Civil, Environmental & Geomatic Engineering, University College London, London, UK

Chris Sampson School of Geographical Sciences, University of Bristol, Bristol, UK; SSBN Flood Risk Solutions, Cardiff, UK

Andreas M. Schaefer Geophysical Institute & Center for Disaster Management and Risk Reduction Technology (CEDIM), Karlsruhe Institute of Technology (KIT), Karlsruhe, Germany

Robert Shcherbakov Department of Earth Sciences, Department of Physics and Astronomy, Western University, London, ON, Canada

Nilesh Shome Risk Management Solutions, Newark, CA, USA

Slobodan P. Simonovic Department of Civil and Environmental Engineering and The Institute for Catastrophic Loss Reduction, The University of Western Ontario, London, ON, Canada

Andy Smith School of Geographical Sciences, University of Bristol, Bristol, UK; SSBN Flood Risk Solutions, Cardiff, UK

David B. Stephenson Department of Mathematics, University of Exeter, Exeter, UK

Kristy F. Tiampo Department of Geological Sciences and Cooperative Institute for Research in Environmental Sciences, University of Colorado, Boulder, CO, USA

Mark Trigg School of Geographical Sciences, University of Bristol, Bristol, UK

Andreas Tsanakas Faculty of Actuarial Science and Insurance, Cass Business School, City, University of London, London, UK

Friedemann Wenzel Geophysical Institute & Center for Disaster Management and Risk Reduction Technology (CEDIM), Karlsruhe Institute of Technology (KIT), Karlsruhe, Germany

Paul Wilson Risk Management Solutions, Newark, CA, USA

Augustin Yiptong Sompo International, Carlisle, MA, USA

Ben Youngman Department of Mathematics, University of Exeter, Exeter, UK

Introduction

Changing climate, skyrocketing exposure along storm- and flood-exposed coastal areas, increasing globalization, and heightened system risk—we might not be living in a safe place. Catastrophe models are the tools that help risk managers and decision-makers to quantify risk probabilities and costs for current events and years to come.

Many of us involved in this volume have spent a significant part of our lives working in numerical risk modeling; a term used interchangeably with catastrophe risk modeling or loss modeling. Every year, our companies spend millions of dollars in licensing fees, modeling specialists, and computer hardware to ensure the availability of the latest tools for managing catastrophe risk. Insurance and reinsurance companies have then built portfolios of tens of billions of premium from the insight obtained. Catastrophe models fashion hundreds of billions in limits and trillions of US dollars in exposure each year. Emergency strategies for millions of people in high-risk areas depend on hazard and risk models, and only climate models and some trading tools in the financial markets might top the money-relevant use of catastrophe risk models. However, despite this extensive usage, little has been published about catastrophe models. The last book dedicated to a similar audience is over 10 years old (Grossi, 2005) and does not cover many of the new ideas discussed here. Most other more comprehensive publications typically concentrate on the regional aspects of risk modeling or concentrate on a single peril, such as earthquake risk, or largely on hazard, leaving the financial aspects barely touched. For example, see Roth and Kunreuther, 1998; Gurenko, 2004; OECD, 2005; Messy and Organisation for Economic Co-operation and Development, 2005; Grace et al., 2003; Froot, 2007; Kleindorfer and Sertel, 2012; Pollner, 2001; Cummins and Mahul, 2009; Daniels et al., 2011; OECD, 2008; Global Earthquake Model, http://www.globalquakemodel.org/; and Global Volcano Model, http://globalvolcanomodel.org/, http://www.gccapitalideas.com/2016/02/11/meeting-the-challenges-catastrophe-modeling/, https://oid.wharton.upenn.edu/profile/kunreuth/.

For more than two decades, catastrophe modeling has been an intrinsic part of the insurance and reinsurance industry. This is despite many major catastrophe events, such as the Newcastle earthquake (1989), Hurricane Andrew (1992), Hurricane Katrina (2005), the Thailand floods (2011) or "superstorm" Sandy (2012), appearing to catch the industry, governments, and society by surprise. Twenty years ago, the industry modeled only a few perils. Today, with strong industry demand for data, and encouragement from rating agencies and regulators, nearly all policies covering natural disasters or man-made risk are subject to numerical modeling.

Catastrophe risk modeling has evolved rapidly as governments, NGOs, (re)insurers, brokers, and private companies have invested heavily to improve their understanding and forecasting of risk. The increased demand for reliable data and risk assessment has spurred the formation of dedicated catastrophe modeling vendors and consulting companies as well. Catastrophe models have evolved into trading platforms and provide a "currency" for highly sophisticated policies, which would be unthinkable without these tools. In addition, the capital markets have become more directly active in the insurance space, investing in insurance-linked securities, such as derivatives, catastrophe bonds, and collateral products, which require similar models. Governments have also begun to follow this market, insuring country-specific disaster risks using the same tools (https://www.gfdrr.org/sites/gfdrr.org/files/documents/Mexico_FONDEN_final_GFDRR.pdf, http://www.proteccioncivil.gob.mx/work/models/ProteccionCivil/Resource/400/1/images/4_%20Catastrophe%20Risk%20Insurance%20Pools_OMahul.pdf; http://www.businessinsurance.com/article/20100113/NEWS/100119962).

The purpose of this volume is threefold. First, to describe risk modeling concepts and use cases, including the insights of model vendors, government research, and (re)insurance practitioners; second, to review current model practice and science; and third, to explain recent developments in the catastrophe modeling space, including general concepts and uncertainty and how new ideas, including "big data," might foster changes in risk management.

The focus is not to provide a detailed discussion of the basic principles that underpin catastrophe models, because authors have successfully tackled this before (see earlier discussion); however, some of the articles discuss general concepts and deemed necessary fundamentals for model creation and functionality, as well as model and parameter uncertainty, at different levels. The purpose is also not to review all possible sorts of catastrophe risk models that have been created, covering a wide range of territories and perils.

This volume, however, covers all major aspects of catastrophe risk modeling, from hazards through financial analysis. Articles include discussion of progress made toward hyperresolution models of global flood hazard (Chapter 9) and model user considerations of uncertainty (Chapters 1 and 4), model robustness, and business decision-making (Chapters 2, 3, 10, 13, and 15). Decision-making includes optimization of deal selection based on the model outcome when considering return maximization versus high-severity risk and capital (Chapter 11). Other questions discussed include how risk model concepts can deviate from a common "long-term average" view of risk to include time-dependent seismicity (Chapters 7 and 8) or the nonstationary nature of the natural environment (Chapter 3). How important are often-neglected indirect losses in the overall assessment of risk (Chapters 5 and 6)? How can modelers use long-term (50 years) historical losses to calibrate models? What common errors might arise if loss inflation does not consider changes in the built environment, construction codes, and so on (Chapters 5 and 6)? What wider data sources are available for creating models (Chapters 1, 5, and 6)? What challenges do remote sensing and other big data sources present if applied within hazard and risk modeling (Chapters 7 and 8)? What is the difference between so-called *parsimonious* models and other loss models, based on much larger numbers of variables (Chapter 2 or Chapter 12) and how can we use risk modeling techniques to, for example, forecast productivity of commodities such as tea (Chapter 12)? Are commonly used stochastic event sets even the most appropriate means for decisions-making (Chapter 13)? Finally, risk often implies consideration of the downside only, underpinned by a concern for maintaining the status quo. Slobodan Simonovic discusses a paradigm shift in risk management and proactive changes toward a more resilient society (see also Taleb, 2014).

This volume aims for a wide audience of those curious to learn how to assess, price, and then hedge the losses from natural and man-made catastrophes. This audience might range from governments to regional risk managers, from scientists to students of environmental and natural sciences, economics and engineering, analysts and practitioners from the financial markets, including fund managers, insurers and reinsurers, brokers, rating agencies, and those from the broader public who have always wondered how insurers manage catastrophes.

Climate change is likely to affect catastrophes in the future—their frequencies and severities as well our exposure to future losses. Climate change is hence a concern for all working in risk management (see also Seo and Mahul, 2009), although it is not a primary focus of this volume. One reason for this is that most insurance hedging takes place on an annual basis, where short-term climate variability is likely to be more important than long-term climate change. The reader may consider climate change to have altered recent risk and data already, making the use of historical data a less valid proxy with which to assess the future. This may surely be true, but is further complicated by the fact that new construction rules and changes in land use and land cover have changed our vulnerability (often down) in most areas. Simply inflating historical losses using climate models along with exposure growth may hence not suffice to explain future losses (Chapter 5). Despite our concern in climate change, recent studies have also suggested that climate change may not be visible in historical loss data available to date (see also Neumayer and Barthel, 2010, or Simmons et al., 2013). Considering climate change in the assessment of insurance risk is, however, a topic of numerous workshops and on the radar screens of many emerging risks managers and chief risk officers.

Hazard has no regional boundaries; however, risk does, given national and regional changes in vulnerabilities and exposure. Most people have exposure to some catastrophe risks and their consequences. Insurance helps because it allows the sharing of losses and hence minimizes the consequences of disasters for all people. Insurance penetration, however, differs hugely around the globe, with high penetration rates for fire and windstorm risk cover (>90%) in most parts of the European Union, the United States, Canada, Japan, Australia, and New Zealand. This contrasts with earthquake risk and/or many areas in Latin America, Africa, or Asia, where insurance penetration is, in general, low, reaching levels as low as 1% or less for homeowners business. This difference in high insurance penetration for many western countries as opposed to low penetration in emerging markets and developing countries is referred to as the "insurance gap" (http://www.bermudareinsurancemagazine.com/article/bermuda-closing-the-insurance-gap), a fact that provides both a thread to all and especially those living in these areas and an opportunity to change. Insurers need capital for large losses that exceed annual net premium. Catastrophe risk drives capital requirements for most insurance companies because catastrophes can affect many, hence defying the general idea of insurance, i.e., the fact that many can pay for the outsized losses of some. In addition, capital tends to chase the same risks, those with the deemed highest returns. This drives capital abundance in the United States and the European Union, whereas other areas suffer a scarcity of insurance capital. One reason for this is that capital has tracked the ability to model catastrophe risk, which in turn has followed the level of recent *insured* losses. This effect continues to increase the concentration of models and availability of data in those areas where capital is high or in excess, rather than supporting a movement of capital and data toward areas where the need for both is greater. It is only recently that this effect has begun to change as investors have started to realize that models might have had a "commoditizing" effect on catastrophe trading, and that there may be a negative correlation between levels of model penetration and available profit. Investors now ask for "nonmodeled" risk from insurance markets to target risk identified by proprietary models rather than that produced by mainstream vendors (such as those described by Chapters 1 and 2). This can include cyber risk, political violence, or natural catastrophe risk in more remote areas.

Catastrophe risk modeling is a youthful science with still only a limited number of involved scientists around the globe. Catastrophe risk modeling is typically an interdisciplinary endeavor, with model creation requiring either a large team of scientists (see, e.g., GEM https://www.globalquakemodel.org/gem/) or demanding scientists who are either generalists or specialists in multiple fields to understand the often-complex processes in hazard, social science, and finance. Risk modeling is also not primarily about understanding and explaining fundamental natural or man-made processes—a task that has excited generations of scientists. Risk modeling is often subject to deep uncertainty because of the nonlinear processes and data limitations involved, and despite risk manager's demand, scientists cannot easily consider model results "precise," given the associated uncertainties involved. Squaring the circle with the need for making detailed decisions and the limitations placed on results by the large uncertainties is a task that keeps risk model developers busy. This, in turn, has fostered a divide in risk modeling approaches between (1) those models based mainly on expert opinion (Chapter 2) and (2) those groups of scientists and practitioners who follow rather parsimonious concepts (Chapter 12). There seems to be a large difference in the value of models for those users who rely on deemed-accurate and precise solutions from approach (1) as opposed to those who follow the rather coarse assumptions inherent in approach (2). Following approach (1) allows deemed-accurate pricing down as far as structured individual risks, with even complex mitigation features attached. Users of approach (1) often assume that the underlying model is deemed valid for making decisions at almost any level of complexity or resolution. This is the opposite to approach (2), which suggests that the user might need to control model risk more effectively by minimizing the number of variables involved. However, behind the scenes, both concepts may be closer than first imagined because both approaches have to make specific assumptions for the idiosyncrasies, deficiencies, or gaps in the case data. Hence, the real difference between the two approaches might be in the confidence that a user places in the model at hand.

Managers and regulators can become obsessed with the downside of risk about which knowledge is often limited. Business is, however, about the upside and making profit and the strategy. Recent changes in regulation (Own Risk and Solvency Assessment, ORSA, https://en.wikipedia.org/wiki/Own_Risk_and_Solvency_Assessment) require companies to conduct and report on its own risk and solvency assessment, including risk appetite and strategy. Model risk and opportunity is part of this. There may be little upside in sharing a risk model with the rest of the market—except perhaps for use as a common trading platform. Yet, despite this, many (re)insurance companies continue to rely on the same few models. This has raised concerns about systemic model risk, along with a growing notion among underwriters that they will miss profitable deals if all share some potentially biased tools.

Loss modeling is considered a *core value* for a (re)insurance company, and management text books argue that such values belong in the business and should not be sourced out to third parties because this might diminish the business franchise value. This is especially true if managers actively use loss models to control both business return upside and downside. The excess capital inflows of recent years along with intermediaries between insurance and reinsurance altering model view might have protected clients against any potential high prices due to a limited number of model vendors that might have hindered business competitiveness. However, proprietary models and numerous smaller modeling companies have recently inundated the market, and most companies now aim to have an "own view of risk" based on their own data and research (Chapter 10). Consulting companies that help the market with this challenge include Karen Clark, the former CEO of AIR (Chapter 13), KatRisk (http://www.katrisk.com/), JBL Consulting (http://www.jbaconsulting.com/), SSBN (http://www.ssbn.co.uk/, Chapter 9), EigenRisk (https://eigenrisk.com/), CoreLogic (http://www.corelogic.com/), and Verisk (http://www.verisk.com/, see also Chapter 2) and major brokers such as AON-Benfield, GC, Willis, and Risk Frontiers (https://riskfrontiers.com/).

The creation of catastrophe risk models is, as can be expected, a business that has made money for some but not necessarily for many. The many reasons for this are outside the scope of this volume; however, three reasons might have played a significant role in this. First, missing trust in model results created without specific knowledge of the markets involved. Second, the accessibility and ease of model application to a specific market place. Third, the already existing spread of a model in the market. The final point also suggests that a majority of companies still feel safer if they know that competitors manage their downside using a similar tool. The cost of setting up a team of model creation specialists may also be a reason; however costs of such an endeavor continue to decrease as processing power become cheaper and tools and models become more readily available. Today, even the creation of a global risk model requires very few people (see also Chapter 9), and risk models are likely to commoditize further. Moreover, what holds for models holds also for the platforms on which these models are deployed. Many large insurers or reinsurers now have their own platforms and/or have access to some of the existing and further commoditized platforms, such as Oasis (http://www.oasislmf.org/), EigenRisk (https://eigenrisk.com/), Impact Forecasting's Elements platform (http://www.insurancebusinessmag.com/asia/news/breaking-news/aon-benfield-releases-enhanced-elements-modelling-platform-49543.aspx), Hazus (https://www.fema.gov/hazus/), AIR (Chapter 2), RMS (Chapter 1), CoreLogic (http://www.corelogic.com/), and/or Cyence (https://www.cyence.net/).

Insurance is less concerned with reading science articles and books, and the catastrophe risk financing market is more about "doers" rather than "technocrats and theoreticians." Small dedicated groups have developed many numerical risk models, albeit often in association with large insurance or reinsurance companies. These small groups might not have the time or the incentive to publish the results of what, ultimately, makes such a difference in our daily lives. The fact that catastrophe risk modeling is often referred to as a "secret sauce" for insurance risk trading might be another reason why concepts and ideas remain private. Most of the authors involved in this volume are experts in or for the insurance market. The three author groups in this volume comprise (1) model and software building vendors and consultants such as AIR, RMS, and Karen Clark, (2) scientists from the United Kingdom, United States, Canada, or Germany who dedicate their lives to modeling, and (3) practitioners and/or people who have worked between the first two groups over their careers.

This volume begins with an introduction to the wider areas of risk modeling, decision-making, and data and concepts and then discusses specific perils such as earthquake and flood and the use of data and modeling of indirect losses in more detail. Also included are two chapters, which introduce the practical everyday use of models and their results in insurance risk management. The volume then reviews innovative ideas, additional concepts, and the use of models for resilience measures, among other topics. The final chapter considers whether the insurance world is prudent in its use of models and uncertainty. Overall, this volume comprises four sections: (I) Catastrophe Models, General Concepts and Uncertainty; (II) Model Creation, Specific Perils and Data; (III) Insurance Use-Cases; and (IV) Model Risk, Resilience and New Concepts.

The panel of authors for Section I includes the head teams for model development from the two main model vendors (Jay Guin from AIR and Nilesh Shome, Mohsen Rahnama, Steve Jewson, and Paul Wilson from RMS). More than any others, these two groups of authors have shaped the current understanding of modeling risk for insurance, governments and/or large multinational organizations such as the World Bank. For over 25 years their ideas, methods, and tools have changed how risk gets traded, whereas their models have become platforms and currencies for the modeled hedging and securitization of progressively more complex products.

Nilesh Shome et al. from RMS focus on quantification of model uncertainty and risk, introducing the reader to the basic concepts of catastrophe models and their component parts, including hazard, vulnerability, and finance. Including Mohsen, the Chief Risk Modeling Officer and General Manager, Models & Data, Nilesh Shome and his team walk the readers through a detailed description of the most important factors to consider when creating an earthquake risk model. They follow this with an introduction to the construction of a typical tropical cyclone model. This includes an emphasis on the different parts of uncertainty involved, namely random (or aleatoric) and knowledge (or epistemic uncertainty). Having created numerous models in the past, Nilesh et al. provide invaluable insights into model sensitivities and what matters most when creating a full-blown catastrophe model, at both an individual component and a whole model level. The only limitation of the ability of Nilesh et al. to create accurate models is the amount of available data.

Jay Guin is the chief research officer of AIR, a company comparable in size and reach to that of RMS. Jay heads up a team of experts and model developers similar to those at RMS. In his paper, Jay introduces current catastrophe modeling practices, with a focus on how a model user may judge the validity and *robustness* of a model. Within this, Jay provides a guide to the different parts of a catastrophe model to familiarize readers with the concept of expert models, those that allow users to assess risk at higher resolution and levels of detail. Jay considers *parsimonious* (see earlier discussion) models as the best practice in testing the results of more complex models.

The next paper is from David B. Stephenson and Ben Youngman, from the University of Exeter, Alasdair Hunter from the Barcelona Supercomputing Center (Spain), and Ian Cook from the Willis Research Network. David Stephenson and his team have been innovators in risk modeling over more than two decades, with many of the tools and models used today, stemming from their research and ideas, including (but not limited to) extratropical storm risk and the clustering of events in time and space around the globe. The paper also reviews how modelers can use the Event Loss Tables (ELTs) produced by catastrophe models to simulate insurance losses due to natural hazards. ELTs are among the most important outputs of catastrophe loss models, and at the core of risk assessment and management for any natural or man-made catastrophe. To make more realistic loss calculations, models must contain a dynamical element, which modulates the parameters. For hurricanes and windstorms, these variables may represent the behavior of large-scale indices, such as the El Nino/Southern Oscillation (ENSO) and the North Atlantic Oscillation (NAO). These variables allow model conditioning to the current state of the climate and further constrain clustering and the impact of climate variability on annual aggregate losses.

Tiziana Rossetto and Ioanna Ioannoua from University College London (UK) are leaders in the field of fragility and vulnerability. Their work has increasingly influenced the ways in which we now translate intensities into losses. Tiziana and her team have become highly influential in the risk and insurance market, following their work with GEM, the Global Earthquake Model (https://www.globalquakemodel.org/), and other models. Their paper looks at empirical modeling approaches to exposure fragility and vulnerability. Rossetto and Ioannoua believe that, in practice, there is no perfect and unbiased event loss data set from which to derive fragility and vulnerability functions. The value of empirical vulnerability and fragility functions hence varies greatly depending on the quality and quantity of data used and the procedures adopted to account for any data bias. This paper presents a summary of their observations from the last 10 years of working on earthquake and tsunami empirical fragility and vulnerability functions. This includes a description of the common biases present in event damage and loss data sets, the consequences of ignoring such biases and possible methods for their management. The paper also includes discussion of the impact of different statistical model fitting assumptions and includes examples drawn from empirical earthquake and tsunami fragility and vulnerability studies to illustrate the key points.

Section II of this volume, Model Creation, Specific Perils and Data, provides the reader with more information on the general use of data, perils, and tools for creating loss models. The Section begins with a paper from James Daniell, Friedemann Wenzel, and Andreas Schaefer from the Karlsruhe Institute of Technology (Germany) about the use of historical loss data for insurance and economic loss modeling. According to the authors, creation of a natural catastrophe risk model requires the following elements: (1) a catalogue of all possible events that can occur within the time considered, (2) an exposure database, and (3) details of the susceptibility of the structures or population at risk to the hazard. This paper focuses on each of these components and highlights recent trends in losses and normalization procedures. The authors show that normalized losses are, in general, decreasing globally, most likely because of increasing hazard mitigation and decreasing vulnerability. The paper ends with an example of an earthquake loss model built from the data discussed therein.

The second paper in this Section, from James Daniell, Bijan Khazai, and Friedemann Wenzel, discusses an indirect loss potential index for natural disasters. The authors separate indirect losses, including losses to infrastructure and the loss of profit, from direct losses using data, literature, and the available methods. The authors present an index for indirect loss potential at both the national and subnational levels using components of the GEM Socioeconomic Indicator Database. The index provides an alternative measure to traditional business interruption models. The index is especially helpful where (the usual) lack of data makes modeling of processes difficult. Indirect losses have become a growing concern more recently because of growing coverage and related loss surge.

In the past, earthquake risk modeling has focused largely on two major concepts, first, an approach referred to as *time-independent* and second, an approach that considers the likelihood of an earthquake, given the length of time that has elapsed since the last major earthquake in the area (see e.g., https://pubs.er.usgs.gov/publication/70033084). The paper from Kristy F. Tiampo, Robert Shcherbakov, Paul Kovacs from the University of Colorado (US), the Western University in London Ontario (Canada), and the Institute for Catastrophic Loss Reduction, Toronto (Canada), provides an overview of the current state of knowledge for time-dependent seismicity-based earthquake forecasting. The focus is on short- and intermediate-term seismicity-based methods. Although many of the techniques employed provide probability gains ranging from 2 to 5000 over time periods of days to years, absolute probabilities remain small. Despite this fact, forecasts, which provide measurable increases (or decreases) in earthquake probability, can provide timely updates for both mitigation and readiness measures. Most of the current risk models do not consider time-dependent seismicity-based earthquake forecasting as described in this paper. However, these measures provide significant insights, meaning that loss models in the future will inevitably embrace these methods. This is especially important where large earthquake losses drive capital needs for insurance companies, such as those in California (US), Japan, South America, New Zealand, or Canada. Even small differences in defined return period losses in the tail of the risk distribution can make a significant difference in capital cost (see Chapter 10).

The following paper, by Kristy F. Tiampo, Seth McGinnis, from Bolder Colorado (US) and Yelena Kropivnitskaya, Jinhui Qin, and Michael A. Bauer from Western University, Ontario (Canada), provides an introduction to the challenges presented by remote sensing and other "big data" sources for hazards modeling. Such data have increasingly become important for loss modeling. This paper discusses some of the big data challenges. Big data are not simply about large volume; the structure, quality, and end-user requirements of big data, including real-time analysis, require new and innovative techniques for data acquisition, transmission, storage, and processing. The authors present two different applications designed to solve the challenges for big data science, from climate models and earthquake hazard assessments.

The paper by Paul D. Bates, Jeff Neal, Chris Sampson, Andy Smith, and Mark Trigg from the University of Bristol (UK) and SSBN (a private company that works on global flood modeling for risk applications) reports on their progress toward the creation of hyperresolution models of global flood hazard. Flood modeling on a global scale embodies a revolution in hydraulic science and has the potential to transform decision-making and risk management in this sector. These authors believe that flood risk might only be understood for a small number of territories, and even where understood, there may be a significant mismatch between national scale model results and long-term loss observations. The paper reviews recent progress in global flood modeling toward hyperresolution simulations using true hydrodynamic equations. The paper further discusses how developments in hydrodynamic modeling can foster a rapid shift from local scale models to global schemes and provides. The final part of their paper discusses the challenges in integrating available global exposure data with hyperresolution flood hazard calculations and addresses the limitations of global flood models, with a specific focus on their use in (re)insurance applications.

Section III of the volume on Insurance Use-Cases presents papers from practitioners in the applied loss modeling and portfolio management areas of the (re)insurance industry. The first paper, by Helena Bickley and Gero Michel from Endurance Specialty Insurance (Bermuda) and Chaucer Syndicates Limited (UK), respectively, discusses the role of catastrophe models in the reinsurance industry. This includes how (re)insurance companies use models; how much companies rely on model results; and the extent to which modern risk takers depend on sophisticated modeling engines, which are paramount for both day-to day underwriting decisions as well as longer-term company strategy. The paper reflects the everyday work of a new generation of actuarial analysts, managers, and statisticians, typically referred to as reinsurance "modelers."

The second paper in this section, by Augustin Yiptong and Gero Michel from Endurance Specialty Insurance (Bermuda) and Chaucer Syndicates Limited (UK) respectively, discusses portfolio optimization and describes how modern insurance companies use catastrophe model results to optimize their risk selection or risk hedging. The paper describes an optimization process based on both, *simulated annealing*, to take care of second-order minima, and the computationally less expensive *hill climbing* method. The paper focuses on a procedure that (re)insurance companies use for everyday decision-making. The description and guidance in the paper may enable readers to reproduce the approach taken. The authors then use the tool to optimize two partly independent portfolios simultaneously, an approach required in reinsurance and retrocessional cases where, for example, the portfolio and/or the return constraints may differ significantly for different portfolios, so allowing for a "mutual arbitrage" or a win–win solution.

The final Section IV, Model Risk, Resilience and New Concepts, looks critically at the use of loss models in the insurance market along with discussing a shift in potential model usage from measuring (and hedging) the status quo to increasing resilience. The first paper, by Patrick McSherry, reminds us to a principle known as Occam's razor that warns against complexity and favors parsimonious models. Transparency and ease of use are also important characteristics of models that will appeal to practitioners. Rather than pursue the holy grail of the illusive single optimal model, an ensemble of diverse parsimonious models may offer the best chance of accurately quantifying risk. Seeking a diverse set of opinions is not only advisable for avoiding surprises but may offer the best means of making predictions. Patrick McSherry then leads us to big data and how crowd sourcing might help finding alternative solutions. Index insurance can provide a powerful and simple means to close some insurance gap in areas where data and insurance penetration is low, e.g., for farmers in developing countries. Patrick McSherry then criticizes complicated black box mentality of vendor CAT models. "By design these CAT models are not transparent due to the need for vendors to protect their intellectual property. While the general recipe for building a CAT model is common knowledge, it is usually difficult to modify any of the ingredients of a commercial model."

Karen Clark from Karen Clark & Company (US), questions the appropriateness of the exceedance probability (EP) loss curve (see also Chapter 3) for risk decision-making. The author explores the idea that the EP curve might not help where decision-making is most important for risk managers, e.g., where a risk taker wants to increase the upside and grow. The paper also questions whether model results really are intuitive. Senior management, no matter whether in the insurance business, in government organizations, or NGOs, may find risk model results rather difficult to understand and hence not necessarily useful for everyday decision-making. The randomly generated loss samples included in catastrophe models may also include events, which are unrealistic and not likely to occur in nature. Many existing models can make it difficult to conduct efficient sensitivity analysis on the underlying assumptions. The paper offers an alternative concept, based on the *Characteristic Event Methodology*, and uses a hurricane model as an example to show how a developer carefully selects the "typical" event characteristics by return period for specific coastal regions and then "floats" this event along the coast at 10-mile increments to provide full spatial coverage and some more intuitive results for senior management.

Slobodan Simonovic then considers some highly promising, practical links between disaster risk management, climate change adaptation, and sustainable development in his essay, *From Risk Management to Quantitative Disaster Resilience: A New Paradigm for Catastrophe*. Simonovic reports on recent noticeable changes in approaches to the management of disasters, which have moved from disaster vulnerability to disaster resilience, with the latter being viewed as a more proactive and positive expression of community engagement with disaster risk management. This paper provides an original systems framework for the quantification of resilience. Different industries can apply the framework that Simonovic reports to quantify and compare different hazard response strategies and to compare the performance between industries under similar hazard conditions. The paper includes railway industry examples, based on the performance of different railway companies under disaster conditions.

The final chapter from Andreas Tsanakas and Laure Cabantous from the Cass Business School (UK) critically assesses the use of risk model results, which may have been caused by the propagation of model flaws and which may lead to poor decision-making. The authors argue that model governance requires a broader conception of model risk to be effective. If, for example, there is a wide range of model specifications, then some of those specifications will fit more naturally than others, with current management expectations, and will increase the potential for bias in decision-making. Tsanakas and Cabantous also report on a Culture of Model Use framework that refines the understanding of how insurance organizations perceive and understand quantitative models. Given that models provide results for some of the most fundamental measures in a company, such as business planning, capital setting, portfolio optimization, risk appetite monitoring, and performance measurement, unbiased usage of the models is paramount. The authors consider questions on how organizations make modeling choices in the presence of substantial uncertainty around key model parameters and expert judgments, which is susceptible to influence from the personal biases of the experts involved and the business incentives on offer to them. For example, underwriters may be inclined to overstate the volatility of their business to justify higher levels of reinsurance purchase or understate it to make the return on capital appear more attractive and to enable them to write more business. Although Tsanakas and Cabantous conclude that practitioners typically have a sophisticated understanding of the model limitations involved, the authors do not fully erase reader concerns over inappropriate model use. Some of these concerns are further addressed in Kunreuther et al. (2013).

Where are we heading to? Undoubtedly, catastrophe risk models will become even more important in daily decision-making across markets. No doubt, also that the currently often painful processes of sourcing of rare data, scarcity of models, or "black box" mentalities restricting how well we may understand model risk and limitations will change toward more complete information and higher transparency. It seems likely that future supply chains will change, and data including weather, climate, real-time event damage, or exposure will come more frequently from crowd sourcing rather than from a limited number of proprietary data sources. Data necessary to feed catastrophe models might hence rather be shared freely than demanding large fees. Models will also cover risk more comprehensively. Modelers have already started to build numerical risk models for pandemic, supply chain, and cyber risk along with fire, leakage, accident, health, space, political violence, and aviation, among others. The term *risk* might well be replaced with resilience, and we might model the upside and compare cost of investments versus the downside over different time scales for the risks we shall model. Future model users will undoubtedly require model accuracy to be specified further herewith, judging model runtime and cost versus utility. Risk is also likely to be shared more completely, and it is only a question of time until insurance (or other forms of risk sharing) penetration rates will be high across the globe.

For many risk managers, however, the answer to the question of whether we will use open-source, open-development, and low-cost platforms to exchange data and risk results widely in the future, or whether we will continue to model risk on a limited number of proprietary and presumably expensive platforms providing "gold standards," remains uncertain. Complex decision-making will continue; however, we may become more aware of the different levels of uncertainty going forward. New decision tools will enter the market, helping risk managers to find appropriate solutions among the numerous potential answers. Rules-based decision-making might change into more principles-based approaches, both internally for companies as well externally, when facing regulations and clients. This, and further advancements in science, will help convert current risk models into short-term forecasting tools, customized for specific needs and single risks, feeding back into global models for portfolio management. Bespoke models will exist alongside platforms, with models becoming the sole judges for commoditized risk business.

I would like to conclude by thanking all of the authors in this volume for their invaluable contributions. As always, it took longer than expected to bring all pieces together but I hope you, the reader, will enjoy this volume as much as we did creating it. Although we may have only scratched the surface of this subject, this volume fills me with confidence that risk modeling is a young science with a great future ahead.

References

Cummins, J.D., Mahul, O., 2009. Catastrophe Risk Financing in Developing Countries: Principles for Public Intervention, 268 pages.

Daniels, R.J., Kettl, D.F., Kunreuther, H., June 7, 2011. On Risk and Disaster: Lessons from Hurricane Katrina, 304 pages.

Froot, K.A., December 1, 2007. The Financing of Catastrophe Risk, 488 pages.

Grace, M.F., Klein, R.W., Kleindorfer, P.R., Murray, M.R., June 30, 2003. Catastrophe Insurance: Consumer Demand, Markets and Regulation, 147 pages.

Grossi, P., 2005. Catastrophe Modeling: A New Approach to Managing Risk. Springer.

Gurenko, E.N., 2004. Catastrophe Risk and Reinsurance: A Country Risk Management Perspective, 322 pages.

Kleindorfer, P.R., Sertel, M.R., December 6, 2012. Mitigation and Financing of Seismic Risks: Turkish and International Perspectives, 286 pages.

Kunreuther, H.C., et al., 2013. http://www.cambridge.org/us/academic/subjects/economics/industrial-economics/insurance-and-behavioral-economics-improving-decisions-most-misunderstood-industry?format=PB&isbn=9780521608268.

Messy, F.A., Organisation for Economic Co-operation and Development, January 1, 2005. Catastrophic Risks and Insurance, Organisation for Economic Co-operation and Development, 421 pages.

Neumayer, E., Barthel, F., 2010. http://eprints.lse.ac.uk/37601/1/Normalizing_economic_loss_from_natural_disasters_a_global_analysis(lsero).pdf.

OECD, January 1, 2005. Terrorism Risk Insurance in OECD Countries, 289 pages.

OECD, September 30, 2008. Policy Issues in Insurance Financial Management of Large-Scale Catastrophes, 312 pages.

Pollner, J.D., 2001. Catastrophe Risk Management: Using Alternative Risk Financing an d Insurance Pooling Mechanisms, 126 pages.

Roth, R.J., Kunreuther, H., July 9, 1998. Paying the Price: The Status and Role of Insurance Against Natural Disasters in the United States, A Joseph Henry Press Book, 320 pages.

Seo, J., Mahul, O., 2009. http://documents.worldbank.org/curated/en/277731468148518556/pdf/WPS4959.pdf.

Simmons, K.M., Sutter, D., Pileke, R., 2013. Normalized tornado damage in the United States: 1950–2011. Environmental Hazards 12 (2), 132–147. http://dx.doi.org/10.1080/17477891.2012.738642.

Taleb, N., 2014. Antifragile: Things That Gain from Disorder.

SECTION I

CATASTROPHE MODELS, GENERAL CONCEPTS AND METHODS

CHAPTER

1

Quantifying Model Uncertainty and Risk

Nilesh Shome, Mohsen Rahnama, Steve Jewson, Paul Wilson

Risk Management Solutions, Newark, CA, USA

INTRODUCTION

Every year different catastrophic events cause significant losses to the insurance industry. For example, natural catastrophic events (III, 2015) caused about $110B overall economic loss and 7700 fatalities in 2014, although the number of events was low compared to other recent years. Among all the different types of catastrophic events that cause loss to the industry, earthquakes and hurricanes generally contribute the most. For example, nine out of top 10 costliest catastrophes in the world since 1950 were earthquakes and hurricanes (III, 2015). The flood risk, however, is becoming to be a significant player in the recent years (e.g., 2011 Thailand flood). Hence, it is of the utmost importance to model these catastrophes accurately, to get an unbiased estimate of loss and reduce the uncertainty in that estimation. This paper focuses particularly on earthquake and hurricane loss modeling and discusses in detail the uncertainties associated with loss estimations.

The primary challenge in developing a robust catastrophe model is the limited availability of data to define the characteristics of events. For example, only limited groundmotion records exist for large magnitude earthquakes in Central and Eastern United States or in regions with similar tectonic characteristics. In addition, there is a limited amount of high-quality location-level loss data from earthquakes in most of the regions in the world as well as from hurricanes in some specific regions, for example, Northeast US, necessary for developing the building vulnerability functions. These data deficiencies drive significant uncertainties in the estimation of risk due to losses for catastrophic events. Thus, catastrophe risk modeling has unique challenges compared to modeling of other sources of insurance loss. For example, the estimation of risk for auto insurance has ample data for developing robust risk assessments following actuarial methods. Catastrophic earthquakes and hurricanes are low-probability, high-consequence events as evidenced, for example, by Hurricane Katrina and the Northridge earthquake. For this reason, risk estimation based on an actuarial approach

Risk Modeling for Hazards and Disasters
http://dx.doi.org/10.1016/B978-0-12-804071-3.00001-X

cannot be carried out for these most severe perils. Hence, instead of following a standard actuarial approach, physical models are developed for these perils based upon scientific research and event characteristic data for estimating risk. These physical models enable us to predict losses that have not been observed before. In fact, even in cases where one or more catastrophe events have been observed before, the quality and the quantity of the data may nevertheless be sufficient to develop an actuarial model that can be generalized over large regions.

The typical loss calculations (performed using products provided by commercial catastrophe modeling companies such as Risk Management Solutions, RMS) generally estimate the mean annual loss or mean loss at specific exceedance probabilities. However, there are significant epistemic uncertainties in the loss results, which users should consider in their decision-making processes. Thus, the loss calculations should consider all types of uncertainties, which include aleatory (also known as random) and epistemic (also known as systematic) uncertainty (Benjamin and Cornell, 1970). Aleatory uncertainty, i.e., the inherent variability in the physical system, is stochastic and cannot be reduced by improving the current approach of hazard analysis. Epistemic uncertainty, on the other hand, is associated with lack of knowledge; it can be subjective and is reducible with additional information. Logic trees can be used to capture the epistemic uncertainty in the estimation of loss. This uncertainty in this approach is represented by separate branches of the logic tree, with one branch for each alternative component model or parameter(s) of a model that a modeler considers reasonable. The modeler assigns each branch either a normalized weight that reflects the modeler's confidence in the choice of the model, or a weight based on the appropriate probability distribution of the parameters.

BACKGROUND OF CATASTROPHE MODELING

Earthquakes, particularly large magnitude ones, can cause significant damage to the built environment over a large region surrounding the fault-rupture zone. The top three costliest earthquakes in the world during 1980–2012 were the Mw 9.0 Tohoku-Oki earthquake (2011), Mw 6.8 Kobe earthquake (1995), and Mw 6.7 Northridge earthquake (1994), causing overall economic losses at the time of occurrence of about $210B, $100B, and $44B, respectively (III, 2015). The structures that are damaged during earthquakes are not only buildings, bridges, etc., but also utilities like water, gas, and sewer lines and lifelines like train tracks and roads. Earthquakes also cause landslides due to the movement of the ground or liquefaction due to loss in the strength of the soil underneath structures. In addition, earthquakes can also cause fires and tsunamis, leading to additional significant damage to structures. Historical losses illustrate that the potential for loss from earthquakes can be profound. Hence it is important to accurately estimate the losses to manage earthquake risks.

Since engineers were primarily interested in designing buildings for life safety until the end of 20th century, postearthquake investigations focused only on the causes of significant damage that led to the collapse of buildings due to earthquakes. As engineers did not collect data on undamaged, slight and moderately damaged buildings from past earthquakes, such data are not available to assess the risk of expected earthquake damage to buildings. In 1985, the Applied Technology Council (ATC) first published a comprehensive report (ATC-13, 1985) to estimate the earthquake losses of existing buildings by developing a suite of vulnerability functions based upon expert opinions. ATC used this approach because there was limited earthquake damage and loss data at that time. The report developed building loss estimates as a function of Modified Mercalli Intensity (MMI) for different facility classes (e.g., low-rise wood frame). The Northridge earthquake (1994), however, provided large number of insurance claims data, which are now used for seismic risk (of loss) assessment of low-rise, single-family, wood-frame buildings in high seismic zones of California, such as Los Angeles and San Francisco (Wesson et al., 2004).

On the other hand, hurricanes cause damage to the built environment due to strong winds, high waves, storm surges, and severe rainfall. Even before hurricanes make landfall the high waves can cause damage to shipping and offshore platforms, and when they do make landfall, hurricanes frequently cause damage to buildings, disruption, and loss of life. However, the risks due to hurricanes are very inhomogeneous in space and time and cannot be well estimated from recent local experience. An individual location may not experience a strong hurricane for many decades, before being suddenly devastated: exactly this happened to parts of Miami in hurricane Andrew in 1992. As a result, hurricane-related risks can only be comprehensively understood using careful analysis of historical records, augmented with computer simulations. This kind of analysis can help us to understand where and when hurricanes are possible and what damage they might do.

The first attempts at a rigorous analysis of hurricane risk were made in the 1960s, by Don Friedman from Travelers Insurance (Friedman, 1972). Fifty years on, and the models used to estimate hurricane risk, known as hurricane catastrophe (cat) models, have become highly complex, involving detailed simulations of thousands of possible future hurricane events using large computers. The starting point for building these models is the set of observational records related to hurricanes. These hurricane records include reports of landfalling storms back to the mid-19th century; measurements of wind speeds, albeit at only a few locations; from early in the 20th century onwards, reports of damage, from newspapers, and, more recently, from insurance company records; aircraft-based reconnaissance from shortly after the second world war; and satellite-based measurements from the 1970s to the present day. Using these various data sources relevant aspects of hurricanes can be studied and modeled. Hurricane cat models then augment this observational record by simulating many thousands of years of surrogate hurricane behavior to represent the whole range of possibilities of what might happen in the next few years. The results from these simulations can be used by anyone with a need to understand hurricane risk, including local government, business, and charities. They are most widely used, however, in the insurance industry, so that insurers can be prepared for whatever payouts they may have to make, and losses they may sustain, as a result of future storms.

Following devastating loss from Hurricane Andrew in 1992 and the 1994 Northridge earthquake, the insurance industry became interested in loss estimation based on physics-based models rather than relying upon the actuarial approach for managing risks due from natural catastrophes. The Federal Emergency Management Agency (FEMA) also first published a comprehensive methodology in 1999 to analytically calculate losses of buildings from earthquakes (HAZUS, 2003). Since there is limited or no data available for developing vulnerability functions for all the different types of construction in different regions, the earthquake loss calculations in the insurance industry are carried out primarily based on analytically developed or engineering judgment-based vulnerability functions leading to significant uncertainty in the loss results. In addition, even before the insurance industry and risk managers ever became interested in estimating risk of losses of buildings from earthquakes, Cornell (1968) developed a methodology to estimate probabilistic seismic hazard. Cornell's methodology soon became standard in the industry for assessing the seismic hazard at building sites. The information of seismic hazard and building vulnerability functions are combined to estimate seismic risk of damage or loss of buildings for insurance companies. The risk assessment provides an estimate of the mean frequency or the probability that a specified amount of loss will occur within a specified period of time.

Hurricane risk is estimated by simulating events by both interpolation and extrapolation of the historical record. They interpolate in the sense that they include hurricanes that lie in between observed hurricanes in terms of landfall locations, strengths, wind speeds, and tracks, essentially filling in gaps in history. They extrapolate in the sense of creating new, but physically possible hurricanes with tracks, strength, and sizes that have not been seen before. The damage caused by hurricanes depends very sensitively on the precise characteristics of the storm, such as forward speed, size, landfall location, and decay rate. By creating a large catalog, or ensemble, of thousands of possible future storms that samples every possibility (within a certain level of discretization), hurricane cat models make it possible to estimate the distribution of possible losses with precision, even at regional and local levels.

However, given the limited historical data available, the limitations of the observations that do exist, and limitations in our current ability to model these observations realistically, there are many challenges involved in the creation of a hurricane cat model, and as a result much uncertainty in the results such models produce. For instance, to account for gaps in data, data must be smoothed, and smoothing assumptions must be made using best guesses based on physical intuition and reasonableness. Such assumptions inevitably lead to uncertainties. In this paper, we give an overview of how hurricane cat models work, and, in particular, we discuss some of the sensitivities and uncertainties associated with such models. We discuss in detail which parts of a hurricane model are well constrained by data, and which parts are not, and which hence lead to a greater level of uncertainty.

As of today, there are three main proprietary catastrophe modeling firms, Risk Management Solutions (RMS), AIR Worldwide, and CoreLogic (also known as EQECat). These firms help to manage risks for a wide range of natural catastrophes such as hurricanes, earthquakes, winter storms, tornadoes, hailstorms, and floods, as well as man-made catastrophes such as terrorism. Insurers, reinsurers, rating agencies, risk managers, and insurance brokers use these models to price and transfer the risks to manage these risks. The following sections provide detailed descriptions of different components of catastrophe modeling, primarily through the example of earthquake and hurricane modeling. The discussion here is reasonably generic and can be applied to any of the major commercial catastrophe models.

DIFFERENT COMPONENTS OF CATASTROPHE MODELS

The catastrophe modeling framework adopted by the firms listed above has three major components or modules, each of which assesses a key element of the risk and engages specific academic disciplines (Grossi and Kunreuther, 2005):

1. **Hazard**: This component defines the frequency, severity, and geographical distribution of events by developing a large catalog of catastrophic events. Additionally, the

component provides an estimate of the intensity i.e., some measure of the severity of the event's impact for each of the events within the affected area. The intensity for hurricanes is typically expressed in terms of wind speed and/or storm surge height and the intensity of earthquakes in terms of peak ground acceleration (PGA) and/or spectral accelerations. The hazard module additionally considers many parameters like the surface roughness (for hurricane hazard) or soil type (for earthquake hazard) to improve the intensity prediction for the events.

2. **Vulnerability**: This component estimates the degree of damage to different classes of buildings by construction and occupancy type, and to their contents, based on the intensity estimated by the hazard component. The model typically assesses the loss to buildings using a vulnerability function (also known as a damage function), which relates the hazard intensity to the distribution of the ratio of the building loss to the building value. Spectral acceleration is often the most common intensity measure for earthquakes, and wind speed for hurricanes.

   Although ground shaking typically causes most of the damage to buildings during earthquakes, damage can also result from secondary hazards like landslides and liquefaction, or from secondary perils like fire following earthquakes, tsunami, and sprinkler-leakage damage. Similarly storm surge can cause significant flood damage during hurricanes.

3. **Finance**: This component estimates the insured losses by applying insurance terms and conditions to either the damage or the ground-up loss estimates derived in the previous step. The financial module typically accumulates the losses through location-level terms, to policy and then program-level conditions applying limits, deductibles, and special policy conditions at each stage to estimate the loss for each of the event defined in the hazard module. The resulting output is an Event Loss Table that provides an assessment of the financial risk exposure to individual events. The financial model then uses this Event Loss Table to calculate the exceedance probability (EP) curve, average annual loss (AAL), and many other loss parameters. The insurance industry uses such loss results extensively for (1) portfolio management, (2) pricing, and (3) capital requirements.

The next section discusses the hazard and the vulnerability components for the earthquake peril in detail, followed by detailed description of hurricanes peril.

EARTHQUAKE HAZARD

The earthquake hazard component estimates the amount of ground shaking at the location of the buildings for a series of potential earthquake events. The event characterization accounts for variations in the location, size, and frequency of events. From this information, the ground-motion model estimates the distribution of shaking intensity of each event, which the model then further uses in the calculation of losses to buildings at different sites. Cornell (1968) developed a comprehensive methodology for the calculation of probabilistic seismic hazard. This article provides an overview of seismic-hazard calculations for different types of sources. It also describes the uncertainty associated with the estimation of the parameters for hazard calculations. This information is then used for assessing the uncertainty in the hazard results.

The sources of seismic hazard fall into two broad categories: (1) fault and (2) area (or distributed). The fault sources include both individual faults and zones of multiple faults where there is potential for future large earthquakes (based on past earthquakes or evidence of seismic activity, such as fault movement). Models typically carry out seismic-hazard calculations for Quaternary faults, i.e., for those active faults that have evidence of movement in the past 1.6 million years. Geologists commonly consider these faults as high risk if there has been movement observed or evidence of seismic activity during the last 10,000 years. These active faults are the source of high potential for causing future earthquakes.

Area or distributed sources, on the other hand, distribute seismicity across an entire region, typically based on historical seismicity. Modelers develop source zones based on large-scale geological structures, tectonic regions, seismogenic faults, observed seismicity (which can be combinations of instrumental macro and micro, preinstrumental, paleoseismological, and many other factors), and geodesy. The distribution of seismicity can be uniform within the zones or can vary in time and space based on the observed seismicity. The *background seismicity* is a special case of an area source, which represents seismicity that cannot be assigned to any known sources.

Components of Earthquake Source Models

The typical earthquake source model comprises four components: (1) fault model, (2) deformation model, (3) earthquake-rate model, and (4) earthquake-probability model. See Field et al. (2014) for details of these components for the California fault model (Uniform California Earthquake Rupture Forecast, version 3, UCERF3).

The *Fault Model* describes the physical geometry of the faults, which includes information on fault trace, upper and lower seismogenic depth, dipping angle, rake, and other relevant information. An example of fault model for California based on UCERF3 is shown in Fig. 1.1. To capture our incomplete understanding of fault geometry, UCERF3 considers alternative fault models to represent the epistemic uncertainties in the fault system geometry.

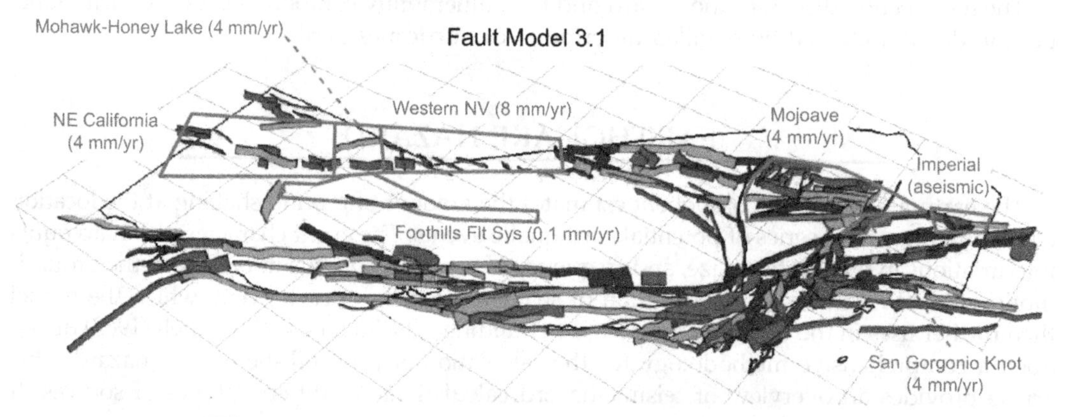

FIGURE 1.1 UCERF3 fault model for California (Field et al., 2014).

The *Deformation Model* estimates the slip rates and aseismicity factors of the fault sections. The slip rates are generally estimated based on observed geologic offsets. Although California is probably the most well-studied region in the world, the slip rates are available for only about 30% of the sections. Nowadays, the rates are also estimated by modeling geodetic measurements such as Global Positioning System (GPS) observations as shown in Fig. 1.2 for California. UCERF3 developed a number of kinematically consistent deformation models to estimate slip rates of known faults, based on geologic and geodetic observation. Fig. 1.3 compares the UCERF3 estimation and the observed slip rates for the southern San Andreas Fault. See Field et al., 2014, for a detailed discussion on different deformation models for California. The deformation models also capture additional information such as the location of observed data, offset features, dating constraints, number of events, and uncertainties, and account for qualitative and quantitative uncertainties in this additional information. Results from all these deformation models capture epistemic uncertainty in the estimated slip rates.

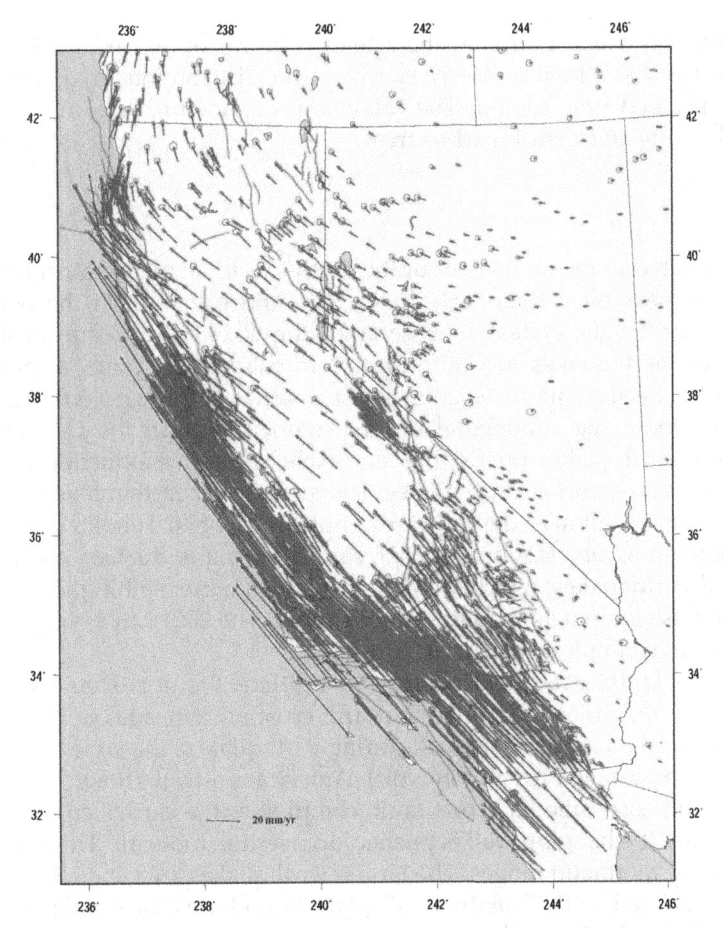

FIGURE 1.2 Distribution GPS velocity vectors in California (Field et al., 2014).

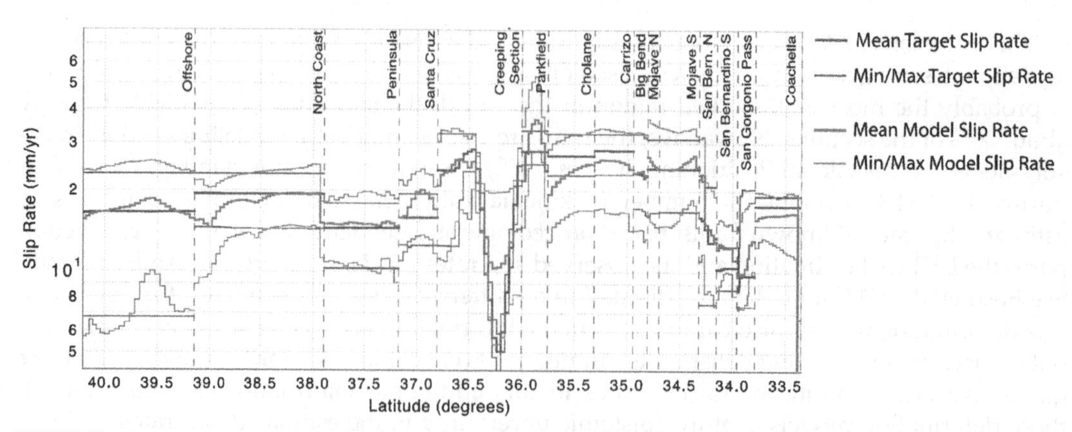

FIGURE 1.3 Comparison of slip rates as observed in paleoseismic data (target) and the average rates estimated by different deformation models for Southern San Andreas Fault (Field et al., 2014).

The *Earthquake Rate Model* estimates the rate of ruptures of all earthquakes throughout a region. The *Earthquake-Probability Model* specifies the distribution of occurrence of earthquakes over a specified time period. The following sections provide an overview of these models for both faults and distributed sources.

Fault Sources

All faults are related to the movement of the Earth's tectonic plates. Most large-magnitude earthquakes occur near plate boundaries (e.g., San Andreas Fault at the boundary of the North American and Pacific Plates). In general, earthquakes either occur on shallow crustal faults (e.g., typical earthquakes in California) or in subduction zones at plate convergent boundaries where one tectonic plate goes under another plate (e.g., oceanic Juan de Fuca plate subducting under the continental North America Plate in the Cascadia subduction zone). The largest earthquakes on Earth occur exclusively at subduction zones and these events are known as megathrust earthquakes. The megathrust earthquakes occur at the interface of two plates at a shallower depth (for example, 2011 M9.0 Tohoku earthquake) and so these are also known as *interface* or *interplate* earthquakes. Subduction zones also generate *intraplate* or *inslab* earthquakes, which are less severe, and occur within the subducting plates at a larger depth (also known as Wadati–Benioff zone). The 2001 M6.8 Nisqually earthquake in Washington is an example of an inslab earthquake.

Different types of faulting mechanisms represent different fault movements: (1) normal, (2) thrust (reverse), and (3) strike-slip. When the earth crust on two sides of faults are stretched, leading to pulling down of a hanging wall during earthquakes, the style of faulting is called *normal*. The Basin and Range Province in North America is a well-known normal faulting region. A reverse fault, also called a thrust fault, compresses the earth's crust on both sides of the fault. As a result, the hanging wall is pushed up over the footwall. This type of faulting is common in uplifting mountain ranges. The largest earthquakes are generally shallow dipping reverse faults associated with "subduction" plate boundaries. Strike-slip faulting, on the other hand, is associated with neither the extension nor the compression of the Earth's crust, but rather both sides of the fault sliding past each other. The San Andreas fault system is a famous example of a strike-slip deformation—part of coastal California is sliding to the northwest relative to the rest of North America. Ground-motion intensity is different for earthquakes with different types of faulting mechanisms. For this reason, accurate modeling of the fault mechanism in hazard computations is important to reduce the uncertainty.

The calculation of probability of fault rupture requires the following information:

1. Rupture area or length
2. Mean magnitude of rupture, based on rupture area
3. Moment rate, based on slip rate or observed rate of occurrence of earthquakes
4. Recurrence rate, based on moment rate and moment magnitude
5. Probability of earthquakes within a time interval

The fault model estimates rupture length/area of faults. This model estimates the magnitude of earthquakes based on the area or length of ruptures of earthquakes from magnitude-scaling relationships. For example, Fig. 1.4 shows an example of the fit of different magnitude-scaling relations to the observed data. The different fitted models in the figure show that the epistemic uncertainty in the estimation of earthquake magnitudes based on fault-rupture area is large for small and large rupture areas (i.e., areas significantly different from the average rupture area of the observed data). The following example of the magnitude-area relationship is based on Hanks and Bakun (2008):

$$\overline{M} = \left(\frac{4}{3}\right) * \log_{10}(A) + 3.07 \quad \text{for } A > 537 \text{ km}^2 \tag{1.1}$$

where A is the area of rupture. The equation shows that the functional form is simple. As Fig. 1.4 shows, the model is quite a good fit to the data. The different color of the dots represents different sources of data. Fig. 1.4 shows the model-to-model variation in the estimation of the magnitude illustrating the epistemic uncertainty in the estimate. The 90% confidence bound

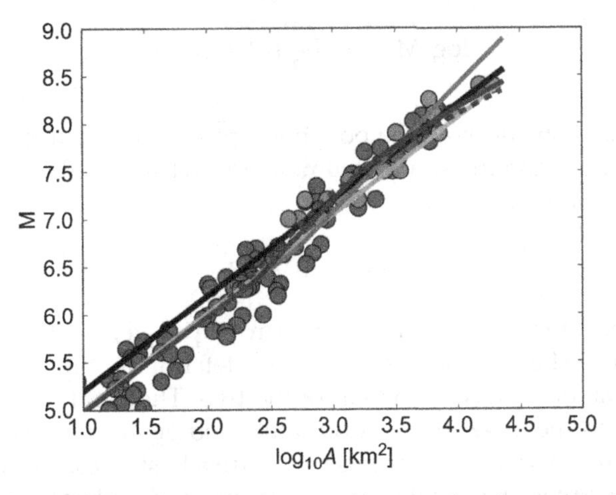

FIGURE 1.4 A comparison between magnitude and area of large earthquakes and the fit of different magnitude-scaling relations to the data (Shaw, 2013).

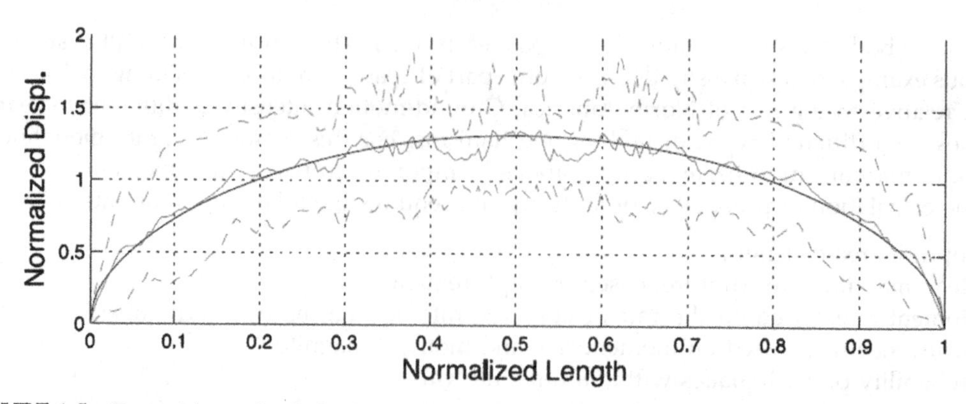

FIGURE 1.5 Tapered (normalized) slip distribution model along the strike of surface ruptures and the one-sigma uncertainty (Biasi et al., 2013).

of the uncertainty in the magnitude is approximately $\overline{M} \pm 0.4$ (Shaw, 2013). Hazard modelers generally assume that multiple models capture half of the uncertainty and that the rest of the uncertainty is aleatory. This is the basis of having ±0.2 epistemic uncertainty in the maximum magnitude in the USGS and other hazard models (e.g., Petersen, et al., 2014).

The next step is calculation of the slip distribution along the strike of ruptures for calculating moment based on average slip for a given rupture. As Fig. 1.5 shows, the slip is generally assumed to taper to the edge of the rupture following the square-root-sine of length of rupture (known as "tapered-slip" model, Biasi, et al., 2013). A simpler alternative to the tapered model, the uniform slip model (or "boxcar" model), is also useful. This is because the end of faults may require a large number of smaller magnitude earthquakes to balance the slip deficits predicted by the tapered-slip model when balancing the moment rates.

The seismic moment M_0 for each event of magnitude M is determined from the moment magnitude relation (Hanks and Kanamori, 1979) as given below:

$$\log(M_0) = \left(\frac{3}{2}\right)M + 16.05 \tag{1.2}$$

The sum of seismic moment of all the possible ruptures on a fault is balanced against the total moment rate to constrain the strain accumulation/release on the fault. The moment rate $(\dot{M}_0)$ is calculated as follows:

$$\dot{M}_0 = \mu \cdot A_s \cdot v \tag{1.3}$$

where A_S is seismogenic area, v is long-term slip rate, and μ is shear modulus ($\approx 3 \times 10^{11}$ dyne/cm^2). The seismic-hazard model determines slip rates from (1) geologic data, (2) geodetic data, or (3) a combination of the two. The slip rate is calculated from the geologic data based on the offset features on faults and age of the feature or from the slip over a period of time observed in paleoseismic trench studies. Geodetic measurements such as GPS are used either to estimate the slip rate or to provide constraints to the geologic slip rates.

Although the stress on faults is released seismically during earthquakes, a significant fraction of stress (varying by fault) is released through aseismic slip (can be interseismic or postseismic), called a *fault creep*. Creep thus reduces the rate of strain accumulation, reducing the slip during earthquakes. Hazard models account for creep by reducing the fault area over which a seismic slip occurs, thus reducing the magnitude of ruptures. In addition, a fraction of total slip, known as *coupling coefficient*, takes part during earthquakes and so can be attributed to seismicity. Thus, the coupling coefficient reduces the event rate. In general, fault creep and the coupling coefficient reduce the moment rate.

The rupture rate (λ) of faults is

$$\lambda = \frac{R \cdot \dot{M}_0}{M_0} \tag{1.4}$$

where R is the reduction in moment rate due to creep and coupling. The rupture rate of faults can also be determined directly from the date of occurrence of historical events and paleoseismically observed dates of events. There is, however, significant uncertainty in the estimation of the recurrence rate of ruptures because of limited historical and paleoseismic data, as well as uncertainty in the data itself. The quantification of the uncertainty in the rate helps modelers to determine the epistemic uncertainty in the estimation of hazard results.

Distributed Sources

This section describes how modelers estimate the rates of earthquakes elsewhere, other than on the known faults. A traditional area source model assumes uniform seismic activity within the source zone. The zones are generally developed based on seismic, tectonic, geological, and geodetic data. The model requires estimates of earthquake rates, the slope of the Gutenberg-Richter model (discussed in the following section), depth distribution of earthquakes, maximum magnitudes, style of faulting, strike and dip directions, etc., to define the distributed earthquakes. All these parameters are defined either based on historical earthquakes in the zone or extrapolated from similar zones. Fig. 1.6 (Woessner et al., 2015) is an example of area sources for the Seismic Hazard Harmonization in Europe (SHARE) project. The shear zones in the UCERF2 model (Field et al., 2009) are another example of distributed sources, but in their case the earthquake rates are calculated based on the rate of deformation across the zone, not based on historical seismicity. Finally, background seismicity accounts for random earthquakes in the model that are not characterized on faults or shear zones.

Smoothing of Seismicity

The frequency of occurrence of earthquakes for distributed sources is calculated from the Gutenberg-Richter (GR) relationship as given below:

$$\log(N) = a - bM \tag{1.5}$$

where N is the number of earthquakes exceeding magnitude M, a is the number of $M \geq 0$ earthquakes, and b is the parameter defining the ratio of larger to smaller magnitude

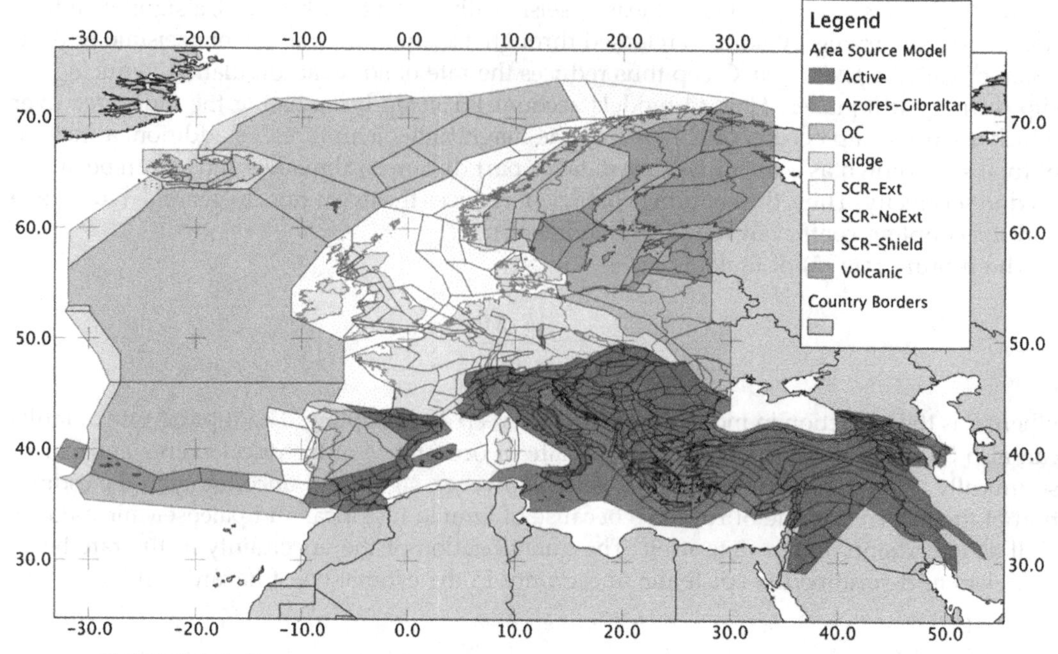

FIGURE 1.6 Tectonic regionalization map overlaying area source zones in SHARE project.

earthquakes. The GR parameters, which are *a* and *b*, are estimated by fitting the historical earthquake catalog to Eq. (1.5). The catalog is generally declustered to remove dependent events such as foreshocks or aftershocks. In addition, modelers assess the completeness of the catalog as a function of location, time, and earthquake size, to prevent the under estimation of earthquake recurrence rates. The larger magnitude earthquakes generally have longer completeness periods. There is uncertainty in the magnitude of earthquakes, particularly those before the 80s, due to the averaging of measurements from seismograph stations. Since different agencies reported different types of magnitudes, e.g., local magnitude (M_L) or body-wave magnitude (M_b), the development of a uniform catalog requires conversion from one magnitude type to another, and this conversion introduces additional uncertainty. The truncation of magnitudes, when reporting the magnitudes, also introduces uncertainties. A detailed discussion on the consideration of all these issues in estimating recurrence rates can be found in the CEUS-SSC report (2012).

Modelers account for such issues in estimating the GR parameters by using a maximum likelihood method (Weichert, 1980). This approach bins the earthquakes into different magnitude ranges to fit the parameters. Gridded sources are used nowadays, for example US National Hazard Map by the United States Geological Survey (USGS), to capture the variation in rates (a) within a zone as an alternative to homogeneous seismicity. The gridded model is developed based on the assumption that the large-magnitude earthquakes are more likely to occur where the smaller magnitude events have occurred more frequently in the past.

Smoothing of Historical Seismicity

To forecast the rates of earthquakes, modelers spatially smooth the rates estimated in the previous section based on historical earthquakes. The approach of smoothing of seismicity falls into two broad categories: (1) *Fixed Kernel* and (2) *Adaptive Kernel*. The 1995 USGS national hazard maps were the first regional scale map, which was developed by smoothing the earthquake rates at regular grids based on a two-dimensional *fixed isotropic Gaussian Kernel* (Frankel, 1995). If there are n_i earthquakes in grid i observed in a declustered catalog, the smoothed value of number of earthquakes, $\widetilde{n}_i$, is the following:

$$\widetilde{n}_i = \frac{\sum n_i \cdot e^{\left(-\frac{\Delta_{ij}^2}{d_c^2}\right)}}{\sum e^{\left(-\frac{\Delta_{ij}^2}{d_c^2}\right)}} \tag{1.6}$$

where Δ_{ij} is the distance between the ith and jth grids and d_c is the correlation distance. The USGS model assumed 50 km correlated distance for the Western US smooth seismicity model. Fig. 1.7A shows the variation of moment rate for the historical catalog in California based on fixed-kernel smoothing (see Field et al., 2014, for details).

The *adaptive-kernel* smoothing approach, based on Helmstetter et al. (2007), varies the size of the kernel based on the nth closest earthquakes. The kernel size can be small in areas where large numbers of historical earthquakes occurred, but large in areas of fewer earthquakes,

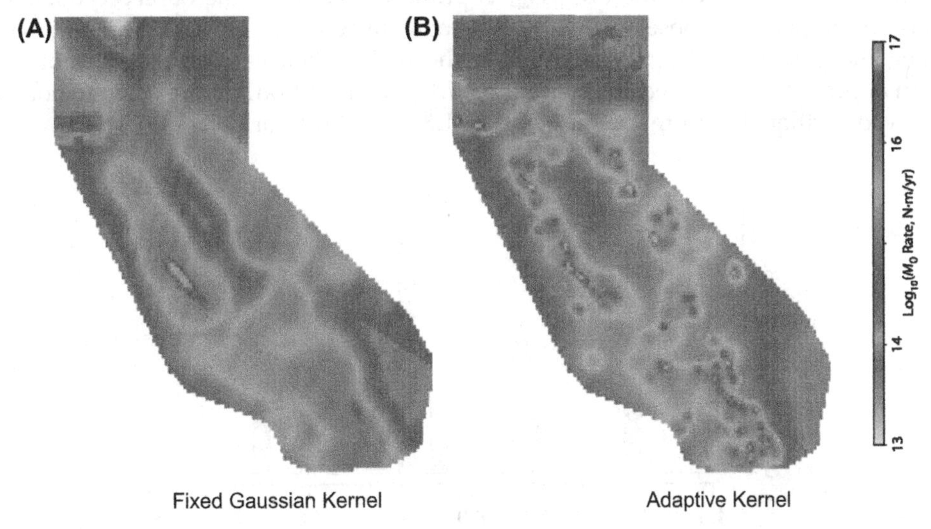

FIGURE 1.7 Comparison of smoothed seismicity maps for moment rates based on different smoothing models in UCERF3 model (Field et al., 2014).

thus spreading historical seismicity over wide areas. The isotropic adaptive Gaussian Kernel, K_d (Helmstetter et al., 2007), is given by the following equation:

$$K_{d_i}(r) = C(d_i)\cdot \exp\left(-\frac{r}{2d_i^2}\right) \quad (1.7)$$

where $C(d_i)$ is the normalizing parameter so that $K_{d_i}(r)$ over an infinite area becomes 1, d_i is the smoothing distance, and r is the distance between an event k and the grid. The smoothing distance d_i associated with an earthquake i is the horizontal distance between event i and the nth closest earthquakes (e.g., UCERF3 has used $n = 8$ for $M \geq 2.5$ earthquakes for smoothing the historical seismicity. See Field et al., 2014, for details). Fig. 1.7B shows the variation in moment rate from fixed-kernel smoothing. The difference in the results between Fig. 1.7A and B from the two different approaches illustrates the epistemic uncertainty in the estimation of background seismicity. Although the smoothing is generally assumed to be isotropic, it is expected that the smoothing should be anisotropic near the faults, introducing additional aleatory uncertainty into the results.

Maximum Magnitude

The calculation of rates based on the GR model requires the information of maximum magnitude (M_{max}) for different seismic zones as the upper truncation point of the GR recurrence curve for the zones. Modelers generally base the M_{max} for a seismic region on observed magnitudes of events in regions throughout the world with similar tectonic characteristics. The maximum magnitude of earthquakes in different nonextended stable continental regions (SCR) as shown in Fig. 1.8 (CEUS-SSC, 2012) shows the variation of M_{max} in a similar tectonic region. Since the recommendation of M_{max} is primarily based on the observed data, which depends on the period of observation, the epistemic uncertainty of M_{max} is quite broad. By assuming the observed M_{max} distribution in the global data as a prior distribution, the Bayesian approach can be used to estimate the M_{max} distribution for a specific region based on observed earthquakes in that region (see CEUS-SSC, 2012; for details).

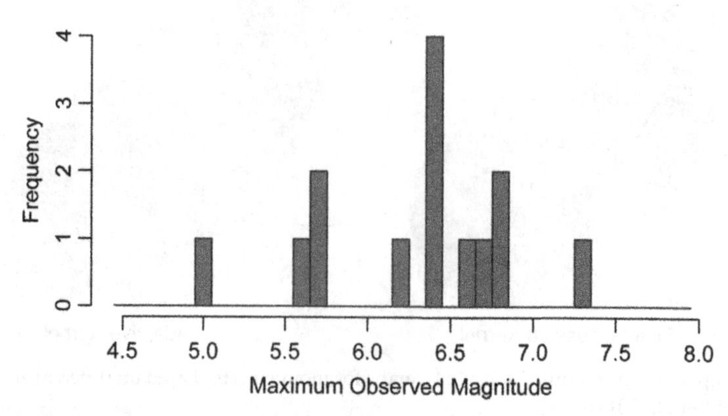

FIGURE 1.8 Distribution of maximum magnitudes as observed in different nonextended tectonic regions in the world (CEUS-SSC, 2012).

Earthquake Probability Model

There are two broad categories of earthquake probability models: (1) time-independent and (2) time-dependent. The time-independent model assumes that earthquakes occur at a constant rate and is represented by a Poisson process. The distribution of times between successive events for a homogeneous Poisson process (random occurrence) is specified by the following probability density function:

$$f_{Exp}(t) = \lambda \cdot e^{-\lambda t} \tag{1.8}$$

where λ is the mean rate of events per unit time (reciprocal of the mean time interval between events). The USGS National Seismic Hazard maps are developed based on this model. The time-dependent model, on the other hand, forecasts relatively large earthquakes on faults at quasi regular intervals based on the elastic rebound theory, as opposed to random occurrence in the time-independent model. Time-dependent models assume that the probability of occurrence of earthquakes on a fault/fault-segment drops significantly after the rupture, following Reid's Elastic Rebound theory. The theory is developed based on the stress renewal model, in which the probability of rupture of a fault drops immediately after the occurrence of an earthquake, because the earthquake releases the tectonic stress on the fault. The probabilities rise subsequently as the stress starts accumulating from tectonic loading. The earthquake rupture occurs when the loading reaches the critical failure thresholds. One of the well-accepted approaches of estimating the interval between the events is based on a physically motivated Brownian Passage-Time (BPT) model (Ellsworth et al., 1999). A detailed discussion on alternative approaches can be found in Fitzenz and Nyst (2015). In the BPT model, the steady rise of the state variable from the ground state, which is the condition just after the rupture of faults, to the failure thresholds is modulated by Brownian motion. The probability density function is defined by

$$f_{BPT}(t) = \sqrt{\frac{\mu}{2\pi\alpha^2 t^3}}\, e^{-\frac{(t-\mu)^2}{2\mu t\alpha^2}} \tag{1.9}$$

where μ is the mean ($=1/\lambda$) and α is aperiodicity of the time interval between earthquakes. The probability density function of BPT for various values of α is shown in Fig. 1.9. It increases to achieve a maximum value at a time greater than the mode of the distribution, and decreases afterward toward an asymptotic value. Thus, a BPT process attains a finite quasi-stationary state in which the failure rate is independent of elapsed time. Since there is limited data rupture sequence for a handful of faults, the epistemic uncertainty in the estimation of μ and α is very high. The uncertainty analysis in hazard or loss calculations should consider this uncertainty.

Ground Motion

Ground-motion models (GMMs) estimate ground-motion intensities for different characteristics of the earthquake sources, propagation path, and site conditions. These models provide the mean and standard deviation of the average of the spectral accelerations, or PGA, or peak ground velocity in two orthogonal directions. The models assume the distribution of

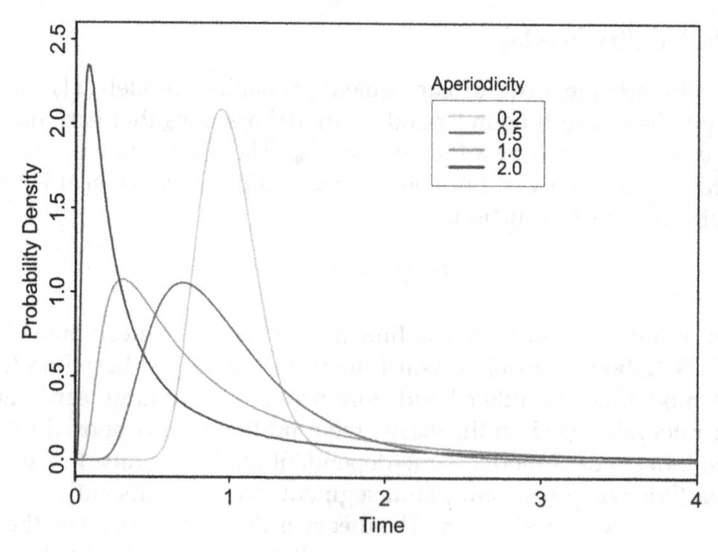

FIGURE 1.9 Probability density functions of Brownian Passage Time model for different aperiodicities.

accelerations to be lognormal. Such models are generally developed using large collections of recorded ground motions, particularly for regions where records are abundant. The models predict ground motion as a function of spectral period, magnitude, source-to-site distance, source characteristics like fault style, site conditions, and many other characteristics. The GMMs fall into three broad categories based on their applicability in different tectonic regions: (1) SCR for shallow crustal earthquakes, such as Central and Eastern United States (CEUS); (2) active continental regions, for shallow crustal earthquakes, such as western United States; and (3) subduction zones (including both interface and intraslab or inslab), such as the Japan Trench. Different researchers have developed a number of models for each of the tectonic regions based on recorded motions where the records are abundant, e.g., in WUS, or based on simulated records from different approaches, e.g., full waveform simulations by Somerville et al. (2001), for regions where records are limited for large-magnitude earthquakes, e.g., in CEUS. These different models predict different ground motions due to differences in the functional forms, regression techniques, data selection, and model parameterizations, thereby capturing epistemic uncertainty in the prediction of ground motion.

The estimation of the intensities varies significantly from model to model as shown in Fig. 1.10A from different models for active regions and in Fig. 1.10B for different tectonic regions. The intensities also depend significantly on the underlying soil conditions. The shear-wave velocity at sites provides an indication of whether the expected shaking from earthquake ruptures can be higher or lower than shaking at bedrock sites. For instance, the average shaking at sedimentary basins (with low shear-wave velocity) is expected to have intense amplification relative to bedrock motion. The shear-wave velocity (V_S) has long been known to be an important parameter for evaluating the dynamic properties of soils. The average shear-wave velocity in the top 30 m, based on travel time from the surface to a depth of 30 m, is known as V_{S30}. Building codes use V_{S30} to separate sites into different

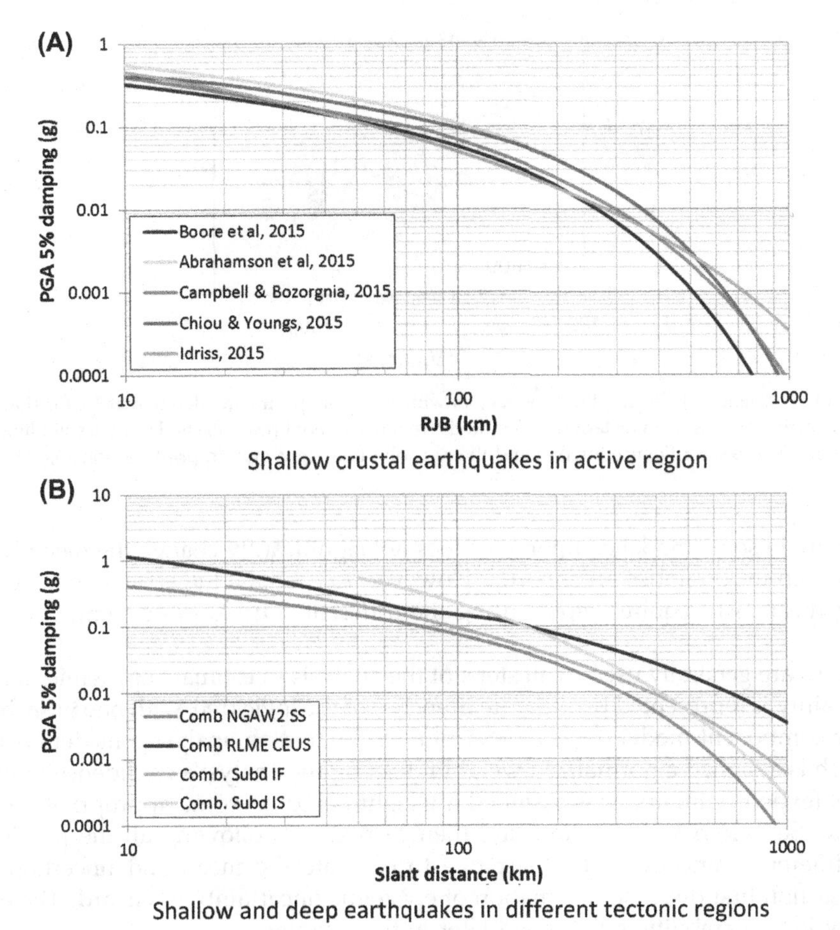

FIGURE 1.10 Comparison of the median peak ground acceleration for M8 earthquakes for different tectonic regions.

classes for engineering design, with the expectation that sites in the same class will respond similarly to a given earthquake. The variation of soil amplification or site factors with V_{S30} is shown in Fig. 1.11. The figure illustrates that the aleatory uncertainty in the amplification factors is high (coefficient of variation ≈ 0.4), but the narrow confidence band suggests relatively small epistemic uncertainty in the estimation of amplification when the shear-wave velocity is known at building sites.

EPISTEMIC UNCERTAINTY IN SEISMIC HAZARD RESULTS

Typical seismic risk calculations estimate the mean risk. Therefore, the calculations do not consider the epistemic uncertainty of modeling different components or the parameters defining those models (i.e., treatment of uncertainty is not exhaustive). As the consideration

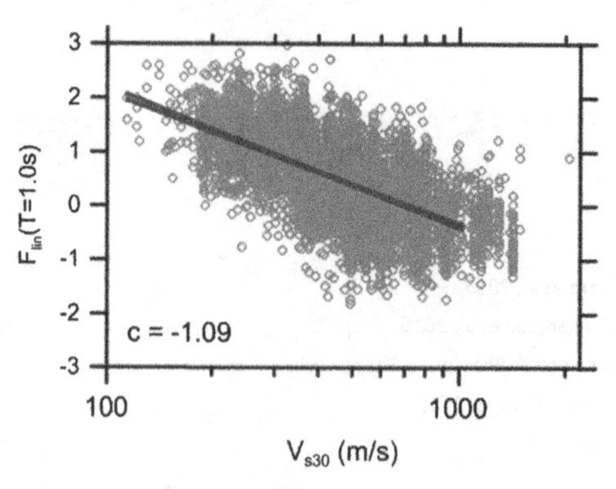

FIGURE 1.11 Variation of site amplification factors with VS30 for spectral acceleration at 1 s for shallow crustal earthquakes in active regions like Western US without the removal of nonlinear effects. The blue solid line (dark gray in print versions) is ±95% confidence interval and the red solid line (light gray in print versions) is the median fit.

of uncertainty in some model components does not significantly change the mean hazard results, thus such uncertainties are ignored. However, to estimate the epistemic uncertainty in estimating mean risk requires representing the different sources of epistemic uncertainties properly.

Logic trees are generally used in all steps of the analysis to estimate epistemic uncertainty. The uncertainty is represented by separate branches of the logic tree, with one branch for each alternative component model or parameter of a model that an analyst considers reasonable. Each branch is assigned a normalized weight reflecting the analyst's confidence in the choice of a model (expert opinion), or is assigned a weight based on some appropriate probability distribution. Seismic risk calculations are then carried out following all the possible paths through different branches of the logic tree to estimate the mean and uncertainty in the risk. This section first discusses estimation of epistemic uncertainty in hazard. The epistemic uncertainty in vulnerability is discussed later in this chapter.

Kulkarni et al. (1984) first introduced logic trees in seismic-hazard analysis to capture the uncertainty associated with the inputs to the analysis, and this has since become a standard feature in PSHA (e.g., Petersen et al., 2014; Stirling et al., 2012). Fig. 1.12 shows an example of the earthquake source-model logic tree that was developed for hazard calculations of California as part of the USGS National Seismic Hazard Model (NSHM). Construction of logic trees entails consideration of different types of epistemic uncertainties: (1) model uncertainty by considering multiple possible models for a hazard component (e.g., magnitude-area relationship) and (2) parameter uncertainty by considering a suite of parameters of any component of hazard model (Bommer, 2012). Different parameter values can be estimated based on expert opinion (e.g., a range of maximum magnitudes for background seismicity). Alternatively, different values may arise from uncertainty in the sample size of data used for estimating a parameter (e.g., uncertainty in the earthquake recurrence rate of ruptures of a fault from a limited number of events based on paleoseismic and/or historical data). Epistemic uncertainty can be represented either through a continuous distribution (e.g., a lognormal distribution of recurrence rate), or through a discrete distribution (e.g., weights of different ground-motion models).

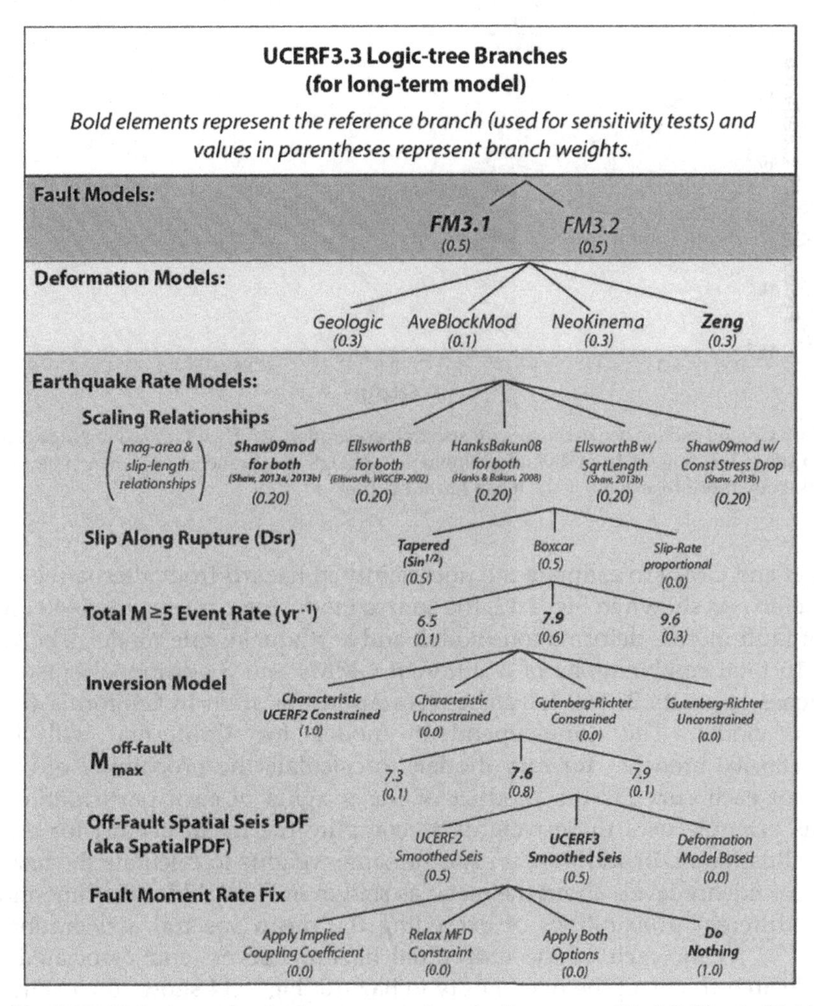

FIGURE 1.12 UCERF3 logic tree for the long-term earthquake forecast model of California (Field et al., 2014).

After establishing branches of a logic tree by identifying competing models or a range of appropriate parameter values, the next step is to assign weights to each branch. As an example, Fig. 1.12 shows the logic tree and the weights for each of the branches of the recent California model. Some researchers consider branch weights to reflect the relative merit of the models or parameters being considered (e.g., Abrahamson and Bommer, 2005). An example would be to assign weights from 1 to 10 to the models and then normalize the weights so that they sum to one. Other researchers may interpret the weights as probabilities (e.g., McGuire, 2004) and so derive weights from a probability distribution.

As a worked example of a logic tree, this section explores the uncertainty in hazard for a site in Los Angeles, California, using the 5-Hz spectral acceleration ground motion with a 2% probability of being exceeded in 50 years (or 2475-year return period). The hazard curves are computed for all the unique combinations of logic-tree branches in the UCERF3

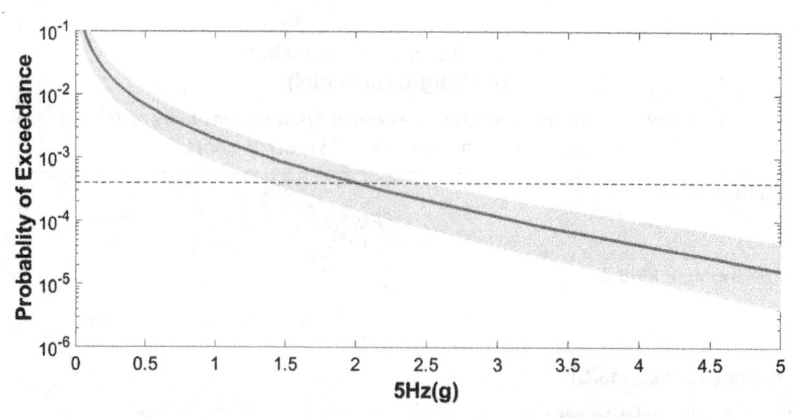

FIGURE 1.13 Ground-motion hazard curves for spectral acceleration at 5 Hz with 5% damping are shown by the cyan lines for a site in Los Angeles from different combination of logic-tree branches as shown in Fig. 1.12. The solid red curve (dark gray in print versions) is the mean hazard curve.

source model and GMM to estimate the uncertainty in hazard from alternative models and their parameters. As shown in Fig. 1.12, the source model logic tree yields 1440 unique combinations of fault model, deformation model, and earthquake rate model. The GMM logic tree yields 15 total combinations of 5 different GMMs and 3 additional epistemic uncertainty branches to create 21,600 hazard curves at any location in California for the time-independent model. The time-dependent model for California will have total $21{,}600 \times 4 = 86{,}400$ branches for four models to calculate the probability of earthquakes. The weight of each curve is the product of the weights of each participating logic-tree branch. This example uses these weights to compute the mean hazard for a site in Los Angeles for illustration. In addition, it uses the same weights to calculate the hazard curves at different confidence levels about the mean as shown in Fig. 1.13 for the time-independent model. The different probabilities of exceeding the mean spectral acceleration at 2% in 50 years ($S_a^{M@2\%/50}$) for each of the individual hazard curves, and associated weight of each, provide an estimate of the uncertainty in hazard. Fig. 1.14 shows the resulting hazard uncertainty for the example site in Los Angeles. The 95% confidence band of the return period of exceeding the $S_a^{M@2\%/50}$ is 1800–4000 years (these are the inverse of exceedance probabilities). The distribution of probability of exceedance is quite smooth, which suggests that a large number of different models influence hazard at this location, with no one model dominating the hazard. This uncertainty should be considered for proper assessment of epistemic uncertainty in seismic risk of buildings.

BUILDING VULNERABILITY

In 1985, Applied Technology Council (ATC) published a suite of vulnerability functions based on expert opinions (ATC-13, 1985). These functions provide loss estimates based on the MMI for different facility classes. Recently, the Global Earthquake Model project also followed a similar approach for developing vulnerability functions (Jaiswal et al., 2013).

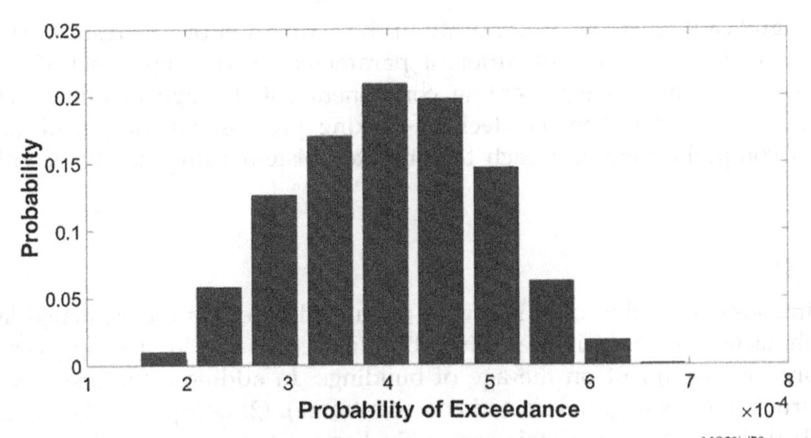

FIGURE 1.14 Uncertainty in the probability of exceeding the spectral acceleration of $S_a^{M@2\%/50}$, which is the S_a at 2%/50 year exceedance probability of the mean hazard curve, for a site in LA.

Although this approach can be used to develop vulnerability functions relatively quickly and easily without any loss data or mathematical model, the results are found to be not very accurate. Nevertheless, this approach can be useful when resource is limited.

Vulnerability functions based on engineering investigations, data collection, and field studies were first developed by Professor Karl Steinbrugge developed for estimating earthquake damage to buildings (Steinbrugge, 1982). These relationships provided to the insurance industry a firsthand insight to the damageability of buildings in the 1990s. Later on, claims data from the Northridge earthquake (1994) helped the industry to develop more robust vulnerability functions for low-rise, single-family buildings for high building seismic zones in California.

The FEMA in 1999 first published a comprehensive methodology to calculate earthquake damage of model buildings analytically, to estimate losses from earthquakes (HAZUS, 2003). Later on, the Consortium of Universities for Research in Earthquake Engineering project (2001) developed an assembly-based vulnerability (ABV) function for single-family, wood-frame buildings. This is a significantly improved analytical approach over the HAZUS. Recently, researchers have focused on developing analytical vulnerability functions based on detailed nonlinear analysis of buildings. For example, the Pacific Earthquake Engineering Research (PEER) center considered a 3D model of 40 + story tall buildings (Shome et al., 2015). In addition, ATC-58 (2011) has developed a number of approaches, simple as well as very detailed and complex, for loss assessment of buildings. Many researchers now use loss assessments by floor (story) within buildings to capture accurately the variation in loss over the height of buildings. This approach is particularly important for mid- to high-rise buildings as building damages vary nonlinearly with height. Researchers analyze the nonlinear response of buildings for large number of ground-motion records to develop the vulnerability functions. To illustrate this approach, this section summarizes a framework developed by Jayaram et al. (2012). This framework assessed losses by story by taking into account the correlation of response, damage, and repair costs of different components of buildings by story height.

There is significant epistemic uncertainty in the estimation of loss results. These uncertainties arise in the estimation of different parameters of damage functions, or due to the use of different models representing components of damage functions. This uncertainty should be considered in the decision-making process. Jayaram et al. (2012) have developed a comprehensive approach to estimate epistemic uncertainty of vulnerability functions.

Methodology

The seismic loss estimation of buildings is calculated based on the estimated lateral load resistance characteristics of buildings. Since building codes are improving over the years, load resistance is dependent on the age of buildings. In addition, the resistance depends on (1) construction class (e.g., reinforced concrete frame), (2) occupancy class (e.g., residential), and (3) region (e.g., low seismic regions in Eastern US). Insurance loss estimation for a portfolio of buildings uses this type of generic information by grouping buildings with similar load resistance capacities. The loss estimation of individual buildings (e.g., ATC-58, 2011), on the other hand, uses the detailed building information to accurately estimate the lateral load resistance capacity for seismic loss calculations. This section discusses the seismic loss estimation approaches that can be easily applied for both portfolios and individual buildings. Although there are many different approaches, the vulnerability functions for buildings derived using an analytical approach are most useful for insurance seismic risk assessments because of the lack of high-quality insurance claims data. Therefore, this section highlights analytical approaches for vulnerability function development.

Earthquakes not only cause damage to structural and nonstructural components of buildings, but also damage contents, such as furniture and movable partitions, and cause the closure of buildings. Such closure incurs losses to the occupants because of temporary rents or hotel stays (more formally called additional living expenses) or loss of income to businesses (called business interruption loss).

Loss Distribution

Since there is significant uncertainty (aleatory and epistemic) in estimating building losses at a given intensity of ground motion, vulnerability functions provide the loss distribution parameters as a function of ground-motion intensities. Modelers generally assumed that a parametric distribution model can adequately represent the loss distribution. Thus, following ATC-13 (1985), a Beta distribution is widely used for building loss calculations. The parameters of the Beta distribution can be estimated from the mean and variance of the sample loss by Method of Moments (Benjamin and Cornell, 1970). The insurance claims data, however, show that combining the Beta distribution with Dirac-delta functions at 0% loss and 100% loss to model the probabilities of zero and complete loss improves the fit to the observed loss distribution. In addition, insurance companies generally pay to completely replace a property (i.e., incur 100% loss) beyond a particular threshold of loss (e.g., 60% loss). Hence, this distribution requires three additional parameters, which are probabilities of 0% and 100% loss as well as the threshold of loss, in addition to the conventional mean and variance for parameterization. Fig. 1.15 illustrates the shape of the resulting five-parameter distribution for losses. This type of distribution may be more appropriate to model the insurance losses.

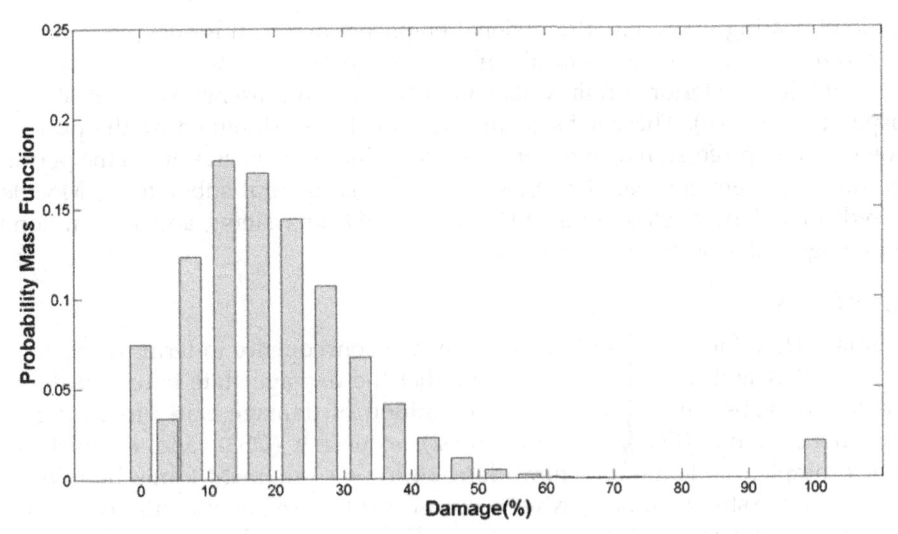

FIGURE 1.15 Five-point beta distribution of building losses at a given intensity of ground motion.

The methodology to estimate the five parameters of the distribution from loss data can be found in Shome et al. (2012a,b).

Analytical Approach of Development of Vulnerability Functions

Since there is limited building damage data from earthquakes, vulnerability functions for a majority of construction classes in most of the regions in the world can be developed only from an analytical approach. Jayaram et al. (2012) have developed a systematic simulation approach based on the PEER Performance-Based Earthquake Engineering (PBEE) loss assessment framework to develop building-specific vulnerability functions. The approach estimates the total building loss by using the building response from detailed nonlinear time-history analysis of buildings. The estimation captures the correlation of losses between different components to improve the estimation of variability in losses. The approach also accurately represents the epistemic and aleatory uncertainties of the random variables, such as ground motions, structural response parameters, loss costs, etc. to quantify the uncertainty in the final seismic loss estimates.

The analytical loss calculation primarily requires estimation of three random variables: (1) repair cost or cost due to demolition associated with a damage state, (2) damage state given a demand parameter (EDP), and (3) EDP given an intensity of ground motion. A brief description of these random variables is given below. Details of these variables can be found in HAZUS (2003).

BUILDING COMPONENTS

To carry out the seismic loss calculations, modelers consider building components into two broad categories or subsystems: (1) structural and (2) nonstructural. This paper illustrates loss estimation for a group of components or subsystems. The structural subsystem (SD), which resists gravity, earthquake, wind, and other types of loads, includes components like columns, load-bearing walls, etc. Professional engineers design this subsystem for building specific conditions. The nonstructural subsystem comprises two categories: (1) drift-sensitive (NSD) (e.g., exterior curtain walls) and (2) acceleration-sensitive (NSA) components (e.g., suspended ceilings). These subsystems are usually not designed by the professionals. The involvement of professional engineers helps to reduce the uncertainty in the performance of structural subsystem compared to those of the nonstructural subsystems. Modelers use drift to predict the damage states of the SD and the NSD subsystems, and acceleration to predict the damage states of the NSA subsystem.

DAMAGE STATES

The damage state for the subsystems represents a consequence in terms of repair and the severity associated with that state. Models predict the damage state based on the fragility functions for the subsystem. The common definition of damage states following HAZUS (2003) are none, minor (DS1), moderate (DS2), extensive (DS3), and complete damage (DS4). For example, moderate structural damage for steel-moment-frame buildings entails the yielding of some steel members, exhibiting observable permanent rotations at connections and major cracks of a few welded connections. The moderate damage of the NSD components or subsystem represents large and extensive cracks of the partition walls, requiring repair and repainting and replacement of gypsum board or other finishes. The moderate damage state of the NSA subsystem or component represents the falling of large number of tiles of suspended ceiling, and disconnection or buckling of the support framing (T-bars) at few locations; leaking of pipes at a few locations for Electrical-Mechanical Equipment; and realignment of elevator machinery.

REPAIR COSTS

The monetary loss from earthquakes is associated with the repair of different subsystems as well as the building replacement cost associated with demolition and collapse. The uncertainty in repair cost arises primarily due to a lack of information about the subsystem. The cost of repair works is generally significantly higher than the cost of new construction, and this is more significant for structural subsystems than the nonstructural subsystems. To estimate the repair cost, models generally assume that the insurer will pay the cost of repair or replacement of the damaged buildings with the materials of like kind and quality, without any deduction for depreciation, which is called the *replacement cost method* for valuation.

PEER-PBEE-Based Analytical Approach

Jayaram et al. (2012) have developed a generic procedure for developing the building vulnerability functions by following a systematic simulation approach. The steps are: quantifying ground-motion hazard using a vector of spectral accelerations; predicting building response parameters such as story drifts, floor accelerations, and residual drifts under the specified hazard; estimating structural collapse and demolition; and calculating story-wise losses based on the responses. The total building loss is the sum of the story losses. The procedure captures the effects of epistemic and aleatory uncertainties in the random variables, such as ground motions, structural response parameters, loss costs, etc. to quantify the uncertainty in the building loss results.

The mathematical framework for the development of the mean vulnerability function based on this approach is given below:

$$E(L|S_a(T_0)) = \int\int\int\int L|dG(L|DM)||dG(DM|EDP)||dG(EDP|S_a(T)||dG(S_a(T))|S_a(T_0)| \tag{1.10}$$

where L is the building loss, $S_a(T_0)$ is the spectral acceleration at a reference period, and $E(L|S_a(T_0))$ is the mean loss conditioned on $S_a(T_0)$; $dG(L|DM)$ denotes the derivative of the probability of exceedance of the building loss given a damage measure (DM); $dG(DM|EDP)$ is the derivative of the probability of exceedance of the DM given an EDP (e.g., story drift ratio); $dG(EDP|S_a(T))$ is the derivative of the probability of exceedance of the EDP given a vector of spectral acceleration, $S_a(T)$); and $dG(S_a(T)|\ S_a(T_0))$ is the derivative of the probability of exceedance of spectral accelerations at multiple periods, $S_a(T)$, given $S_a(T_0)$. The reference period, denoted here T_0, is not necessarily the fundamental period of the analytical model of a structure. The fundamental period is a good choice when the structural performance is reasonably elastic and dominated by the first mode (i.e., for low-rise buildings), but it is not a good choice for tall buildings whose performance is nonlinear and dominated by multiple frequencies. Although the equation predicts structural response based on a vector of spectral accelerations, the final vulnerability is a function of a single intensity measure, which ensures simplicity of use without sacrificing any accuracy.

The variance of the vulnerability function can be estimated as follows:

$$Var(L|S_a(T_0)) = \int\int\int\int (L - E(L|S_a(T_0))^2|dG(L|DM)||dG(DM|EDP)||dG(EDP|S_a(T))| \times |dG(S_a(T))|S_a(T_0)| \tag{1.11}$$

The above integrals are evaluated by using Monte Carlo simulation. The Monte Carlo simulation approach involves simulating all the random variables in Eqs. (1.11) and (1.12) (which are $S_a(T)$, EDP, DM, and L) and then computing the mean and the variance of L for a wide range of $S_a(T_0)$ values. The Monte Carlo simulation approach provides flexibility and allows consideration of accurate but complex models for the different random variables.

The EDP variable values required in this approach for loss calculations are the interstory drift ratio (SDR), peak floor accelerations (PFA), and the residual drift ratio (ResDR). The model uses the SDR values to predict the damage to SD and NSD components at each story level, the PFA values to predict those for the NSA components, and ResDR to determine if a building would have to be demolished after an earthquake. The EDPs are first calculated based on nonlinear time-history analysis of buildings for a suite of ground-motion records.

During earthquakes, there is a possibility of collapse of structures due to excessive lateral drifts and deterioration of strength leading to loss of the gravity load resistance of structures. To evaluate this possibility, this model uses a collapse probability curve, which provides the probability of collapse of buildings as a function of $S_a(T_0)$. The collapse probability curve is developed using a logistic regression (see Shome and Cornell, 2000, for details) and can be expressed as follows:

$$\ln\left(\frac{p}{1-p}\right) = \beta_0 + \beta_1 \cdot S_a(T_0) + \varepsilon \tag{1.12}$$

where p = probability of collapse; β_0 and β_1 are the regression coefficients; and ε is the regression error.

Jayaram et al. (2012) showed that a more efficient and accurate loss calculations involves estimating the mean and the standard deviation of the EDPs at each story by fitting the nonlinear analysis results to a vector of spectral accelerations $S_a(T)$ than simulation of buildings responses by nonlinear structural analyses of buildings for different ground-motion records. This method involves jointly simulating EDPs for the entire building based on the correlation between different EDPs at the same story and between EDPs at different stories. The extent of these correlations is computed using the regression residuals. It is generally assumed that the EDPs follow the multivariate lognormal distribution.

Damage Measure and Repair Cost

To estimate losses of different subsystems of buildings, the framework presented in the previous section to estimate the repair costs associated with different damage states from the fragility functions. In this approach, the repair costs conditioned on the DM variable in Eq. (1.11) are modeled as a multivariate lognormal random variable with the median and dispersion and correlations of repair costs between different damage states over the height of buildings. The repair costs for different subsystems across different damage states are expected to be correlated because of common factors involved in repairs, such as the same contractor hired for the repairs and the same material used for repairs of different subsystems across different stories.

The analytical model then calculates the total loss by summing up the sample losses of different subsystems at each story and normalizing the resulting total by the building value to get the damage ratio. To illustrate the resulting range of damage ratios, Fig. 1.16 shows a scatter plot of damage ratio results as a function of $S_a(T_0)$ based on simulation of a 20-story modern steel building in Los Angeles, California. The final step in creating the analytical building vulnerability functions is to calculate the mean and CoV of the damage ratios by binning the data points into $S_a(T_0)$ bins. By estimating these values a number of times by first sampling the building parameters from the epistemic uncertainty, the modeler can then derive the overall epistemic uncertainty in building vulnerability.

SECONDARY HAZARD AND SUBPERILS

In addition to shake damage, earthquakes can also cause landslides due to ground movement or liquefaction due to the loss of strength in the soil underneath structures. These phenomena are secondary hazards associated with earthquakes. Sometimes these secondary hazards can contribute significantly to the overall losses for an earthquake, e.g., damage from liquefaction in the 2011 Christchurch earthquake. These secondary hazards are

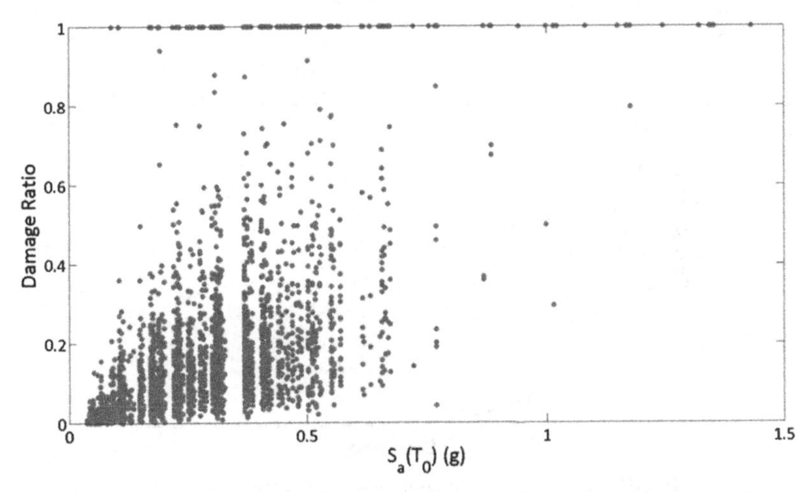

FIGURE 1.16 Simulated damage ratios of a modern 20-story SMF in Los Angeles as a function of $S_a(T_0)$.

high-gradient hazards, i.e., hazard changes rapidly for certain parameters (e.g., slope of ground for landslide or soil strength for liquefaction) over short distances. Since there is limited site-specific information for insurance loss calculations, the uncertainty in loss estimation is even higher when the significant fraction of losses is from secondary hazard.

Earthquakes can also cause fires and tsunamis, resulting in significant fire damage or water damage due to tsunamis. These earthquake subperils need to be considered to improve the seismic loss estimation. The severity of subperil is dependent on some of the characteristics of earthquakes, which may not be very important for shake losses. For example, PGA is important for number of ignitions for fire damage or slip distribution on fault during earthquakes for tsunami. In addition, the loss from subperils depends on many factors other than primary building characteristics, for example, availability of fire engines for fire damage following earthquakes or presence of levees or other defense systems for tsunami. The coefficient of variation of damage (CoV) for these subperils is very high and so when these damages are added to the shake damage, the CoV for total loss increases significantly, particularly when the contribution to total loss is significant as observed in the 2011 Tohoku earthquake. Since insurers do not capture all the required information for accurate loss calculations for secondary hazards or subperils, the epistemic uncertainty in insurance loss for both secondary hazards and subperils are high.

METHODOLOGY FOR LOSS CALCULATIONS

As discussed before, the distribution of a building loss for a given event is commonly modeled using the two parameter Beta distribution or using the improved five-parameter distribution as illustrated in Fig. 1.15. In addition, modeling the loss distribution of a portfolio of buildings also requires evaluating the spatial correlation of loss between two locations. The correlation model can be a function of the separation distance between the two locations and may also depend on other parameters. Ignoring or inaccurately representing the correlation of damage between

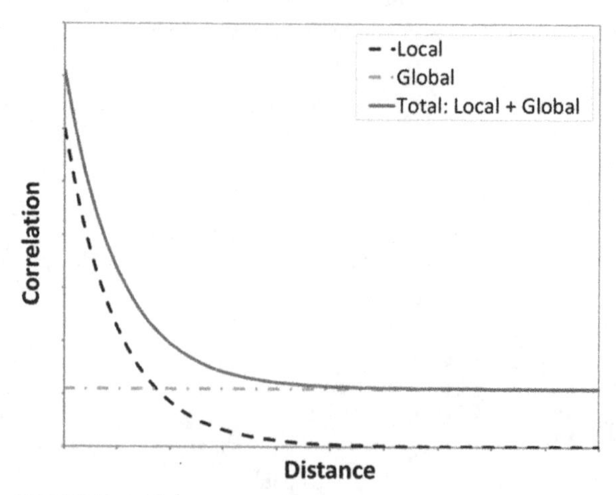

FIGURE 1.17 Different components of correlation model.

locations has a significant impact on the standard deviation of loss of portfolio of buildings, and thus affects the losses at different return periods. However, current industry practice is to either completely ignore these correlations or use a simplified correlation model.

A generalized correlation model can have two components (Fig. 1.17): (1) a constant global component and (2) a distance-dependent local correlation component. The distance-dependent local component can be observed in the claims data and reflects the fact that buildings closer to each other suffer similar deviations in the losses from their predicted means (i.e., if one building has above-average loss, its neighboring building tends to have above-average loss due to similarities in construction, site conditions, environmental conditions, etc.). The global component arises from loss deviations that are common for all the sites during a given event (e.g., higher stress drop).

To carry out a loss assessment, it is necessary to first simulate catastrophe events that can be expected in the future together with their annual rates, and secondly estimate the losses due to each of these simulated events. The analytical approach provides an estimate for the first two moments of the losses. Since the loss distribution is not normal, an analytical solution for the estimation of aggregate loss distribution for portfolio of buildings is not possible. Therefore the analytical framework for catastrophe risk assessment makes an assumption of aggregate distribution of loss (e.g., Beta distribution). In contrast, a simulation approach can provide an accurate estimate of the distribution of losses for the portfolio of buildings.

Catastrophe risk assessment considers the possibility of different event characteristics (size, fault mechanism, etc.) that may occur in the future at different locations. Each event of different size (e.g., category 1 hurricane, magnitude 5 earthquake, etc.) occurs with a specific frequency. The smaller/minor events are typically more frequent than the larger/major events. While developing catastrophe models, a large set of diverse events is first compiled based on the random variables or random fields representing different characteristics of a catastrophe (e.g., central pressure for hurricane or magnitude for earthquakes). The list of events is known as the stochastic catalog. Each of these events occurs at an annual rate based on empirical evidence and analytical models.

LOSS RESULTS AND THE UNCERTAINTY

The vulnerability functions developed as described in previous sections are used to carry out seismic risk assessments for buildings. Models integrate these vulnerability functions with hazard curves to compute the uncertainty in AAL and the return period losses. As discussed, the methodology developed by Jayaram et al. (2012) can be used to account for epistemic uncertainties in the vulnerability functions by developing a suite of functions. The losses are then sampled from these functions to compute the epistemic uncertainty in loss costs and many other loss parameters. Fig. 1.18 shows, for illustration, the normalized 64% confidence interval of the AAL and the 500 yr return period loss of a modern 20-story steel moment-frame building in Los Angeles (see Jayaram et al., 2012, for details). The relative extent of the contributions of the epistemic uncertainties in vulnerability and hazard to the total epistemic uncertainty depends on whether the loss assessment is carried out for a single building or a portfolio of buildings. The result shown here is an example of building-specific seismic risk assessment based on analytical vulnerability functions. For a single building, as seen in the figure, the vulnerability function uncertainty has a larger contribution to the total uncertainty.

In contrast, the impact of vulnerability function uncertainty for portfolio of buildings depends on how the buildings are distributed geographically. If the portfolio of buildings is distributed over a large geographical region, and the characteristics of the buildings are different (e.g., different construction class), the uncertainty in building vulnerability across the portfolio of buildings will be mostly independent and average out. However, the hazard uncertainty is highly correlated across multiple locations and will not average out. Therefore, for a portfolio, hazard uncertainty will have a higher contribution to the total uncertainty. The sensitivity of losses of a well-distributed portfolio of buildings in Los Angeles, California, to different ground-motion models relative to the results based on the average Next Generation Attenuation for Western US version 1 (2008) relations (NGA-W1) (http://peer.berkeley.edu/products/nga_project.html) is shown in Fig. 1.19 (Molas et al., 2012). The results show that the change in loss results from the base model is within 10% from the average model, while the changes for the models with additional epistemic uncertainty in ground motion range from 35% to 70%. The subsection on ground-motion discusses the drivers of differences in the results from different models and the additional epistemic uncertainty.

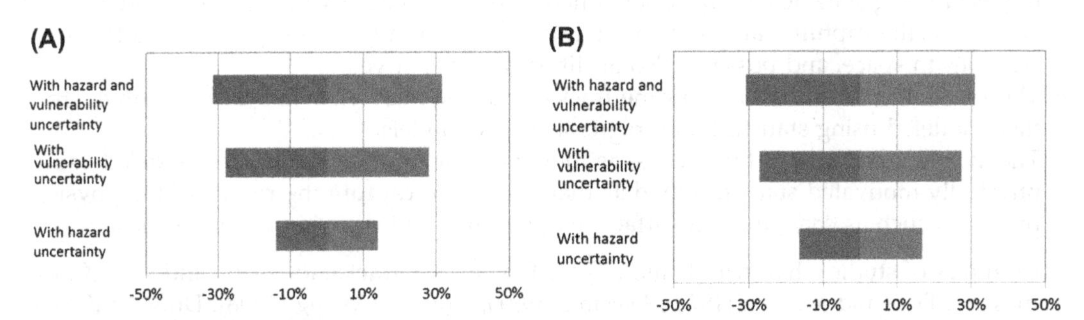

FIGURE 1.18 Impact of different sources of epistemic uncertainties in the loss results of a modern 20-story steel moment frame building in Los Angeles (Jayaram et al., 2012). (A) Average annual loss and (B) 500-year loss.

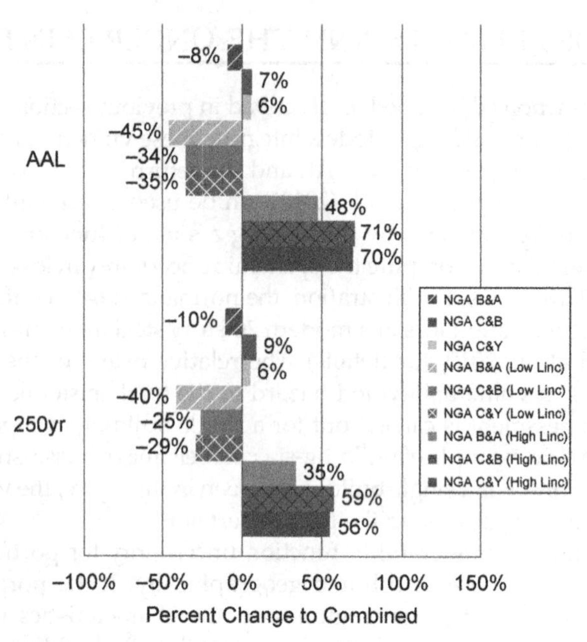

FIGURE 1.19 Sensitivity of the AAL and 250-year loss for a well-distributed portfolio in Los Angeles county for different NGA-W1 ground-motion models (Molas et al., 2012).

OVERVIEW OF A TYPICAL HURRICANE CATASTROPHE MODEL

A typical state-of-the-art hurricane catastrophe model consists of a large number of components that model all aspects of hurricanes that are relevant to predicting likely damage distributions, from the early stages of their life cycle over the ocean, to their landfall, impact on property, and triggering of insurance payouts. We now describe each of the components in turn:

- Locations of hurricane genesis (or formation) over the Atlantic Ocean are typically modeled using statistical models, calibrated to historical observations. Such models would usually capture variations in the probabilities of hurricanes forming at different locations in space, and possibly also at different times of year.
- The speed and direction of hurricane tracks, and how they vary in space and time, are then modeled using statistical autoregressive type models.
- The intensity and size of hurricanes, as they move along their tracks, are modeled using physically motivated statistical models that attempt to capture the results of the physical processes such as changing sea surface temperatures and their impact on hurricanes.

Examples of studies that include models for the genesis, tracks, intensity, and size of hurricanes are: Fujii and Mitsuta (1975), Darling (1991), Chu and Wang (1998), Drayton (2000), Vickery et al. (2000) and Hall and Jewson (2007).

- The transitioning state of hurricanes is modeled to determine whether storms remain purely tropical or whether and when they gradually start to change into extratropical disturbances, due to interactions with midlatitude weather systems (Loridan et al., 2015a,b).
- The landfall rates and intensities may be adjusted, in some way, to match historical landfalling rates and intensities on a regional basis.
- The filling rate of hurricanes over land is modeled using models that capture the probability that storms will maintain their intensity far inland, fill rapidly, and decay, or something in between.
- Lysis models determine whether and when storms die, either over the ocean, or over land.
- Overall frequencies of hurricanes are often modeled with the Poisson distribution, but other distributions are also sometimes used, such as negative binomial.
- Effects related to the clustering of hurricanes in space and time, as seen in certain seasons, particularly 1968 and 2004, are sometimes included in models, although the largest annual losses are generally driven by single large hurricane losses rather than by such clusters, and so such effects are not of primary importance.
- The frequency of hurricanes can be modeled as constant in time. Such models are known as Long Term Rates (LTR) models, and in these models the frequencies are typically estimated using all available data back to 1900 for landfalls or 1950 for basin hurricanes. Alternatively the frequencies can be modeled as varying in time to capture the decadal time-scale fluctuations that are observed in hurricane numbers. These models are known as Medium Term Rates (MTR) models, and are based on analysis of these observed fluctuations, and their relationships with varying sea surface temperature and climate change (Jewson et al., 2007).

Combined together, these components can then be used to simulate a large number of years of hurricane activity[1].

A typical simulation might consist of 100,000 years of hurricanes. With hurricanes making landfall at a rate of roughly 1.7 storms per year on average in the US, 100,000 years simulation of US risks would contain roughly 170,000 storms. This many years of simulation, or similar, is typically considered necessary to fill in the gaps in the historical hurricane record in a reasonably complete way, to achieve relatively smooth results in space, and to achieve relatively good convergence of the loss results from the model at regional level.

Having generated a large set of simulated storms, a hurricane cat model then generates hazard values for each of the perils related to hurricane: wind, waves, surge, rainfall, and flood.

To model wind damage, wind-field models create plausible overland wind fields as a function of the simulated track, location, forward speed, size, and intensity. These wind fields are then adjusted to account for boundary layer effects and surface roughness, based on a representation of the land surface.

[1]One might wonder whether some of all of these steps could be achieved together in a single simulation of a numerical model. In fact, the current generation of numerical models is still several orders of magnitude too slow for such simulations to be useful in this respect.

To model waves and surge, over-ocean wind fields are created and used to drive hydrodynamic models of the ocean that can simulate wave heights and storm surge. Over the ocean, these are used to calculate wave damage to offshore platforms. At the coast, models of sea defenses and hydrodynamic models of inland inundation due to storm surge are used to calculate overland water depths.

To model rainfall, physically motivated statistical rainfall models are used to create realistic rainfall footprints based on the storm characteristics, including a number of the physical processes that affect rainfall such as interaction with orography, rain-drift, moisture supply, and so on (Grieser and Jewson, 2012).

To model flood, simulated rainfall is used to drive rainfall-runoff models, and hydrodynamic models for inundation.

This modeling process creates, for each simulated storm, wind, surge, wave, rainfall, and flood hazard footprints, discretized at an appropriate resolution. The resolution used would typically be higher for surge and flood than for wind, to capture the greater variation of flood depths over small distances.

These hazard footprints, whether for wind, surge, wave, rainfall or flood, are then used to calculate damage via vulnerability curves. These curves model a distribution of possible damage percentages as a function of the hazard. Different vulnerability curves are used for different buildings, typically at a great level of granularity. In the RMS model, for instance, buildings are primarily classified as a function of construction type, year built, number of stories, and occupancy. They are then subsequently subclassified, using "secondary modifiers", according to attributes such as roof type, roof age, and floor area. Each combination of primary classification and subclassification is given a different vulnerability curve. This ultimately leads to many thousands of vulnerability curves for different building types. Percent damage values from the vulnerability curves are converted into actual damage using estimates of the location, characteristics, and value of each building.

As the damage as a function of given hazard is modeled probabilistically (since vulnerability curves give a distribution of possible damages, not a single damage value), each hurricane can be simulated multiple times, to give different losses each time. This captures the fundamental randomness involved in the processes that lead to building damage. Each simulated loss is known as a sample. The uncertainty related to the number of years of simulation, and which storms occur, is known as primary uncertainty, while the uncertainty related to the number of samples simulated for each storm is known as secondary uncertainty.

The damages calculated for each storm and each building can be processed in various ways to give estimates of, for instance, the AALs for each building or region, and the distribution of possible losses for each building or region (typically shown using EP curves) and these metrics are the final output of the modeling process.

COMPONENT LEVEL UNCERTAINTIES

Each of the modeling steps listed above contains uncertainty to a greater or lesser extent. At this point, a full detailed analysis and comparison of all the uncertainties in a hurricane cat model has not been attempted and would be very challenging. However, significant work can and has been done that gives reasonable insights into where the uncertainties lie. We select a number of key examples for discussion.

Sensitivity to Numbers of Years and Numbers of Samples

As discussed above, a typical hurricane cat model might consist of 100,000 years of simulation, and each year can be repeated with different random samples for the determination of simulated losses. This raises the following convergence-related questions: how well converged are the results after 100,000 years of simulation, do we need more years of simulation, and does using extra random samples for each storm contribute materially to the level of convergence achieved? These issues have been discussed in some detail in Kaczmarska et al. (2017), in a study on a flood cat model. Assuming the same conclusions apply similarly to hurricane, as they are likely to, the answers to the above questions seem to be:

- For national level results, 100,000 years of simulation is enough to achieve a high level (roughly 1%) of convergence on AAL. EP metrics converge slightly less well than AAL, but still reasonably well (roughly 2%). At these high levels of convergence the simulation errors are likely much smaller than other sources of uncertainty in the model, suggesting further years of simulation may not be needed.
- For national level results, running extra samples of simulation does not increase the level of convergence in a material way. This is because the variations in damage from one simulated sample to another are much smaller than the variations in damage from 1 year to the next, or from one storm to the next.
- Moving to higher resolutions than national, the convergence gets progressively worse. At the highest resolution possible, which is individual buildings, convergence of AAL is often 5% or worse.
- At the highest resolutions, running extra samples does contribute materially to convergence in some cases.

Sensitivity to Choices for LTR Models

There are a number of sensitivities and uncertainties related to the modeling of LTRs for hurricanes. The most obvious is to consider the parameter uncertainty related to the estimation of the overall number of hurricanes. A back-of-the-envelope estimate is a good place to start: since there are roughly 1.7 hurricanes per year making landfall in the US, and we have roughly 100 years of hurricane landfall data, the standard sigma-over-root-n formula for the estimate of the standard error of the mean number of hurricanes gives roughly 0.12 hurricanes, or 7%. This then translates to a standard error of also roughly 7% on estimates of the AAL. The impact is less than 7% on the tail of EP curves, since large EP losses are generally driven by the occurrence of large single events rather than the annual frequency of storms.

Another uncertainty related to LTRs is whether to use distributions other than the Poisson for modeling hurricane frequencies. The obvious alternative is the negative binomial, which allows for overdispersion. The overdispersion of overall hurricane numbers is, however, low, and so using a negative binomial turns out to have little effect, less than the parameter uncertainty considered above.

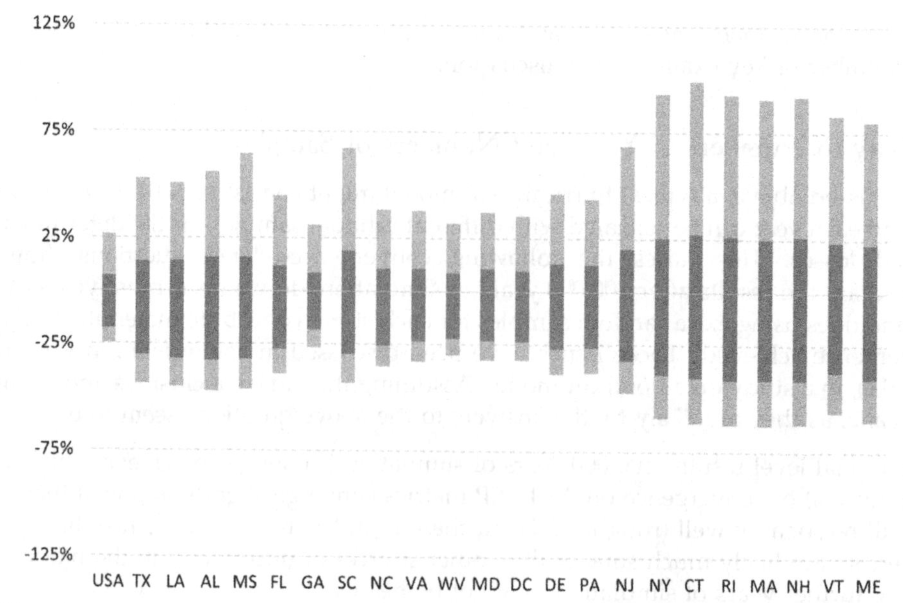

FIGURE 1.20 By state (& all US) percentage of difference in the AAL relative to the median for the 5th, 25th, 75th, and 95th percentile of a 1000 member bootstrap (with replacement) of the historical record after application of RMS landfall smoothing logic.

A third uncertainty related to LTRs is the parameter uncertainty related to landfalling numbers in individual regions, rather than just overall. In general, the regions with the largest numbers of hurricane landfalls, such as Florida, have the best estimated hurricane frequencies, and the regions with the fewest, such as the US Northeast, have the least well-estimated frequencies. This then feeds through into uncertainties on estimates of AAL in these regions, and Florida has lower percentage uncertainty than the Northeast.

To illustrate how the accuracy of AAL estimates varies by region, Fig. 1.20 shows the percentage difference in the AAL by state relative to the median for the 5th, 25th, 75th, and 95th percentile of a 1000 member bootstrap (with replacement) of the historical record after application of RMS landfall smoothing logic, which smooths the historical record along a sequence of landfall gates defining the US coastline.

Sensitivity to Choices for MTR Models

During the period 2005–15, MTR forecasts for future numbers of hurricanes were significantly higher than LTR forecasts, consistent with the perception that this was a period of raised hurricane activity. For instance, the RMS MTR forecasts were predicting roughly 2.0 landfalls per year during this period, versus LTR predictions of roughly 1.7. The choice of whether to use LTRs or MTRs was therefore a critical one. The latest evidence now suggests that we may be coming to the end of this active period of hurricane activity, and that MTR predictions may be similar, or below, LTR predictions in the future, although this is not yet settled.

TABLE 1.1 RMS MTR Forecasts Made With 13 Different Models in 2015

	Category 1–5	Category 3–5
Model	**Rate**	**Rate**
Direct main development region (MDR) sea surface temperature (SST)	2.05	1.02
Indirect MDR SST	2.09	0.88
Direct MDR + Indo-Pacific (IP) SST	1.97	0.92
Indirect MDR + IP SST	2.04	0.84
Long term	1.70	0.62
Direct MDR + IP SST shift	1.23	0.61
Indirect MDR + IP SST shift	1.83	0.62
Direct shift	1.71	0.70
Indirect shift	1.64	0.71
Active baseline: direct MDR + IP SST	1.31	0.65
Active baseline: indirect MDR + IP SST	1.92	0.67
Active baseline: direct	1.80	0.74
Active baseline: indirect	1.72	0.79

If one chooses to use MTR forecasts, then there is further sensitivity related to the choice of which MTR model to use. As an example, Table 1.1 shows the RMS MTR forecasts made with 13 different models in 2015, based on different but reasonable decisions about how to predict sea surface temperatures and future hurricane numbers.

Furthermore, there are choices about how to regionalize MTR forecasts and distribute them across hurricane categories, which can have a significant influence on loss. To illustrate the impact of choice of MTR model (including regionalization model), on regional losses, Fig. 1.21A shows the change in loss-cost (average annual losses normalized by exposure) by zipcode across the US using the highest individual rate model relative to a weighted average "best" model combination of all 13, while Fig. 1.21B shows the same but for the lowest individual rate model.

Sensitivity to Choices for Track, Intensity, Size, Transitioning, and Filling Rate Models

For a hurricane model in which landfalling rates are calibrated against historical landfalling rates, the local results will not typically be highly sensitive to details of how track and intensity models are constructed, since the landfall calibration process to some extent overwrites the landfall rates produced by the track and intensity models. The sensitivity to regional landfall rate data (discussed above) is then more important. Track and intensity

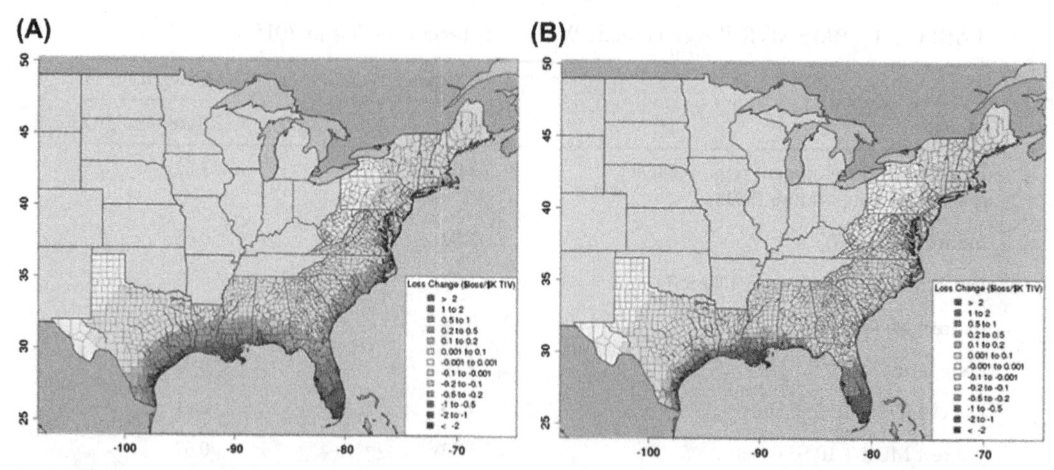

FIGURE 1.21 Change in loss-cost (AAL normalized by exposure) for the highest (A) and lowest (B) individual rate model relative to a weighted average "best" model combination of all 13 RMS MTR forecast.

models do have a big influence, however, on the level of correlation between regions in such models, e.g., the correlation between the Caribbean and the US.

The size of hurricanes, however, is not typically calibrated against historical landfall sizes. Furthermore size is generally rather poorly observed and recorded. As a result, hurricane size models contain significant uncertainty, and this feeds thru into uncertainty in loss results.

Transitioning is also poorly observed, and the wind fields of transitioned storms are difficult to model. Information for how to model such storms comes as much from numerical model simulations (such as those cited above) as it does from observations. As a result all aspects of transitioning storms are uncertain. This has the most impact on the uncertainty in loss results from hurricane cat models in the NE US, where transitioning storms form a larger proportion of the total number of storms.

Filling rate is generally modeled as a function of latitude and hurricane characteristics. Filling rate models are calibrated against wind and surface pressure observations over land. These observations are, however, rather limited, and as a result filling rate models also tend to be rather uncertain. As with transitioning, our knowledge of filling rate, and what affects it, is significantly enhanced by the use of numerical model studies (Colette et al., 2010). Significant uncertainties remain, however, and small changes in the filling rate can have large impact on inland losses.

Sensitivities to Overland Wind-Field Models

Overland wind fields, and the losses they cause, are influenced by a number of modeling choices such as which wind-field structure model to use, which land-surface data sets to use, how to convert land-surface information to roughness estimates, and how to convert those roughness estimates to impacts on the winds. The spatial resolution at which these calculations are done is also important. Of these various model components, tests at RMS suggest that the wind-field structure model is not particularly important, for a given size and strength. All aspects of the treatment of land-surface, however, and its influence on wind speeds, are critical. This includes uncertainties are in the identification in the land-use classes, how those classes are converted to roughness, and in the models used to convert those roughness estimates to wind impacts.

As an example, consider a hypothetical location situated 500 m from the coast where the upstream overland portion is urban and then infinitely open-water. If, for simplicity, we assume the roughness of the open-water has been set equal to 0.001 m following the work of Powell et al. (2003), and the roughness for the urban region is sampled from a range of roughness lengths of 0.7–1.5 m based on the range defined as "regularly-built town" by Wieringa (1993), after the appropriate averaging time adjustments we can calculate a set of corrections using the model described by Cook (1986). Each correction corresponds to a sampled roughness length and adjusts the modeled open-water wind field to one representing the real overland roughness, in this case a coastal town. The variation on the correction is +/−2%; applying these to a typical two-story, shingled gable roof, wood-frame wall structure vulnerability damage curve (HAZUS) can lead to significantly larger uncertainties of loss, as high as +/−13%, due to the nonlinear effect of the damage curve.

Sensitivities to Vulnerabilities

The vulnerability curves that convert hazard values to loss have a critical influence on the final loss results from hurricane cat models. Some of these vulnerability curves are reasonably well constrained by data, such as those for residential buildings in Florida, for which there is data from the hurricanes of the last 20 years, in particular from hurricanes Charley, Frances + Jeanne Ivan, Rita, and Wilma.

Others, such as those for buildings in the US NE, where there have been no recent hurricanes, with the exception of "superstorm" Sandy, are much less well estimated. To capture these uncertainties, RMS produces low and high versions of all vulnerability curves, based on expert judgment of the level of uncertainty. This uncertainty varies by region.

Another approach to understanding vulnerability uncertainty is to follow the approach described in Shome et al. (2012a,b). If building loss data for a number of hurricanes in recent years are available, then the empirical vulnerability functions can be developed and a mixed-effects regression model can be used to quantify the epistemic uncertainty in the loss estimate due to different sources of uncertainties like "interevent," "intercompany," etc., in the loss results. The intercompany uncertainty explains the differences in the loss results across different companies and this uncertainty is attributable to the differences in the underwriting practices, differences in the claim settlement practices, and other company-specific factors. The interevent uncertainty explains the differences in the loss results between different storms at the same wind speed due to the unique characteristics of that storm that are not captured by the model, including differences in the duration of the storms or the rainfall intensity in a storm. This more rigorous way of calculating the epistemic uncertainties in the loss results provides a sanity check on the high and low sensitivities presented in models.

Shome et al. (2012a,b) quantified the uncertainty in the average annual loss and the losses at different return periods for a portfolio of buildings in Miami. The one-sigma uncertainty in the AAL and the losses at the three return periods are shown in Fig. 1.22. The results show

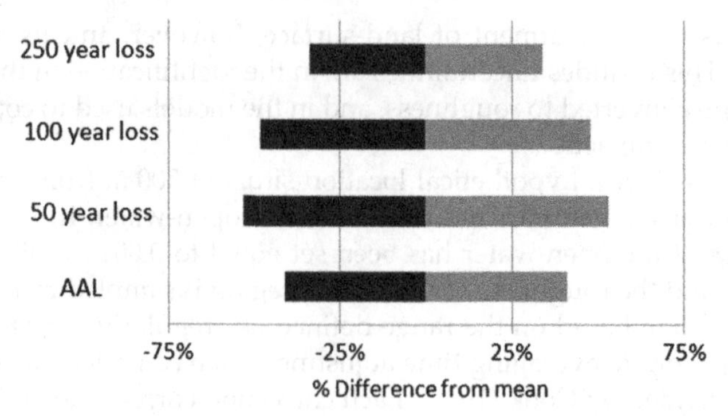

FIGURE 1.22 One-sigma uncertainty in the AAL and the losses at different return periods for epistemic uncertainties only in vulnerability (Shome et al., 2012a,b).

that the uncertainty in loss estimates decreases with the increase in the return period. The figure shows that the uncertainty in AAL is about 30%. It is observed that the intercompany uncertainty changes the variance in the losses, but does not alter the mean value of the loss itself. The interevent uncertainty, on the other hand, alters the mean value because this is constant across all the sites, but does not add uncertainty to the loss estimates.

Other Wind Uncertainties

In addition to the discussion above, there are various other uncertainties that arise in the estimate of hurricane wind losses. One particularly important issue is the question of how exactly to model postevent loss amplification (PLA), which can arise for a number of reasons, but which is principally due to claims inflation that occurs due to shortages in materials and labor following large hurricanes. The exact level of likely PLA depends on a number of economic and political factors.

Surge Uncertainties

The modeling of surge is influenced by a number of significant uncertainties, even if state-of-the-art hydrodynamic surge models are used. We will discuss and give examples of the impact of three particular sources of uncertainty.

Our first example is the modeling of elevation: very small changes in elevation can determine whether a building is hit by a storm surge or not, and the elevations of most buildings are not known precisely enough to determine borderline cases very accurately. Catastrophe loss models may allow users to specify a building elevation, or they may derive building elevation from digital elevation models. The resolution of such models can lead to significant uncertainty in loss.

The second is the vulnerability. As discussed above there are significant uncertainties with respect to how buildings will respond to the hazard at a location and the same logic applied to surge modeling. To capture these uncertainties, RMS produces low and high versions of all vulnerability curves, based on expert judgment of the level of uncertainty.

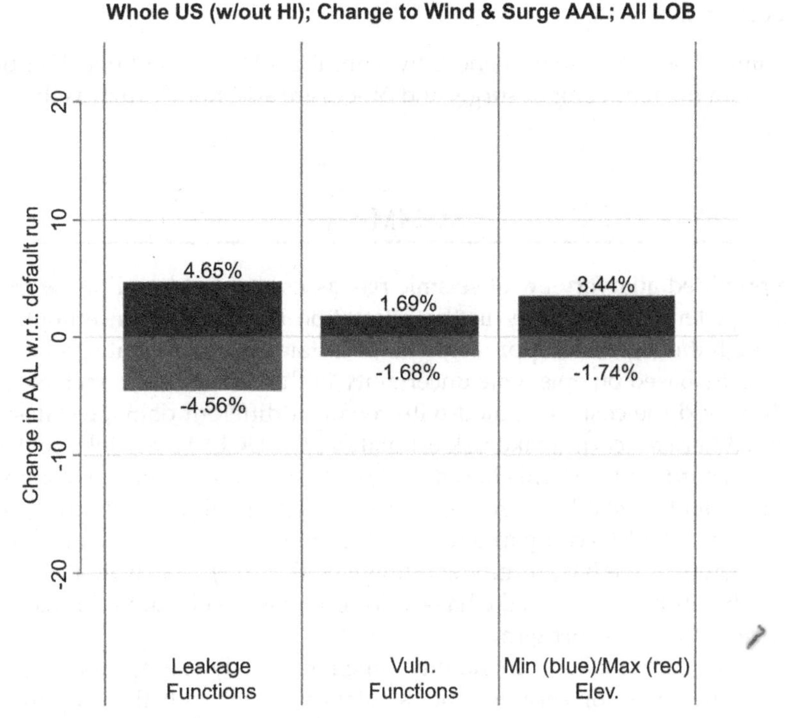

FIGURE 1.23 Percentage of change in the combined wind and surge average annual losses for all LOB assuming; on the left, high, and low leakage assumptions based on expert judgment and the variation in leakage observed across multiple historical storms; in the middle, RMS's high and low vulnerability sensitivities applied only to the surge component of loss; on the right, assuming all properties lie at the minimum or maximum elevation based on a 10 m-high resolution elevation model overlaid on RMS surge hazard grid. Blue (dark gray in print versions): High, Red (light gray in print versions): Low.

The third, relevant only to the insurance industry and the calculation of possible insurance payouts, is the extent to which damage due to wind is inadvertently or deliberately ascribed to surge, or vice versa, when an insurance claim is made. This is called policy leakage. It often occurs when a policy covers only wind or surge, and when there is damage due to both on the same building. Assumptions that are made about the future policy leakage rates can make a large difference to the estimates of future wind and surge losses, although do not affect estimates of wind plus surge losses together, since leakage only involves a transfer of loss from one peril to the other.

Fig. 1.23 below contrasts these three sources of uncertainty showing the percentage change in the combined wind and surge average annual losses for all LOBs assuming: on the left, high and low leakage assumptions based on expert judgment and the variation in leakage observed across multiple historical storms; in the middle, RMS's high and low vulnerability sensitivities applied only to the surge component of loss; and on the right, assuming all properties lie at the minimum or maximum elevation based on a 10 m high resolution elevation model overlaid on RMS surge hazard grid. Not unexpectedly the uncertainty in the financial payout due to surge losses is larger than the physically based uncertainty.

Flood Uncertainties

The modeling of flood is also influenced by a number of significant uncertainties. Many are similar to those in the modeling of surge, and others are additional, such as the uncertainty in river defense levels.

SUMMARY

We have provided an overview of seismic risk assessment of buildings with an emphasis on capturing epistemic uncertainty in the estimation of risk. The uncertainty in hazard is captured through the logic-tree approach. The uncertainty in earthquake vulnerability is estimated analytically based on epistemic uncertainty in the different parameters of building capacity, fragility, and the cost associated with repair of different damage states of buildings.

As illustrated here for earthquake risk estimation, the UCERF3 model calculates epistemic uncertainty in hazard for California based on 1440 logic-tree branches. In contrast, the previous UCERF2 model had 480 logic-tree branches. In fact, the uncertainty in the estimation of hazard increases in UCERF3 compared to UCERF2, with the increase in our knowledge about different parameters of the hazard model (Shome et al., 2014). The authors expect the uncertainty in the estimation of earthquake hazard for other less well-studied regions will also increase significantly in the short term.

For developing earthquake vulnerability functions, analytical approaches have become standard in recent years for seismic risk calculations. Although these approaches require large numbers of parameters, researchers are actively estimating those based on laboratory experiments and observed damage data (see, for example, ATC-58, 2011, for a suite of recently developed fragility functions or recent experiment results in NEES Nonstructural Grand Challenge Project; http://nees-nonstructural.org/documents/index.html). Significant uncertainty remains in the estimation of those parameters with regard to vulnerability function development. The uncertainty, however, will decrease over the years as researchers continue to gather additional information from experimental or observed data.

Regarding the approach for estimating the uncertainty in the estimation of earthquake risk, the logic tree has become a popular approach over the years, primarily for earthquake hazard models. Although it is commonly followed, populating logic trees with all the publicly available models does not necessarily improve estimation of epistemic uncertainty. Correlation between the models arises because of some common features in the models or due to the use of a similar set of data, and this correlation results in inaccurate estimates of the epistemic uncertainty. The accurate estimation of uncertainty requires that the suite of models in the logic tree for any particular component should be collectively exhaustive. However, achieving such an exhaustive suite of models is even more difficult because of incompleteness in the collective scientific knowledge. The authors expect that the framework for estimating uncertainty in risk and in the scientific understanding of different components of risk models will improve over the years, helping to improve the accuracy in the results.

Hurricane catastrophe models have also been improved significantly over the last 50 years. They have transformed the way that the insurance industry understands the hurricane risks they face. This has made the industry more resilient in the face of natural catastrophes, which is good for their customers and for society in general. However, significant uncertainties still remain in the models, many of which cannot easily be eliminated.

The main limiting factor in reducing the hurricane uncertainties is the amount of observed data available. Those regions that experience the most hurricanes and the most hurricane losses, such as Florida, are also those regions where we have the most data. This applies both to climate observations, which give us information about the frequency and characteristics of hurricanes, and to damage observations, which give us information about the effect of hurricane hazards on buildings. These regions are therefore where the risks are most easily and most accurately quantified. For those regions with fewer hurricanes, such as the NE US, much less data are available, and risk estimates, although much lower, are also much more uncertain.

The key to improving models over time, and reducing uncertainties, is therefore mostly to do with the continued collection and archiving of as much data as possible, for all aspects of hurricane behavior, both from a climate perspective, and a loss and damage perspective. Given that data, modelers will be able to further test and refine their models.

Although this paper focused only on the uncertainty in medium- to long-term risk, assessing short-term risk (from days to few weeks) is also important in the insurance industry, particularly for Insurance Linked Securities and catastrophe bonds, to respond to impending hurricanes or aftershocks of earthquakes. Measurements from different sensors of reconnaissance aircraft, buoys, Doppler radar, and many other measurements can be combined with the detailed storm track to develop accurate storm footprint for loss calculations from hurricanes to reduce the uncertainty in the results (http://www.hwind.co/Blog/How-Hurricane-Technology-Can-Reduce-Risk-Uncertainty). Similarly, for earthquakes there is lack of information in the quantification of increased seismicity from aftershocks (for example, as observed in 2011 Tohoku earthquake leading significant increase in the insurance premiums). Recently USGS has initiated Operational earthquake forecasting for providing real-time forecasts of earthquakes, which will help the insurance industry to estimate seismic risk accurately over very short time period (Field et al., 2016).

References

Abrahamson, N.A., Bommer, J.J., 2005. Probability and uncertainty in seismic hazard analysis. Earthquake Spectra 21 (2), 603–607.

ATC-13, 1985. Earthquake Damage Evaluation Data for California. Applied Technology Council, Redwood City, CA.

ATC-58, 2011. Guidelines for Seismic Performance Assessment of Buildings (75% Draft). Applied Technology Council, Redwood City, CA.

Benjamin, J.R., Cornell, C.A., 1970. Probability, Statistics, and Decision for Civil Engineers. McGraw-Hill, New York.

Biasi, G.P., Weldon, R.J., Dawson, T.E., 2013. Appendix F—distribution of Slip in Ruptures. Uniform California Earthquake Rupture Forecast, Version 3 (UCERF3)—The Time-Independent Model. U.S. Geological Survey Open-File Report 2013–1165.

Bommer, J.J., 2012. Challenges of building logic trees for probabilistic seismic hazard analysis. Earthquake Spectra 28 (4), 1723–1735.

CEUS-SSC, 2012. Technical report: Central and Eastern United States Seismic Source Characterization for Nuclear Facilities. EPRI, Palo Alto, CA, U.S. DOE, and U.S. NRC.

Chu, P., Wang, J., 1998. Modelling return periods of tropical cyclone intensities in the vicinity of Hawaii. Journal of Applied Meteorology.

Colette, A., Leith, N., Daniel, V., Bellone, E., Nolan, D., 2010. Using mesoscale simulations to train statistical models of tropical cyclone intensity over land. Monthly Weather Review.

Cook, N.J., 1986. Designer's Guide to Wind Loading of Building Structures Part 1. Butterworth-Heinemann.

Cornell, C.A., 1968. Engineering seismic risk analysis. Bulletin of the Seismological Society of America 58 (5), 1583–1606.

CUREE, 2001. Improving loss estimation for woodframe buildings. In: Porter, K.A., Beck, J.L., Seligson, H.A., et al. (Eds.), CUREE-Caltech Woodframe Project- Element 4, Economic Aspects. Consortium of Universities for Research in Earthquake Engineering, Richmond, CA.

Darling, R., 1991. Estimating probabilities of hurricane wind speeds using a large-scale empirical model. Journal of Climate.

Drayton, M., 2000. A stochastic basin-wide model of Atlantic hurricanes. In: 24th AMS Conference on Hurricanes and Tropical Meteorology.

Ellsworth, W.L., Matthews, M.V., Nadeau, R.M., Nishenko, S.P., Reasenberg, P.A., Simpson, R.W., 1999. A Physically-Based Earthquake Recurrence Model for Estimation of Long-Term Earthquake Probabilities. U. S. Geological Survey, OFR 99–522.

Field, E.H., Arrowsmith, R.J., Biasi, G.P., Bird, P., Dawson, T.E., Felzer, K.R., Jackson, D.D., Johnson, K.M., Jordan, T.H., Madden, C., Michael, A.J., Milner, K.R., Page, M.T., Parsons, T., Powers, P.M., Shaw, B.E., Thatcher, W.R., Weldon II, R.J., Zeng, Y., 2014. Uniform California earthquake rupture forecast, version 3 (UCERF3) — the Time-independent model. Bulletin of the Seismological Society of America 104 (3), 1122–1180. http://dx.doi.org/10.1785/0120130164.

Field, E.H., Dawson, T.E., Felzer, K.R., Frankel, A.D., Gupta, V., Jordan, T.H., Parsons, T., Petersen, M.D., Stein, R.S., Weldon II, R.J., Wills, C.J., 2009. Uniform California earthquake rupture forecast, version 2 (UCERF 2), 2009. Bulletin of the Seismological Society of America 99, 2053–2107. http://dx.doi.org/10.1785/0120080049.

Field, E.H., Jordan, T.H., Jones, L.M., Michael, A.M., Blanpied, M.L., Operational Earthquake Forecasting Workshop Participants, 2016. The potential uses of operational earthquake forecasting. Seismological Research Letters 87 (2A). http://dx.doi.org/10.1785/0220150174.

Fitzenz, D.D., Nyst, M., 2015. Building time-dependent earthquake recurrence models for probabilistic risk computations. Bulletin of the Seismological Society of America 105 (1), 120–133. http://dx.doi.org/10.1785/0120140055.

Frankel, A., 1995. Mapping seismic hazard in the central and Eastern United States. Seismological Research Letters 66 (4), 8–21.

Friedman, D., 1972. Insurance and the natural hazards. ASTIN Bulletin.

Fujii, T., Mitsuta, Y., 1975. Synthesis of a Stochastic Typhoon Model and Simulation of Typhoon Winds. Technical report. Kyoto University Disaster Prevention Research Institute.

Grieser, Jewson, 2012. The RMS TC-rain model. Meteorologische Zeitschrift.

Grossi, P., Kunreuther, Howard (Eds.), 2005. Catastrophe Modeling: A New Approach to Managing Risk. Springer Science + Business Media, Inc., Boston.

HAZUS-MH MR3, 2003. Multi-hazard Loss Estimation Methodology: Earthquake Model. Department of Homeland Security, Washington, D.C.

Hall, T.M., Jewson, S., 2007. Statistical modeling of North Atlantic tropical cyclone tracks. Tellus 59A, 486–498. http://dx.doi.org/10.1111/j.1600-0870.2007.00240.x.

Hanks, T.C., Bakun, W.H., 2008. M-log A observations of recent large earthquakes. Bulletin of the Seismological Society of America 98 (1), 490–494.

Hanks, T.C., Kanamori, H., 1979. A moment magnitude scale. Journal of Geophysical Research 84, 2348–2350.

Helmstetter, A., Kagan, Y.Y., Jackson, D.D., 2007. High-resolution time-independent grid-based forecast for M≥5 earthquakes in California. Seismological Research Letters 78 (1), 78–86.

Insurance Information Institute, III, 2015. http://www.iii.org/fact-statistic/catastrophes-global.

Jaiswal, K.S., Wald, D.J., Perkins, D., et al., 2013. Estimating structural collapse fragility of generic building typologies using expert judgment. In: 11th International Conference on Structural Safety and Reliability (ICOSSAR), New York, USA.

Jayaram, N., Shome, N., Rahnama, M., 2012. Development of earthquake damage functions for tall buildings. Earthquake Engineering and Structural Dynamics 41 (11), 1495–1514.

Jewson, S., Bellone, E., Khare, S., Laepple, T., Lonfat, M., Nzerem, K., O'Shay, A., Penzer, J., Coughlin, K., 2007. 5 Year prediction of the number of hurricanes which make us landfall. Book chapter. In: Elsner, J. (Ed.), Hurricanes and Climate Change.

Kaczmarska, J., Jewson, S., Bellone, E., 2017. Quantifying the Sources of Simulation Uncertainty in Natural Catastrophe Models. http://dx.doi.org/10.1007/s00477-017-1393-0.

Kulkarni, R.B., Youngs, R.R., Coppersmith, K.J., 1984. Assessment of confidence intervals for results of seismic hazard analysis. In: 8th World Conference on Earthquake Engineering, San Francisco, CA.

Loridan, T., Scherer, E., Dixon, M., Bellone, E., Khare, S., 2015a. Cyclone wind field asymmetries during extratropical transition in the western North Pacific. Journal of Applied Meteorology and Climatology.

Loridan, T., Khare, S., Scherer, E., Dixon, M., Bellone, E., 2015b. Parametric modeling of transitioning cyclone's wind fields for risk assessment studies in the western North Pacific. Journal of Applied Meteorology and Climatology.

McGuire, R.K., 2004. Seismic Hazard and Risk Analysis. MNO-10. Earthquake Engineering Research Institute, Oakland.

Molas, G.L., Aslani, H., Shome, N., Cabrera, C., Taghavi, S., Rahnama, M., Bryngelson, J., 2012. Impacts of earthquake hazard uncertainties on probabilistic portfolio loss risk assessment. In: 15th World Conference on Earthquake Engineering (WCEE), Lisbon, Portugal.

Petersen, M.D., Moschetti, M.P., Powers, P.M., et al., 2014. Documentation for the 2014 Update of the United States National Seismic Hazard Maps. U.S. Geological Survey Open-File Report 2014–1091. http://dx.doi.org/10.333/ofr20141091.

Powell, M.D., Vickery, P.J., Reinhold, T.A., 2003. Reduced drag coefficient for high wind speeds in tropical cyclones. Nature 442, 279–283.

RMS, 2015. North Atlantic Hurricane Models RiskLink 15.0 (Build 1625). Florida Commission on Hurricane Loss Projection Methodology. https://www.sbafla.com/method/ModelerSubmissions/CurrentYear2013ModelerSubmissions/tabid/1512/Default.aspx.

Shaw, B.E., 2013. Appendix E—Evaluation of Magnitude-Scaling Relationships and Depth of Rupture. Uniform California Earthquake Rupture Forecast, Version 3 (UCERF3)—The Time-Independent Model. U.S. Geological Survey Open-File Report 2013–1165.

Shome, N., Jayaram, N., Krawinkler, H., et al., 2015. Loss estimation of tall buildings designed for the PEER tall building initiative project. Earthquake Spectra.

Shome, N., Powers, P.M., Petersen, M.D., 2014. Comparison of epistemic uncertainty in hazard in California between UCERF2 and UCERF3. In: SSA 2014 Annual Meeting, Seismological Research Letters, March/April 2014, vol. 85, pp. 365–558. http://dx.doi.org/10.1785/0220140014.

Shome, N., Jayaram, N., Rahnama, M., 2012a. Uncertainty and spatial correlation models for earthquake losses. In: 15th World Conference on Earthquake Engineering (WCEE), Lisbon, Portugal.

Shome, N., Jayaram, N., Rahnama, M., 2012b. Hurricane loss assessment of building portfolios considering epistemic uncertainties. In: 13th International Conference on Wind Engineering (ICWE13), Amsterdam.

Shome, N., Cornell, C.A., 2000. Structural seismic demand analysis: consideration of collapses. In: 8th ASCE Specialty Conference on Probabilistic Mechanics and Structural Reliability, St Luis, USA.

Somerville, P., Collins, N., Abrahamson, N., Graves, R., Saikia, C., 2001. Ground Motion Attenuation Relations for the Central and Eastern United States—Final Report, 2001 (Report to U.S. Geological Survey).

Steinbrugge, K.V., 1982. Earthquakes, Volcanoes and Tsunamies – an Anatomy of Hazards. Skandia America Group, New York, NY.

Stirling, M., McVerry, G., Gerstenberger, M., et al., 2012. National seismic hazard model for New Zealand: 2010 update. Bulletin of the Seismological Society of America 102 (4), 1514–1542. http://dx.doi.org/10.1785/0120110170.

Vickery, P., Skerlj, P., Twisdale, L., 2000. Simulation of hurricane risk in the US using an empirical track model. Journal of Structural Engineering.

Weichert, D.H., 1980. Estimation of the earthquake recurrence parameters for unequal observation periods for different magnitudes. Bulletin of the Seismological Society of America 70, 1337–1356.

Wieringa, J., 1993. Representative roughness parameters for homogeneous terrain. Boundary-Layer Meteorology 63, 323–363.

Wesson, R.L., Perkins, D.M., Leyendecker, E.V., et al., 2004. Losses to single-family housing from ground motions in the 1994 Northridge, California, earthquake. Earthquake Spectra 20, 1021–1045.

Woessner, J., Laurentiu, D., Giardini, D., Crowley, H., Cotton, F., Grünthal, G., Valensise, G., Arvidsson, R., Basili, R., Demircioglu, M.B., Hiemer, S., Meletti, C., Musson, R.W., Rovida, A.N., Sesetyan, K., Stucchi, M., The SHARE Consortium, 2015. The 2013 European seismic hazard model: key components and results. Bulletin of Earthquake Engineering 13, 3553–3596. http://dx.doi.org/10.1007/s10518-015-9795-1.

Further Reading

Bozorgnia, Y., Abrahamson, N.A., Atik, L.A., Ancheta, T.D., Atkinson, G.M., Baker, J.W., Baltay, A., Boore, D.M., Campbell, K.W., Chiou, B.S.-J., Darragh, R., Day, S., Donahue, J., Graves, R.W., Gregor, N., Hanks, T., Idriss, I.M., Kamai, R., Kishida, T., Kottke, A., Mahin, S.A., Rezaeian, S., Rowshandel, B., Seyhan, E., Shahi, S., Shantz, T., Silva, W., Spudich, P., Stewart, J.P., Watson-Lamprey, J., Wooddell, K., Youngs, R., 2014. NGA-West2 research project. Earthquake Spectra 30 (3), 973–987. http://dx.doi.org/10.1193/072113EQS209M.

Knutson, T.R., McBride, J.L., Chan, J., Emanuel, K., Holland, G., Landsea, C., Held, I., Kossin, J.P., Srivastava, A.K., Sugi, M., 2010. Tropical cyclones and climate change. Nature Geoscience 3, 157–163.

PEER-2011/05 Moehle, J., Bozorgnia, Y., Jayaram, N., et al., 2011. Case Studies of the Seismic Performance of Tall Buildings Designed by Alternative Means. Task 12 Report for the Tall Buildings Initiative. Pacific Earthquake Engineering Research Center Report, Berkeley, CA.

Petersen, M.D., Frankel, A.D., Harmsen, S.C., Mueller, C.S., Haller, K.M., Wheeler, R.L., Wesson, R.L., Zeng, Y., Boyd, O.S., Perkins, D.M., Luco, N., Field, E.H., Wills, C.J., Rukstales, K.S., 2008. Documentation for the 2008 Update of the United States National Seismic Hazard Maps. U.S. Geological Survey Open-File Report 2008–1128.

Russell, L.R., 1968. Probability Distribution for Texas Gulf Coast Hurricane Effects of Engineering Interest (Ph.D. thesis). Stanford University.

Scherbaum, F., Kuehn, N.M., 2011. Logic tree branch weights and probabilities: summing up to one is not enough. Earthquake Spectra 27 (4), 1237–1251.

Shome, N., 2015. Seismic loss assessment. In: Beer, M., Kougioumtzoglou, I.A., Patelli, E., Au, I.S.-K. (Eds.), Encyclopedia of Earthquake Engineering. Springer-Verlag Berlin Heidelberg. http://dx.doi.org/10.1007/978-3-642-36197-5_257-1.

Vickery, P.J., Masters, F.J., Powel, M.D., Wadhera, D., 2009. Hurricane hazard modeling: the past, present, and future. Journal of Wind Engineering and Industrial Aerodynamics. http://dx.doi.org/10.1016/j.jweia.2009.05.005.

Yue, L., Ellingwood, B.R., 2006. Hurricane damage to residential construction in the US: importance of uncertainty modeling in risk assessment. Engineering Structures 28, 1009–1018.

CHAPTER

2

What Makes a Catastrophe Model Robust

Jayanta Guin

AIR Worldwide, Boston, MA, USA

INTRODUCTION

Today, 30 years after the first catastrophe models were introduced to the industry, it might seem reasonable to argue that catastrophe modeling is a mature field. Certainly, models have changed the way catastrophe risk is managed—profoundly. Yet 30 years later we are seeing a growing number of new initiatives and new approaches across the private and public sectors, as well as in academia.

This is not altogether surprising given the tremendous complexities of the phenomena we are trying to capture in our models. Weather and climate are known to exhibit chaotic behavior but our understanding of it remains incomplete and certainly the degree of uncertainty in earth processes is even higher—so from a modeling point of view, we observe nature, generating events as a number of stochastic processes. Witness the sheer velocity with which scientific research is being conducted throughout the world, leading to new knowledge as reflected in thousands of publications, unprecedented levels of data being collected every second, and new ways of observing the same phenomenon.

So from a scientific perspective, it seems that we are still in the early innings of the still maturing field of catastrophe modeling. The explosion of data and compute power today has and will continue to give us new ways of answering the same question that modelers answered three decades ago, and each new attempt will give us new answers and make us feel that we are smarter than we were yesterday. The reality is that, as science continues to evolve, so will the answers we get from our favorite model(s). Therefore the quest for that elusive "truth" will continue, and for very good reason.

Still, as a practitioner in this field—whether you are a model developer or a model user—the last 30 years of experience provides a great opportunity to reflect on how we evaluate models and whether we can identify the models that are more robust. With so many versions of models across so many modelers it is an opportune time to reflect on how we go about

Risk Modeling for Hazards and Disasters
http://dx.doi.org/10.1016/B978-0-12-804071-3.00002-1

finding the "useful" models in the sense the great statistician George Box used that term when he said, "All models are wrong but some are useful." To that end, let us revisit the main goals of a catastrophe model and its value proposition.

First and foremost, a catastrophe model is designed to provide estimates of financial risk and it should be unbiased across the entire spectrum of risk. It should be robust enough that it brings (price) stability to the view of risk within the market (the so-called insurance value chain). It should give decision makers a view of risk appropriate to the time horizon over which they run their business.

Now, these are lofty goals and ones the model developer can never lose sight of, even as new versions of a model are released. They can be supported as long as the model (and its subsequent versions) can be validated—and that notion brings us to the crux of this chapter. Can a catastrophe model really be validated? Or is it more a combination of validation and checking for robustness from several perspectives. There are many models being developed in the scientific community, some simple, others complex, but more often than not they are designed to answer a narrow set of questions related to a complex phenomenon.

Numerical weather prediction models are extremely complex, for example, but they are typically used only to make short-term weather forecasts. A catastrophe model dealing with an atmospheric phenomenon, on the other hand, has to address—*probabilistically*—the frequency and severity of future weather extremes over large geographies, incorporate impact assessment on a wide range of properties exposed to extreme weather, and translate all of that to monetary outcomes before and after policy conditions are applied.

This complex series of nonlinear processes throughout the entire model leads to a system in which the end results are quite sensitive to individual parameters in various model components and therefore it poses many challenges to model calibration and validation. The fact that a model is designed to quantify *extreme* risk (including outcomes for which there is no historical precedent) and its corresponding estimates of uncertainty would suggest that model robustness is a better framework to think about the endeavor of testing models rather than absolute validation. Therefore, this chapter attempts to provide a framework on how a model can be evaluated for robustness in light of limited historical data that greatly varies in quality as we go back in time, an incomplete but rapidly evolving state of science and understanding, and widely varying quality of model inputs (so-called "exposure data" in cat modeling terminology).

But this chapter is not a recipe for mechanistic model testing. Instead, it attempts to answer questions like these: What does it mean to validate the tails of the exceedance probability (EP) curves that the model produces (losses that have no historical precedence)? How much effort should be spent in validating each component of the model individually as opposed to doing a holistic evaluation? Is there a natural bias of the model evaluator that impacts the final conclusion?

It is assumed that the reader is familiar with the essential building blocks of how most catastrophe models are built, with the concept of simulation, and with the statistical terms that are typically associated with cat model output and risk metrics. There is also an assumption (as one engages in the exercise of testing model performance with past loss experience) that the inputs into the model are of the quality that the model expects and we will not dwell on this point in the rest of this chapter.

PROMISE OF THE STOCHASTIC CATALOG

The stochastic catalog of events is a critical component of any cat model because it reflects many fundamental and important assumptions. As we make new observations in nature, the catalog is the point of entry that allows us to go back and critique our model, hopefully leading to largely favorable conclusions if the model is robust.

In theory, the stochastic catalog holds the promise of providing the complete universe of plausible events that have either occurred historically or could occur in the future—each associated with the correct degree of relative probability. "In theory" is emphasized because while a model may arguably be robust in its formulation, limitations imposed on its implementation in practice may render it deficient. The limitations stem from the fact that compute power and time are limited and therefore modelers are required to make simplifications in order for the model to be deemed practically useful. For example, theory might suggest that in order to represent the complete universe of events for a particular peril, the stochastic catalog should have 10 million (or even tens of millions) unique realizations; however, that would impose a computational demand that would make the model unusable, particularly when today's decision makers expect an answer in near real time.

In reality, therefore, the stochastic catalog is a representative sample from the universe of plausible events, which means that with all else equal, the robustness of the catalog results from how the sampling is done and the scale at which the model results are being evaluated. Let us imagine we are interested in understanding risk from hurricanes in the northeastern United States. Since New England does not experience hurricanes frequently, the limited sample size might provide a sufficiently stable view of risk for an entire portfolio, but model results might be less stable if the goal is to understand risk for a single property. This leads us to the concept of convergence in results, which means that given a catalog size, model results are more robust at larger spatial scales. Using optimization techniques the results at smaller scales can be improved over a simple random sample but, nevertheless, there is a compromise that has to be made. The good news is that with ever increasing compute power and advances in parallelization, this limitation can be largely overcome with larger catalogs; moreover, any such limitation is *not* a true reflection of the robustness of the model's formulation.

The more interesting aspect of the model evaluation process lies in understanding the extent to which a model has contemplated events that have not yet occurred but are plausible. Staying with our example of Northeast hurricane risk, a number of questions emerged after Hurricane Sandy in October of 2012. There were many unusual aspects of this storm, one of which was the westward turn Sandy made as it approached the coast of New Jersey. The historical record of landfalling Northeast hurricanes (see Fig. 2.1A) prior to this storm shows the tendency of storms to head in a northeasterly direction at these latitudes, which is consistent with our meteorological understanding of atmospheric steering currents and interactions with other large-scale pressure systems as storms get closer to the continental United States. As we all know, Sandy defied the historical norm and took a westerly turn as shown in Fig. 2.1B.

Notwithstanding how the operational forecast models fared with this prediction, for a catastrophe model, the obvious question is whether the stochastic catalog in use prior to

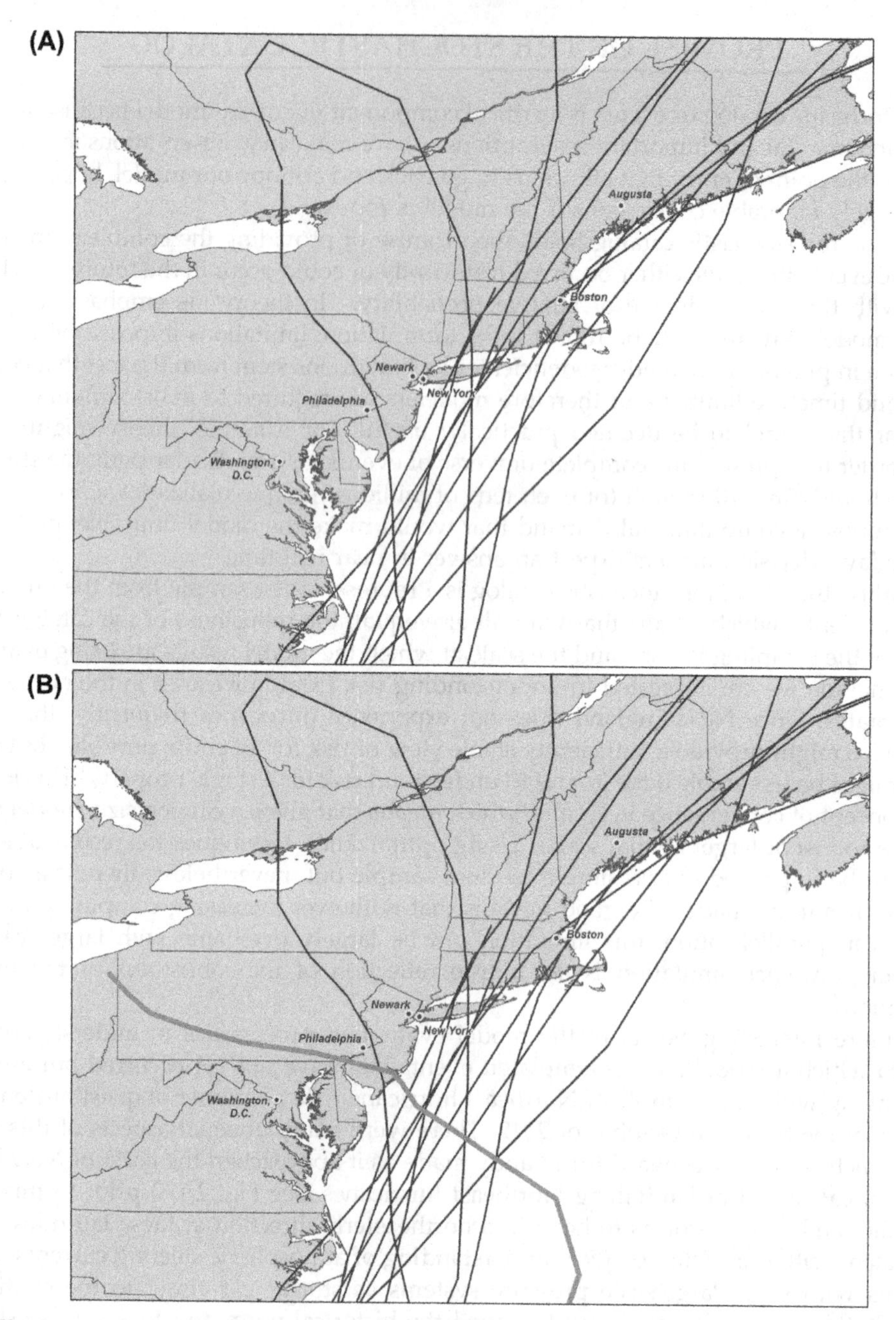

FIGURE 2.1 (A) The tracks of hurricanes making landfall in the Northeast since 1900. Note the dominant tendency of storms heading in a northeast direction. (B) The track of Superstorm Sandy (2012) added; the westward movement of the storm as it made landfall in New Jersey is unusual.

Sandy accounted for tracks similar to Sandy's. With respect to this specific question, a model is robust if, given a large enough sample size, it can simulate storm tracks with this westerly turn off the coast of New Jersey. However, if the model is formulated in such a way that this is simply not a scenario that can be generated, then one can conclude that, at least in this aspect, the model needs improvement. We are not even talking about the relative frequency with which a model represents such storm tracks; that is an even harder question to evaluate and one that involves subjective judgment. But it is fair to say that, as we make new observations, if we can find similar analogs in a large enough sample of simulated events from the underlying model we can take comfort in the robustness of the stochastic catalog. Of course, it doesn't always happen this way and our three decades of modeling experience have given us numerous surprises. Before we take a deeper look at such unwelcome surprises, let us also look at what other criteria events in a stochastic catalog should meet.

Simulated events as defined by a set of parameters should be physically realistic and the correlation across the set of parameters should be adequately captured. The parameter space should be scientifically justified and scientifically bounded. We need to ensure that an earthquake catalog represents the full range of magnitudes that is plausible for a particular tectonic environment with known geological features. But we ought not to take an overly conservative approach and simulate earthquakes of magnitude 9 when there is no physical justification. These two competing facets of model building forces modelers to use the best available data, scientific research, and scientific judgment while accounting for uncertainty due to the inherent randomness in nature or lack of knowledge (Guin, 2010).

Likewise, in meteorological models we need to evaluate whether the assumptions about extremes are scientifically justified. Let us say the lowest observed central pressure for hurricanes in the open Atlantic is 882 mbar (Hurricane Wilma, 2005), while the lowest recorded central pressure at US landfall is 892 mbar (Labor Day Hurricane, 1935). Given that we are attempting to simulate the complete universe of all plausible events, should a model include events with a landfall central pressure of 882 mbar or lower and how would that change as we consider higher latitudes along the US coastline? Or are there other physical considerations that would argue that since most storms tend to weaken as they enter the shallower waters near the coastline, the lower limit for central pressure at landfall should be higher than the lower limit in the open ocean. This is just one example of the many questions that the model developer has to answer using statistical extrapolations, an understanding of physics, and scientific judgment. More importantly, the model evaluator has to form an opinion on these because these assumptions are invariably quite different across models. This is what makes catastrophe models distinctly different from "data-driven" models; the catastrophe model is based on a hybrid approach of blending statistical modeling techniques with physical modeling and scientific judgment, again with the intention of making sure the simulated events are physically realistic.

Since the first, purely statistical catastrophe models were introduced 30 years ago, we have come to realize that there is an equally important place for physics and physical modeling to ensure model robustness. However, since our physical understanding of natural processes is not yet (and may never be) complete, we should not risk over confidence in our model predictions based solely on physics. It is well accepted, for example, that climate is nonstationary and therefore as the earth warms, it is not prudent to use the past as the sole good predictor of the future. However, the limited past experience is extremely valuable in evaluating our model predictions. For example, following the very active 2004–05 Atlantic hurricane seasons, which resulted in an unprecedented 10 landfalling hurricanes in the United States, it was tempting to conclude that we had entered a new climate regime as evidenced by physics. However, there is value in first challenging the existing model to explain past variability in hurricane occurrence, as one is surer of the observed record than of future predictions. A careful analysis of the past historical record of landfalling hurricanes did not show a strong signal due to warmer than average conditions in the Atlantic sea surface temperatures (Dailey et al., 2009). Certainly the 10 additional years of experience since the 2004–05 hurricane seasons have shown that the rush to implement physical assumptions around climate change was not warranted—another lesson that our knowledge of all the underlying physical processes is limited. Ten years later, the conversation has now turned to understanding hurricane "droughts," with Florida recently experiencing the longest period of quiescence in the historical record.

If we acknowledge that in a stochastic catalog we are trying to simulate how events occur in nature, then we are immediately drawn to questions around the temporal characteristics of how events occur in nature versus in our simulations. Is there a tendency for natural events to occur in clusters, for example, and if so, what tests might we conceive to determine whether the model captures that tendency? Although not guaranteed, most models do reasonably well when evaluated with measures of central tendency (mean, median), but since we are in the business of risk analysis where measures of variability are more important, it is always a challenge. In general, nature tends to exhibit more variability (in space and in time) than models and that is what a model evaluator should focus on. We may experience seasons where a particular region gets battered by a train of storms (clustering in space and time within a season) or we may see consecutive years of heightened activity followed by consecutive years of calm (interannual variability). Does a model's stochastic simulation represent such variability adequately? There are a set of criteria against which the model evaluator can assess adequacy. While establishing a truly exhaustive set of tests is unlikely, a practical approach can expose major model limitations.

Coming back to the notion of model formulation and its limitations, the Tohoku earthquake of 2011 is a salient example of nature reminding modelers that we might still be in the early innings of a maturing field. It is a fact that most models, including the most widely used catastrophe models and the official model of the government of Japan, did not anticipate a magnitude 9 earthquake on the Japan Trench, off the Pacific coast of northern Japan. This failure of the models is attributed to the (mis)interpretation of the nearly 500 years of historical data during which the region had exhibited large earthquake ruptures in the order of magnitude 8 and lower, but never a magnitude 9 (see Fig. 2.2). (Note: a magnitude 9 earthquake releases roughly 30 times more energy than a magnitude 8 earthquake.) This is also a failure of scientific judgment in the model-building process and ignoring the findings of other physical models (Stein and Okal, 2011) that did postulate the possibility of a magnitude 9 earthquake in the region. Newer, physically based models (McCaffrey, 2008; Loveless and Meade, 2010) that have used earth observations using GPS, especially after the 2004 Boxing Day tsunami in Indonesia, gave enough evidence to infer the possibility of a magnitude 9 earthquake in the region of Japan Trench. In hindsight, we now have a better physical understanding of long-term strain accumulation on faults and rupture processes, and newer generations of models are accounting for them.

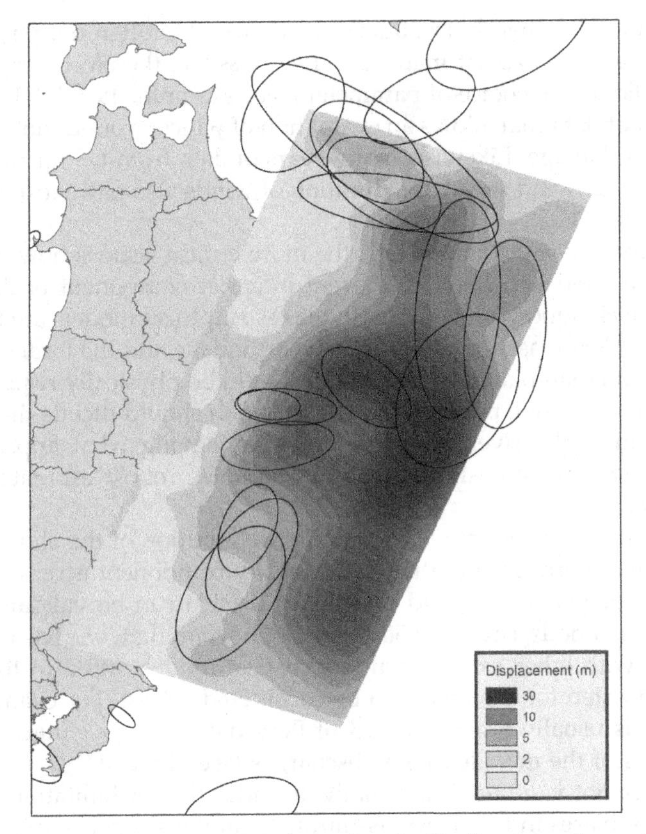

FIGURE 2.2 Estimated slip in the 2011 M9.0 Tohoku earthquake. Shading shows the varying degree of lip across the rupture plane. The ellipses show the extent of slip of previous large earthquakes, all of which are considerably smaller than the 2011 event.

Nevertheless, the need for scientific judgment cannot be eliminated because the model building process is not simply to account for extreme events in all locations but rather to bring to bear the latest physical models to complement the historical data and apply learnings from one region of the world to other regions on a scientific basis. Coming back to testing the robustness of a model and having observed Tohoku, one has to ask questions regarding the basis for the extreme events in the model and their relative frequency—yet another task in which the model evaluator has to form her own opinion and take informed decisions.

SIMULATION OF EVENT INTENSITIES AT LOCATION

In a catastrophe model, losses are estimated for each simulated event. The first step in that process is to estimate a spatial field of hazard parameters that are used to estimate damage. The selection of those hazard parameters can be tricky and not all models agree on which parameters describe damage the best. An engineering understanding of damage mechanisms is often the foundation for this choice, but we recognize it is often not possible to model all damage mechanisms in an explicit manner. When possible, therefore, empirical evidence is also required to validate the choice of parameters. For example, the 2011 Tohoku earthquake generated a rich set of data that allows an evaluation of which ground motion parameters are the best predictor of damage. Likewise, many years of data from US hurricanes have shown that, in addition to peak wind speed, the duration of winds also partly explains the degree of damage at a location.

Beyond the choice of hazard parameters, the more critical issue is how they are modeled. For some perils like wind and flood, this event intensity component of the model can use physical modeling techniques, while for earthquakes empirical models are the norm because of the immense complexity associated with each earthquake and the local geology. Nonetheless, the local intensities simulated by the model should be physically realistic and bounded. For example, in a flood model, principles of conservation should dictate the volume of water moving downstream in the river network. However, having what appears to be a good model for simulating local intensities does not guarantee robust estimates of probabilistic hazard (PH) at a location.

The PH at a location is essentially a probability distribution of the chosen hazard parameters and is obtained by integrating the local intensity component across all the events in a stochastic catalog. For perils like wind and flood, the PH can be validated with observed data for lower return periods, say up to 50 years. But beyond that, one has to resort to reasonability and testing with other extreme-value statistical models fitted to the observed data. This is an important step for the model to be considered robust. For example, in an inland flood model, there is usually a good record of flow data in river gauges that can be used to validate quantiles of the modeled peak discharges (see Fig. 2.3).

If the model does not validate, it is usually an indication of limitations in the intensity-simulation model or biases in frequency assumptions in the stochastic catalog. Interestingly, this step of validation is very challenging for earthquakes, because we do not have a long time series of ground motion observations at a fixed seismic station (as opposed to weather stations that have been around for more than 100 years in some places and are exposed to events relatively frequently). However, in certain landscapes that are seismically active, geological artifacts provide clues about the reasonability of a modeled estimate. Such a line of thinking was used when observing that in the deserts of California, nature has preserved precariously balanced rocks. That is, the scientific consensus of PH was revealed to be overestimated because, had it been accurate, the rocks should have toppled long since. Indeed, newer estimates of PH from the consensus United States Geological Survey model are, in fact, lower.

Going back to the theme of variability and of nature exhibiting more variability than models actually capture, it is relevant when we think about modeling local intensities. If we have the luxury of having a dense set of high-quality observations, whether it is for windstorms or earthquakes, we observe tremendous variability. This is physically often understood, but in a model (and because it *is* a model), not all variables are captured, so we must simulate the variability stochastically such that the footprints of simulated intensities look realistic compared to observations and have the right spatial correlation at various scales. For example, simulation of rainfall is often the starting point for a flood model. The "texture" of simulated rainfall is very different if the model is resolving precipitation at 1,

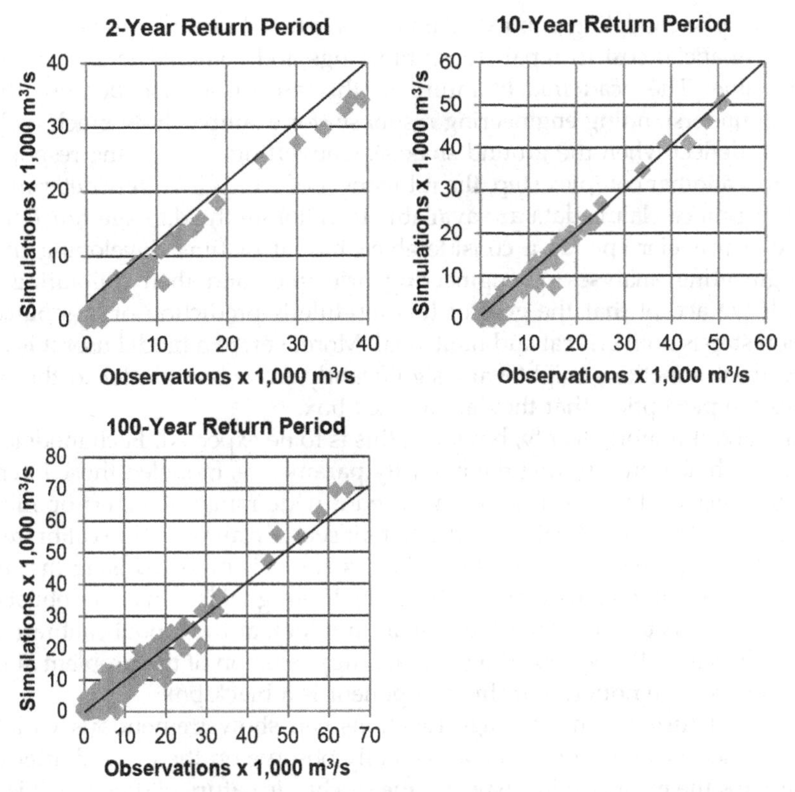

FIGURE 2.3 Modeled and distribution-fitted quantiles of observed annual peak discharges for the 2-, 10-, and 100-year return periods.

10, or 100 km scales. This is quite intuitive but critical in the evaluation process because other components of the flood model are impacted by this very first choice. We expect that the simulated fields at high resolution should have the texture of what we observe in radar, with much higher variability as compared to smooth fields in a lower resolution model.

To conclude this section, there are a number of decisions a modeler has to make with respect to the spatial scale at which the hazard parameters are estimated. This is critical for the next component of the model, namely the damage estimation module, as the two are closely coupled. The model evaluator therefore needs to be aware of the modeling choices that were made and the rationale for those choices. Only then can the overall model be evaluated for robustness.

MODELING DAMAGE

This component of the catastrophe model comprises a family of damage (or vulnerability) functions, which represent a statistical relationship between varying degrees of damage and their associated probabilities and varying levels of local intensity. The damage functions are often perceived to be the "black box" in a catastrophe model.

In the catastrophe modeling context, damage, whether to buildings or their contents, is defined as the financial cost to repair those buildings and contents as a fraction of the full replacement value. The academic literature in this area is sparse because the focus in academia is on understanding engineering response (for example, how much will a building at the top floor deflect when the ground shakes). Converting engineering response to financial loss requires another complex step, the robustness of which is largely determined by how much actual insurance claims data are available to calibrate the damage functions.

In practice, a modeler spends a considerable amount of time developing the functions based on engineering analyses or engineering principles and then calibrating them with claims data. If we accept that the goal of this module is prediction of loss outcomes, then this calibration step is both critical and nontrivial. Moreover, to a model user it is not obvious how or why these functions might vary significantly from one model to the next, which further drives the perception that they are a black box.

If you think about it more deeply, however, this is to be expected. Each modeler has made a different set of choices in selecting the intensity parameters, modeled those intensities with different approaches at different resolutions, used a wide range of expert opinion in developing damage functions, and, finally, used different sets of claims data to calibrate the model. The final results are quite sensitive to these choices and it is not surprising that the damage functions across modeling teams vary significantly. If our goal is to have a robust model that predicts losses, one has to probe into these areas and look at the model calibration and validation more holistically. Proper due diligence and appreciation of the problem at hand, is the only way to get past the notion that this component is a black box.

To explore a bit further why damage functions can show tremendous variation across modeling teams, let us start with what we actually observe on the ground after a major catastrophe. Perhaps the most striking aspect is the stochastic nature of damage. It is more often the case than not that in a single neighborhood you will observe a wide range of damage to seemingly similarly constructed properties. If we imagine hurricane damage, this wide variation could come from real differences in the construction quality and maintenance of the properties, differences in the strongest winds and the direction from which they came, randomness in the airborne debris that breaks windows, either the shelter offered by trees or damage due to trees that have been uprooted, and the list goes on. For some very localized perils with extreme intensity—imagine the core area impacted by an EF-5 tornado—this variation in damage is limited, but such examples are more the exception than the rule.

Keeping this picture in mind and go now to the modeling world, we recognize that a model tries to capture damage with one, or two, or three hazard intensity parameters. For example, if the hurricane model estimates damage using maximum wind speed and wind duration, then losses due to water infiltration, tree damage, and damage due to debris are either not modeled or they are captured indirectly. If we make a philosophical choice of not attempting to capture the other sources of damage then we will have a model that will underestimate losses when compared with real loss data. Moreover, the modeler's choice is also limited by the inputs that are expected by the model. For insurance risk modeling, the physical assets are typically described by a small set of characteristics and this largely limits how well damage estimates can be resolved. An excellent discussion on this topic and other key factors that are important for damage estimation can be found in Jain (2010).

Given this complex set of challenges how can we test robustness? One practical approach is to compare relative vulnerabilities across asset classes and then evaluate the absolute vulnerability for assets for which more claims data are available. In general, loss data are more abundant for residential properties (larger population) and are very limited in the case of complex, more heterogeneous commercial and industrial structures. The Maule earthquake in Chile (2010) caused severe damage to several high-rise buildings but those are a small fraction of all the high-rise buildings in the region affected. With detailed knowledge of an individual building, engineers can explain the cause of failure, but in the absence of such information, the modeler has to estimate the probability that a small fraction in a population of tall buildings will experience significant damage—and that requires judgment and a number of assumptions in the absence of data. Checking the model performance at an aggregated level (for example, total event losses or losses for a county), assuming the portfolio has several tall buildings is a prudent way of validating the model. The modeler does not have enough data to validate all damage functions and neither does the model evaluator. So focusing on the overall process and the underlying engineering assumptions is a sensible approach.

Another way of appreciating why model calibration is critical for the robustness of a model starts with understanding the goals of engineering construction. Buildings are designed to withstand a certain level of intensity and, in recent times, to achieve a certain level of performance. In an actual event, when the local hazard exceeds that design threshold, damage is likely to occur and can often lead to rapid deterioration of the entire building. Imagine what happens to a wood home under strong winds once the windows are breached. Strong winds can quickly tear apart the rest of the structure if the framing is not strongly connected. So in reality we observe many examples of this rapid acceleration towards large failures. In the modeling world, the damage functions reflect the same behavior, which means they are quite sensitive to changes in the local intensity parameters over a critical range of values. This sensitivity in a model, while not desirable, is an essential feature and having slight biases in the model components become easily apparent in an uncalibrated model. The scarcity of observation data and the complexity of the phenomena being modeled make it difficult to assess and therefore eliminate those biases. Moreover, it can happen that we make many more observations in one component of the model than the other. Such a situation necessitates a recalibration as well. When new equations for ground motion prediction were introduced in 2008 as a result of a large concerted effort to develop a clean dataset (the New Generation Attenuation, or NGA, equations), earthquake damage functions had to be recalibrated.

FINANCIAL MODEL AND VALIDATING LOSSES

A catastrophe model generates a large array of output, both financial and nonfinancial, but the former is the most widely used for insurance purposes. Financial losses are expressed as a probability distribution of losses at various scales of aggregation—starting with losses to a single building all the way to a complete portfolio of assets. There are several opportunities for validating loss metrics derived from these probability distributions and significant amount of time should be dedicated to this effort in order to evaluate model robustness.

By definition, since the problem is one of extremes, these probability distributions are extremely skewed and fat-tailed. With a limited history of actual loss outcomes, it is challenging to validate such distributions. Certainly one can evaluate the so-called frequency part of the distribution, meaning the portion that estimates losses characterized by a higher probability. For example, if there is a fairly complete loss record for 30–40 years, it is a simple matter to estimate statistically the uncertainty bounds of losses at shorter return periods like 2-, 5-, 10-, and 20-year return periods. If the cat model's results are outside these bounds, one has to have good justification. Because the distributions are much skewed, validating the mean (referred to as average annual loss [AAL]), is quite difficult, although understandably tempting. For more frequent atmospheric perils, it can be reliably done (with caution), but for infrequent perils like earthquakes, it is nearly impossible.

While not strictly validated, there should also be enough time spent on studying the "right-tail" of the distributions where we have the most extreme simulated events. They have to be examined for physical reasonability, beyond validation of individual model components. It is particularly prudent to evaluate these extremes to see how the model behaves when all model components are assembled together. Validating a cat model involves a bit of art as well as science because the evaluator's personal biases come into play. The biases can be due to degrees of risk aversion; the assumer of risk will always have a more conservative expectation than one who is ceding risk. The modeler, who is not a player in risk transfer, should not have such biases and rather focus on ensuring the model components are unbiased.

Occurrence of extreme events also provides an opportunity to validate an existing model. The edge conditions can be tested more readily than others. A model is clearly deficient if the actual loss from an extreme event exceeds the far right tail of the model's distribution. Occurrence of multiple extreme losses in short window of time can also lead to questions about a model. For example, if two large losses occur within a 10-year period and they are estimated in the model to have an annual probability of occurrence of 0.5%, it is very likely the model has limitations—or at least a close examination is called for.

For insurance applications, a catastrophe model has to go well beyond estimating the total cost of repair (financial damage) and there is usually a financial modeling component that translates damage to insurance payouts. Insurance (and reinsurance) policy structures vary from a simple deductible on a home all the way to very complex terms. Accurate modeling of these terms and conditions probabilistically is a challenge but validating them becomes an even bigger challenge. Recent history has shown that models have better skill in estimating losses for portfolios of risk with simple "primary" insurance terms, and the skill deteriorates when subjected to estimating losses for "excess and surplus" insurance terms. A practical discussion of the complexities involved in estimating losses for commercial portfolios including policy conditions can be found in Jain (2011).

From a theoretical point of view (a mathematical discussion of this is outside the scope of this chapter), this is to be expected, but the question is how one should test the robustness of the model when it comes to testing such complex portfolios. This can be accomplished by looking at the model performance over a larger set of such portfolios and if a bias is apparent, it is usually a sign of deficiency in how the financial model treats the uncertainties in damage predictions and how probabilistic loss estimates are aggregated. This alone is a fascinating area of catastrophe modeling that merits a dedicated chapter. It is fair to say that in order to have the confidence that a model is robust; at least one-third to one-half of the total model development time should be dedicated to model validation.

CLOSING THOUGHTS

As with any other model, a catastrophe model attempts to capture almost infinitely complex, nonlinear processes and therefore, by definition, will always have limitations when compared to reality. However, well-constructed models that have been tested for robustness are the best tool for practitioners who are involved in pricing and managing catastrophe risk. There is no even close alternative. Model users have to spend time to understand a model and the critical assumptions that underlie it. Since a catastrophe model has many parameters, proper due diligence takes significant time and should not be compared with the level of effort involved in a purely statistical model with a few parameters. However, from a practical point of view, time may be limited and, under such circumstances, it is better to devote the majority of the time in evaluating loss estimates and the stochastic catalog. These two anchors at the two ends of the model capture the essence of frequency and severity assumptions—the most critical aspects of the risk that are being evaluated. The time requirement will also depend on the scale at which risk is being evaluated. For example, if the subject of interest is a single large risk, much more due diligence is needed in order to gain confidence in the model results.

The case can be made that a parsimonious model (a model with few explanatory variables) often is equally powerful as compared to a complex catastrophe model. There is a complementary role for parsimonious models in the overall validation process. In its simplest form, for perils that have a few decades of reasonably credible loss history, one can fit a statistical model to construct a loss EP curve with associated confidence bounds and use it to benchmark the EP curve from a cat model. Caution needs to be exercised in studying the type of events reflected in that limited loss history because adding another year of data that includes an extreme loss occurrence can significantly impact the results from the statistical model, a problem that a cat model avoids. Other parsimonious models could consider simpler formulations of each of the main components of a cat model and provide equally credible validation of losses at certain "scales." However, such a model may be suitable only for estimating risk at a coarser resolution. A well-constructed, cat model on the other hand can provide insights into risk at a much higher level of granularity along with internal consistency across the seemingly infinite space of model parameters. Nevertheless, there is value in postulating simpler models and testing the results of a more complex model and should be considered as best practice.

When actual events occur, users naturally gravitate towards looking for a "similar" event in a model's stochastic catalog. While tempting, one has to avoid simple conclusions about the overall robustness of the model; remember, an event is a single realization of an infinite number of possibilities. It is also important to understand the composition of the actual loss in the context of which components of loss are being modeled. Finally, equal effort needs to be paid to regions that have not had an extreme loss in recent history. Human psychology leads us to spend more time on what we have recently witnessed (recency bias) than on scenarios that are generated by computers. Practitioners should reflect on how much time they have spent on studying recent disasters as opposed to the disasters that have not occurred in recent history but have already happened in their computer simulations. An appendix has been provided to summarize a set of guidelines for the practitioner on how they can evaluate a catastrophe model, assuming they have the necessary resources.

The hope is that this chapter provides enough motivation for model developers, model users, and model evaluators to reexamine their models and have a better appreciation of the complexities of catastrophe risk analysis.

Acknowledgments

The author would like to thank his longtime colleague, Beverly Porter for her insightful comments, suggestions, and editorial help.

References

Dailey, P.S., Zuba, G., Ljung, G., Dima, I.M., Guin, J., 2009. On the relationship between North Atlantic sea surface temperatures and U.S. Hurricane landfall risk. Journal of Applied Meteorology and Climatology 48, 111–129.

Guin, J., 2010. Understanding Uncertainty. http://www.air-worldwide.com/Publications/AIR-Currents/2010/Understanding-Uncertainty.

Jain, V.K., 2010. Anatomy of a Damage Function: Dispelling the Myths. https://www.air-worldwide.com/Publications/AIR-Currents/attachments/AIR-Currents–Anatomy-of-a-Damage-Function/.

Jain, V.K., 2011. Uncertainty in Estimating Commercial Losses—And Best Practices for Reducing It. http://www.air-worldwide.com/Publications/AIR-Currents/2011/Uncertainty-in-Estimating-Commercial-Losses%e2%80%94and-Best-Practices-for-Reducing-It/.

Loveless, J.P., Meade, B.J., 2010. Geodetic imaging of plate motions, slip rates, and partitioning of deformation in Japan. Journal of Geophysical Research 115, B02410.

McCaffrey, R., 2008. Global frequency of magnitude 9 earthquakes. Geology 36 (3), 263–266.

Stein, S., Okal, E.A., 2011. The size of the 2011 Tohoku earthquake need not have been a surprise. EOS 92 (27).

APPENDIX A

This section is not intended to be a complete and formulaic approach to model evaluation but provides a set of practical guidelines to form an opinion on the robustness of a model. Many of the suggestions below have been discussed in the body of this chapter, which provides better context to the task at hand. Note that there are situations when a catastrophe modeler incorporates a well-known, published model; for example, the hazard component of US earthquake models is usually based on the national hazard model of the United States Geological Survey (USGS). In that case, the user should engage in model verification to ensure the model accurately, or at least reasonably, reflects the USGS model. Of course, if instead the goal is to validate the USGS model, that is an entirely different matter. From a decision-making point of view, one can have an opinion on the USGS model, but we should not confuse verification of the catastrophe model with validation of the USGS model. However, it is still prudent to compare the modeled frequency of earthquakes with the historical record, in regions that have sufficient data.

Below is a set of checks that should be performed, organized by major model component. If resources are limited, one should spend time on evaluating the stochastic catalog and overall losses.

Stochastic Catalog

1. Compare modeled frequency of events with historical records across the spectrum of severity parameters and by region. In regions where the historical data are sparse or the model deviates significantly from history, evaluate the physical explanation for such divergence. Usually there is limited data in the extremes and so the physical basis for the estimates has to be understood.
2. For extreme events, frequency measures in the model should have smooth continuity across regions in the model domains if the physics justifies it. Discontinuities should be supported by a physical basis.
3. Evaluate the temporal characteristics of the occurrence of events in the stochastic simulation. For atmospheric perils, which in some cases exhibit clustering in space and time, evaluate characteristics of interarrival times and compare with the historical record.

Local Intensity

1. Evaluate whether the model employs the appropriate set of intensity measures.
2. Evaluate the validation of intensity measures against historical observations.
3. Evaluate the spatial extents of the event footprint compared to historical events.
4. Understand how the model treats spatial variability of the intensity measures in its simulations.
5. Understand the spatial resolution of the model and evaluate against the use cases. Increasing spatial resolution is not a sufficient condition to ensure model robustness. A model with a coarse resolution could provide robust loss estimates at an aggregated level but may not be suitable for pricing risk for an individual industrial plant.

Damage Functions

1. Evaluate the modeling and validation approach and check for relativity in vulnerability across the many asset classes that the model supports.
2. Evaluate the validation of the damage functions with historical loss experience and develop an appreciation of the extent of claims data that has been used in the validation process.
3. Understand all model parameters and check for reasonability of changes in vulnerability as the parameters are varied.

Financial Calculations

1. Understand the approach of propagating uncertainty through the financial engine.
2. Evaluate the actuarial formulae that are used for applying the policy conditions and ensure reasonableness of results. Tests can be created to evaluate how the model behaves as financial terms are varied.

3. Evaluate model behavior for edge conditions. For example, if a simple policy is deductible and has a limit, then under the highest levels of severity, the policy should not payout more than the limit.

Overall Model

1. Evaluate model performance for past historical events at the level of granularity that reported historical data allows. This is best done with a company's actual exposure and claims data concurrent with the time the event occurred. Test this across as many events as possible to check for any biases in the model. This step should also be done at the highest level of spatial aggregation of losses, i.e., for all properties impacted by the event, often referred to as "the industry."
2. Evaluate for reasonability the probability of occurrence of historical loss levels as estimated by the model. For atmospheric perils, there is often loss data for the past 10 or 30 years that can be used to evaluate the frequency component of the model's results (the exceedance probability curve). A lack of good agreement can imply problems in several components in a model. For low frequency phenomena, like severe earthquakes, the occurrence probability needs to be explained in the context of the entire model. This requires careful consideration of the full range of events that are considered in the model, the spatial distribution of exposures, and any peculiarity of a historical event that might make it an outlier, and so on. Considerable judgment is involved in this step.
3. Ask questions regarding "what-if" scenarios and whether a plausible event is represented in the stochastic catalog. This step requires good domain knowledge of what constitutes a plausible event; however, an examination of past historical events is a good place to start to inform future scenarios by simply perturbing certain parameters or relocating them spatially, within reasonable limits.
4. Check for internal consistency in the overall model. For example, are losses for historical events and stochastic events modeled with the same consistent approach? If there are differences, one should understand them and evaluate the implications for model results, if any.
5. Evaluate the shape of the loss distribution. Catastrophe models generate loss distributions of various shapes and characteristics, which are largely a function of the frequency of the underlying phenomenon and event severity characteristics. Many popularly used loss metrics are moments of these loss distributions and one has to carefully evaluate them. In pricing, the mean or average annual loss (AAL) is often used and there needs to be an evaluation of how to benchmark the AAL with past historical experience. Because loss distributions of catastrophes are usually fat-tailed, the tail contribution to the AAL can be substantial and can pose challenges when compared to a short historical record.
6. Evaluate modeled events that drive the tail of the modeled distribution. This requires domain expertise to form an opinion on whether the simulated events are physically realistic and plausible. By studying these events, the user of the model can develop a great intuitive sense of what drives risk for the particular peril.

CHAPTER

3

Towards a More Dynamical Paradigm for Natural Catastrophe Risk Modeling

David B. Stephenson[1], *Alasdair Hunter*[2], *Ben Youngman*[1], *Ian Cook*[3]

[1]Department of Mathematics, University of Exeter, Exeter, UK; [2]Barcelona Supercomputing Centre, Barcelona, Spain; [3]Willis Towers Watson, Willis Research Network, London, UK

INTRODUCTION

This chapter reviews how Event Loss Tables (ELTs) produced by catastrophe models are used to simulate insurance losses due to natural hazards. A formal statistical interpretation is presented in terms of a static mixture of compound processes and implications for clustering of losses are discussed. The consistency of this approach is then critiqued on the basis of physical understanding of natural hazard processes. A more dynamic nonstationary approach is demonstrated using various idealized dynamical mixing models and results are compared for an artificial ELT typical of windstorm losses. The tail of the aggregate loss distribution is found to be highly sensitive to the modeling choices. The chapter concludes with a summary of main findings and some ideas for further developments.

LOSS SIMULATION

The most direct way to estimate insurance loss distributions would be to use historical claims data. However, such an approach is not reliable due to the rarity and quality of the loss event data, the shortness of the historical records, and the nonstationarity of the losses (e.g., due to technological and economic trends). An alternative approach known as catastrophe modeling (Friedman, 1972) relies upon developing a model to simulate a large sample of hazard events, and then for each of these synthetic hazards, vulnerability and exposure

Risk Modeling for Hazards and Disasters
http://dx.doi.org/10.1016/B978-0-12-804071-3.00003-3

information are used to calculate property damage and subsequent insurance losses. Cat(astrophe) models have become the *de facto* approach to quantify natural hazard risk after the unprecedented losses in the early 1990s; examples include the cluster of windstorms that affected Europe in 1990, Hurricane Andrew in 1992, and the Northridge earthquake of 1994 (Grossi and Kunreuther, 2005). Many cat models use a discrete event simulation approach whereby random numbers for hazard variables are drawn from a simplified parametric stochastic model of the hazard process that has been fitted to past data. Alternatively, a large sample of plausible hazard events can also be generated *ab initio* by direct numerical simulation from a complex physical model of the hazard (e.g., a regional climate model). However, such numerical simulation is often computationally expensive and requires ad hoc bias correction to adjust for model misrepresentation of extreme events. This chapter will focus on the more common stochastic simulation approach.

Event Loss Table Definition

For a given insurance portfolio (i.e., spatial vulnerability and exposure information), most probabilistic cat models produce an ELT, which can be used to create plausible simulations of losses for the forthcoming insurance period (hereafter, assumed to be the forthcoming calendar year). An ELT typically has thousands of rows each of which represents the properties of a certain possible type of hazard, namely, its rate of occurrence ρ and a set of parameters θ that describe the distribution of losses that would be incurred if such an event were to happen. The distribution is bounded from above by the maximum exposure and represents uncertainty in losses for this given hazard event due to imperfect knowledge of vulnerability and exposure in the portfolio (so called *secondary uncertainty*).

Fig. 3.1 shows a scatter plot of rate and the loss expectation (i.e., the mean of the loss distribution) for each of $I = 55{,}000$ rows from an artificial ELT created by simulation from a bivariate lognormal distribution fitted using maximum likelihood to mimic ELTs typically generated by European windstorm catastrophe models. Rates and losses vary over many orders of magnitude with rows having larger mean losses tending to have smaller and less uncertain rates.

An ELT is essentially a very large set of model parameters for a finite mixture model of compound Poisson processes that can then be used to simulate losses. Losses in a year (or other desired time period) are assumed to be a mixture of independent losses drawn from distributions specified by parameters given by each row of the ELT. The $N = \sum N_i$ losses in a year consist of $N_i \sim Poi(\rho_i)$ losses drawn independently from a known cumulative distribution function $F(x;\theta_i)$ for each row $i = 1,2,\ldots,I$ in the table. Because the total rate is usually small and there are many rows in the ELT, most of the N_i counts are 0, a few are 1, and very few are greater than 1.

In the simplest case, each type of hazard in the ELT leads to an exact loss, in other words, there is no additional so-called secondary uncertainty caused by imperfect knowledge of vulnerability and exposure. In such cases, losses from row i of the ELT are always identical fixed values equal to the expected loss $\theta_i = E(X_i)$ and the cumulative distribution function $\Pr(X \leq x) = F(x;\theta_i)$ is a step function equal to 0 when $x \leq \theta_i$ and 1 when $x > \theta_i$. More generally when there is secondary uncertainty, this can be represented either by specifying parameters of distributions (e.g., the 3-parameter Beta distribution) for each row in the ELT, or by

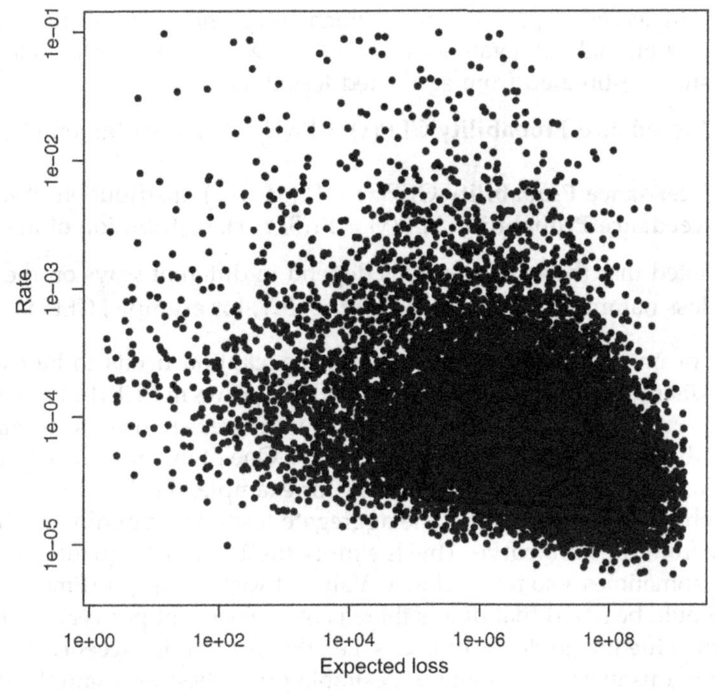

FIGURE 3.1 Scatter plot of rate against expected loss (on logarithmic axis scales) for the 55,000 rows of the illustrative ELT typical of European windstorm losses. Negative association due to smaller rates for higher losses is typical of most hazards.

replicating each row in the ELT and dividing its rate among a discrete set of fixed loss values according to their probabilities (J.C. Rougier, personal communication).

To simulate losses, one could loop over all the rows in the ELT. However, it is much faster to first find the total number of events in a year by drawing from a Poisson distribution $N \sim Poi(\rho)$, where $\rho = \sum \rho_i$, and then draw N losses from distributions having θ sampled N times by replacement from the θ_i each with probability ρ_i/ρ (for a proof see Rolski et al., 2009; Section 4.2.2). For example, in the case of no secondary uncertainty, this can be done using the following R language code:

```
rhosum <- sum(rho)
N <- rpois(rhosum)
losses <- sample(x=theta, size=N, replace=TRUE, prob=rho/rhosum)
```

where `rho` and `theta` are the column vectors of rates and expected losses, respectively, in the ELT.

Loss Distribution Functions: CEP, OEP, and AEP

Decisions about risk capital, portfolio management, and premium pricing are based on summary measures of distributions of various simulated loss variables. Following common practice, consider the time period of interest to be 1 year. Suppose that in year k there are

$N_k > 0$ single event losses $\{X_{jk}; j = 1,\ldots,N_k\}$, which have annual maximum loss $M_k = \max\{X_{jk}; j = 1,\ldots,N_k\}$ and annual aggregate loss $S_k = X_{1k} + X_{2k} \ldots + X_{N_k k}$. The following distributions are then usually estimated from simulated loss data:

- **Conditional Exceedance Probability** $\mathrm{CEP}(x) = \Pr(X_{jk} > x)$: distribution of single event loss;
- **Occurrence Exceedance Probability** $\mathrm{OEP}(x) = \Pr(M_k > x)$: distribution of maximum loss;
- **Aggregate Exceedance Probability** $\mathrm{AEP}(x) = \Pr(S_k > x)$: distribution of aggregate loss.

It should be noted that these distributions depend in different ways on the joint distribution of rate and loss parameters represented by the ELT, for example, $\mathrm{CEP}(x) = \rho^{-1} \sum_{\theta_i > x} \rho_i$ in the special case of no secondary uncertainty. Hence, caution needs to be exercised not to modify the joint distribution when merging and deleting rows in ELTs but to increase computational speed (so-called *boiling down*). Various risk measures are used to summarize exceedance probability distributions (Embrechts et al., 2005). One of the most widely used measures is the return value for a specified return period, for example, the T = 200-year return value, $\mathrm{AEP}^{-1}(1/\mathrm{T})$, is obtained by reading off the aggregate loss corresponding to the 1/T exceedance probability from the AEP curve. This is simply the $1 - 1/\mathrm{T}$'th quantile of the aggregate loss distribution sometimes also referred to as Value-at-Risk $\mathrm{VaR}_{1-1/\mathrm{T}}$ (Embrechts et al., 2005; Section 2.2). It should be noted that unless there is only one event per year, $\mathrm{CEP}^{-1}(1/\mathrm{T})$ is not the T-year return value for single event losses, i.e., the loss that is exceeded on average every T years. However, this metric has some undesirable properties: for example, it is not subadditive, i.e., the quantile of a sum of losses from two portfolios can exceed the sum of the quantiles of each individual portfolio. Subadditivity is one of the most important axioms of coherence for diversification as it guarantees that an aggregate portfolio made up of smaller portfolios is not exposed to greater risk than the sum of the constituent portfolios (Dowd and Blake, 2006). A more coherent measure is provided by *Expected Shortfall* ES_p—the expectation of losses that exceed the p'th quantile (sometimes also referred to as Tail Value at Risk TVaR and Tail Conditional Expectation). Expected shortfall can be obtained either by Monte Carlo simulation or by integration of VaR quantiles over the tail of the loss distribution:

$$ES_p = \frac{1}{1-p}\int_p^1 \mathrm{VaR}_q dq = E(X|X > \mathrm{VaR}_p)$$

Expected shortfall ES_p is never less than VaR_p, and when $P = 0$, is simply the expectation $E(X)$, i.e., the mean loss. Both VaR and ES risk measures will be used here to summarize loss.

CLUSTERING

Scientific Background

Unlike losses simulated from an ELT, natural hazard events are not generally independent; background conditions can cause hazards to arrive in close succession, which enhances the risk of large aggregate losses, e.g., the winter 2013/14 cluster of European windstorms

Christian, Xavier, Dirk, and Tini caused insured losses of 1.38, 0.96, 0.47, and 0.36 billion USD that in total exceeded 3 billion USD.

By fitting stochastic models to historical storm track data, new scientific discoveries have been made in recent years about the clustering of windstorms. Notably,

- there is more clustering than expected due to chance, i.e., there is significant overdispersion in storm counts to the south and north of the Atlantic storm track region and over West Europe (Mailier et al., 2006);
- clustering can be largely explained by dynamic modulation of Poisson rates by the North Atlantic Oscillation and other large-scale climate modes (Mailier et al., 2006);
- clustering increases for more intense windstorms especially over North Europe and Scandinavia (Vitolo et al., 2009), which is most likely related to the fact that rate and intensity of windstorms are positively correlated and increase together over large parts of North Europe (Hunter et al., 2016).

Similar findings have been reproduced in storms simulated by climate models (Kvamstø et al., 2008; Pinto et al., 2013; Economou et al., 2015). These results are stimulating new research into mechanisms for clustering of extreme storms (e.g., Rossby wave breaking; Pinto et al., 2014). Furthermore, many of these aspects of clustering also apply to other natural hazards such as hurricanes (Mumby et al., 2011) and floods (Villarini et al., 2013).

Overdispersion in Counts

A simple way to characterize clustering is to assess the variance in counts of events. If the events occur randomly in time with no clustering, then the number of events N in a fixed time period will be Poisson distributed with a variance equal to the constant mean rate. If clustering is present, this will be associated with additional variation that causes the variance to exceed the mean counts—a phenomenon known as *overdispersion.* Clustering is sometimes more narrowly defined to be additional variance in counts to that accounted for by Poisson sampling AND known variations in the rate, i.e., more variance than can be explained by a nonhomogeneous Poisson process. Overdispersion can be assessed by the statistic $\phi = Var(N)/E(N) - 1$, which indicates clustering when positive. This statistic has been extensively used in the studies mentioned in the previous section.

To understand how clustering behaves for more extreme events, it is useful to consider overdispersion in the number of events $N(u)$ that have losses/intensities exceeding threshold u. Vitolo et al. (2009) used such a quantity to show that there was increased clustering for windstorms with more extreme vorticities on the northern flank of the Atlantic storm track that extends into Scandinavia and North Europe.

For losses generated from an ELT, the Laws of Total Expectation and Variance can be used to show that the expectation and variance of $N(u)$ are given by

$$E(N) = \sum_{i=1}^{I} \rho_i(1 - F(u;\theta_i))$$

$$Var(N) = \sum_{i=1}^{I} \left(\rho_i(1 - F(u;\theta_i)) + \rho_i(1 - F(u;\theta_i))^2\phi_i\right)$$

For generality to allow row counts to be negative binomially as well as Poisson distributed, it is assumed that the row counts N_i have expectation $E(N_i) = \rho_i$ and variance $Var(N_i) = \rho_i(1+\phi_i)$, where ϕ_i is a prescribed overdispersion for counts generated by row i. By taking the ratio of these expressions, the overdispersion in $N(u)$ is given by

$$\phi(u) = \frac{\sum_{i=1}^{I} \rho_i (1 - F(u;\theta_i))^2 \phi_i}{\sum_{i=1}^{I} \rho_i (1 - F(u;\theta_i))}$$

which leads to various interesting deductions. Firstly, $\phi(u)$ is strictly positive only if one or more of the rows has overdispersion (i.e., $\phi_i > 0$) and so if the rows all have Poisson counts with $\phi_i = 0$ then there is no overdispersion in the total counts above any threshold. Secondly, in the presence of secondary uncertainty as $u \to \infty$, $1 - F(u;\theta_i) \to 0$ and so $\phi(u) \to 0$, i.e., overdispersion and clustering vanish in total counts of extreme losses even if row counts are overdispersed (e.g., negative binomial). Finally, in the special case of no secondary uncertainty, $1 - F(u;\theta_i)$ is only either 0 or 1 and so overdispersion simplifies for independent row counts to

$$\phi(u) = \frac{\sum_{\theta_i > u} \rho_i \phi_i}{\sum_{\theta_i > u} \rho_i},$$

i.e., a weighted average of the overdispersion in each of the rows with losses θ_i that exceed the threshold u. When all the rows have the same dispersion $\phi_i = \phi$, this leads to overdispersion in the total counts that is independent of threshold, i.e., $\phi(u) = \phi$ but this is a singular result valid only when there is absolutely no secondary uncertainty.

Building Clustering into ELT Calculations

Scientific evidence for storm clustering has led to several recent attempts to build clustering into cat model loss simulations. The simplest approach is to assume that row counts are independently distributed as negative binomial rather than as Poisson distributions. From the arguments in Section Overdispersion in Counts above, this will result in overdispersion of losses above low thresholds that unfortunately vanishes for more extreme losses. Khare et al. (2015) describe a more sophisticated approach that splits the ELT into $K > 1$ independent clusters/groups of mutually dependent rows that have negative binomial counts with different shape parameters (overdispersion) for each cluster. Such an approach will also result in vanishing overdispersion in total counts above sufficiently high thresholds. Furthermore, there is little scientific evidence for believing that there are a finite number of independent groups of windstorm type—a continuum is perhaps more justifiable for describing vortex features in a turbulent atmosphere. An alternative approach for creating clusters of losses is to create a parent loss using Poisson sampling from the ELT and then generate offspring losses by making repeated draws from the secondary uncertainty distribution of the parent row. Such a Neymann-Scott process has been implemented on the financial platform at Willis Re but it is difficult with this approach to get overdispersion to increase for more extreme losses. Furthermore, it could be argued that such a process is nonphysical in that it creates clusters of losses for a given hazard rather than creating losses for a cluster of hazards as occurs in the real world.

DYNAMICAL MIXTURE MODELS

Concept

Fig. 3.2 shows a simple graphical model for how ELT simulations can be made more dynamical. Random (mixing) variables Z and Z' can be introduced to represent time-varying drivers, which modulate the rate and loss distribution parameters (e.g., the North Atlantic Oscillation and Scandinavian Pattern indices for European windstorms; see Hunter et al., 2016). The drivers can be either independent or correlated with one another and can be specific to the spatial location of the portfolio.

As a first step, one can ignore modulation of losses and replace the static ρ_i rates in an ELT with dynamical rates $\lambda_i(Z)$ that depend on a time-dependent random mixing variable Z(t), which modulates the hazard rate. Such an exogenous mixing approach is already a common practice in credit risk modeling (Crouhy et al., 2000; Frey and McNeil, 2003). Credit risk portfolios take a similar form to ELTs; a portfolio will consist of a bank's loans each of which is characterized by a probability of defaulting and a corresponding loss. Although loans are

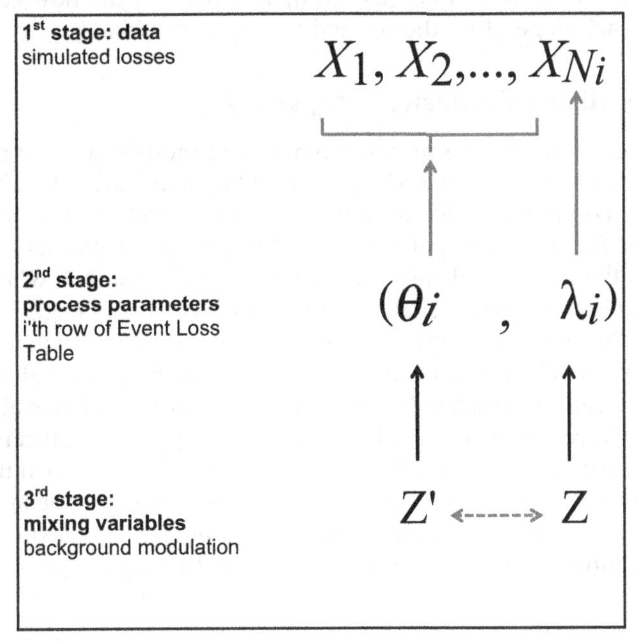

FIGURE 3.2 Schematic graph showing the hierarchical relationship between variables used to simulate losses from an ELT. *Blue arrows* (light gray in print versions) denote dependency, i.e., random variables at the tip are drawn from distributions controlled by variables at the base, for example, $X \sim F(\theta_i)$ and $N_i \sim Poi(\lambda_i)$. *Black arrows* denote deterministic functions and the *dashed arrow* represents association that may exist between variables that drive rates and those that drive hazard intensity (e.g., for windstorms over Scandinavia; see Hunter et al., 2016).

assumed to default independently of each other, they can be affected by the same set of background factors such as the rate of inflation, interest rates etc. To include these drivers in credit risk models, probabilities of loans are regressed on a common set of random variables that represent these factors.

Adopting a Generalized Linear Model approach, the counts from row i of the ELT can be modeled as $N_i \sim Poi(\lambda_i)$, with the rate depending on the mixing variable via $g(\lambda_i(Z)) = \beta_{0,i} + \beta_{1,i}Z$, where $g(.)$ is a monotonic function known as a *link function.* When $g^{-1}(\beta_{0,i} + \beta_{1,i}Z)$ is Gamma distributed, this gives a Gamma-Poisson mixture for counts, which is equivalent to assuming negative binomial counts. The simplest dynamical model is obtained by considering an identity link function and a mixing variable that can take only binary values, 0 and 1, each with probability 0.5, i.e., $Z \sim Ber(0.5)$. Binary mixing is a crude coin-toss approach for modeling years that can have either low or high states (i.e., $Z = 0$ or $Z = 1$) of some climate mode such as the North Atlantic Oscillation. A more continuous approach is to represent the climate index using a standardized Gaussian distributed variable, i.e., $Z \sim N(0,1)$. Following usual practice in Poisson regression, the link function $g(.)$ is then chosen to be the *canonical* logarithmic link, which prevents having negative rate parameters. The rates of each row are then log-normally distributed and analytic expressions can be used to find the first and second moments for each row. Since the lognormal distribution is often quite similar to the Gamma distribution, this gives counts for each row that are close to having a negative binomial distribution. However, it should be noted that the counts in all the rows (and not just groups) of the ELT are now correlated with one another since they each depend on the mutual mixing variable Z.

Estimation of the Mixing Parameters: $\beta_{0,i}$ and $\beta_{1,i}$

Because of the lack of driver information from hazard models, it is not possible to regress ELT rates on known drivers such as leading modes of climate variability. Therefore, alternative approaches are required for estimating the mixing parameters. A simple method of moments approach for finding these parameters is to impose constraints on the expectation $E(N_i) = E(g(\lambda_i))$ and the variance of row counts $Var(N_i) = (1+\phi_i)E(N_i)$, where overdispersion $\phi_i = Var(g(\lambda_i))/E(g(\lambda_i))$. To ensure that mean behavior is similar (e.g., expected loss), it is natural to assume that the expectation of the row counts should equal that of the static approach i.e., $E(N_i) = \rho_i$. Overdispersion can be set to nonnegative values $\phi_i \geq 0$ for each row to reproduce desired clustering behavior in the losses such as a believable overdispersion profile $\phi(u)$. These two constraints for each row give two equations that can be solved to find the beta parameters, e.g., $\beta_{1,i} = \sqrt{\phi_i \rho_i / VarZ}$ and $\beta_{0,i} = \rho_i - \beta_{1,i}E(Z)$ when the link function is the identity function $g(\lambda_i) = \lambda_i$. For the special case where the overdispersion of row i is chosen to be $\phi_i = \phi\rho_i/\rho$ and the rates are modulated by a binary $Z \sim Ber(0.5)$ mixing variable via an identity link function (i.e., binary mixing), beta estimates are given by

$$\beta_{1,i} = 2\rho_i\sqrt{\frac{\phi}{\rho}}$$

$$\beta_{0,i} = \rho_i - \frac{\beta_{1,i}}{2}$$

When $\phi_i = \phi\rho_i/\rho$ and the rates are modulated by a Gaussian $Z \sim N(0,1)$ mixing variable via a logarithmic link function (i.e., lognormal mixing), beta estimates are given by

$$\beta_{1,i} = \sqrt{log(1+\phi/\rho)}$$

$$\beta_{0,i} = log\rho_i - \frac{log(1+\phi/\rho)}{2}$$

Note that the $\beta_{1,i}$ slope parameters increase monotonically with the clustering parameter ϕ, i.e., increased clustering is associated with greater dependence of the rates on the mixing variable. The above expressions assume for convenience that rates increase with the mixing variable, but negative relationships (e.g., the effect of ENSO on US hurricanes) can also be easily represented, e.g., by reversing the signs of the $\beta_{1,i}$ slope parameters and for binary mixing setting $\beta_{1,i} = \rho_i + \beta_{1,i}/2$. Since rates can never be negative, $\phi > \rho$ is nonadmissible for the binary mixing model, whereas any value $\phi \geq 0$ is allowed when using the logarithmic link function.

Choice of Row Overdispersion: Scaling Properties

The special case $\phi_i = \phi\rho_i/\rho$ can be proven to give dynamical rates for each row that scale similarly with the mixing variable, i.e., $\lambda_i = \rho_i f(Z)$ where $E(f(Z)) = 1$ and $Var(f(Z)) = \phi/\rho$. Conversely, rates scaling as $\lambda_i = \rho_i f(Z)$ with $E(f(Z)) = 1$ implies $\phi_i = \phi\rho_i/\rho$ where $\phi/\rho = Var(f(Z))$. This powerful equivalence holds for all possible distributions of mixing variable and choices of link function. An important consequence of such scaling is that the CEPs (and the mean loss per event) are independent of Z and so the mixing variable only affects the rates but not the distribution of the single event losses (see Hunter, 2014; Appendix C). Finally, such scaling guarantees that $Cov(\lambda_i, \lambda_j) = \phi\rho_i\rho_j/\rho$, and hence overdispersion in the number of all losses $\phi(0) = \phi$ when the link function is the identity.

The scaling choice, $\phi_i = \phi\rho_i/\rho$, generally leads to $\phi(u)$ decreasing from ϕ to zero with increasing threshold u. However, one may wish to create dynamic mixing models that can reproduce counts that have different clustering profiles, for example, overdispersion that stays constant with threshold, e.g., $\phi(u) = 0.3$ for all u. For the simplest case with no secondary uncertainty, this can be achieved by iteratively estimating the beta parameters starting at the row in the ELT that has the largest mean loss $\theta_{(1)}$. The dispersion parameter for this row can be estimated by setting it equal to the dispersion of counts exceeding $\theta_{(1)}$, i.e., $\phi_{(1)} = \phi(\theta_{(1)})$. The dispersion parameters for rows with progressively smaller θ_i can then be found iteratively by rearranging expressions for $\phi(u)$ such as this one is valid for linear link functions when there is negligible secondary uncertainty:

$$\phi(u) = \frac{\sum\limits_{\theta_i>u}\sum\limits_{\theta_j>u}\sqrt{\phi_i\phi_j\rho_i\rho_j}}{\sum\limits_{\theta_k>u}\rho_k} \tag{3.1}$$

See Section 5.3.4 of Hunter (2014) for more details.

Models and Simulations

To illustrate these approaches, simulations have been made with five different ELT simulation models:

- M1: static sampling with no mixing, i.e., constant rates giving $\phi(u) = 0$
- M2: dynamic sampling with a binary mixing and $\phi_i = \phi\rho_i/\rho$ with $\phi = 0.3$
- M3: dynamic sampling with lognormal mixing and $\phi_i = \phi\rho_i/\rho$ with $\phi = 0.3$
- M4: dynamic sampling with binary mixing designed to give $\phi(u) = 0.3$
- M5: dynamic sampling with lognormal mixing designed to give $\phi(u) = 0.3$.

For each model, losses were simulated by the following algorithm:

```
Make J=100000 random draws of Z to represent varying annual conditions;

For j in years 1,2,...,J
{
  Calculate new rates λi(zj) for each row in the ELT;
  Draw annual counts Nj~Poi(λ(zj)) with total rate λ(zj) = Σ_i λi(zj);
  Draw Nj losses from loss distribution having parameters θ sampled Nj times by
  replacement from the θi row parameters with probability λi(zj)/λ(zj).
}
```

Results

Clustering

Fig. 3.3 shows how the overdispersion of the annual counts of the simulated losses depends on the threshold for each of the models. The static model, M1, is simply a mixture of compound Poisson processes and so has no dispersion for all threshold choices. Models M2 and M3 have decreasing dispersion with threshold, which closely follows $\phi = \mathrm{CEP}(u)$, which is obtained by substituting $\phi_i = \phi\rho_i/\rho$ into Eq. (3.1). Models M4 and M5 have succeeded in generating losses that have constant dispersion of 0.3 in counts above any threshold value, although the binary model shows a drop off due to the nonadmissibility condition mentioned earlier.

Aggregate Exceedance Probability

Fig. 3.4 shows AEP curves for simulated losses from the models M1–M5. Little difference can be seen between the curves for models M1, M2, and M3. Models M4 and M5 with constant overdispersion have AEP curves that differ from those of the other models for aggregate losses that are rarer than the 0.05 exceedance probability (20 year return period).

Constant overdispersion has substantially increased the risk of extreme aggregate loss as can be seen in risk measures shown in Table 3.1. The value-at-risk $\mathrm{VaR}_{0.995}$ for the models M2 and M3 are not significantly different from that of M1, but $\mathrm{VaR}_{0.995}$ for M4 and M5 are 18% and 32% greater than that for model M1. The difference between the expected shortfalls $ES_{0.995}$ is even more pronounced, with M5 having a shortfall that is 50% greater than that of M1. All models give the same expected annual aggregate loss, with a

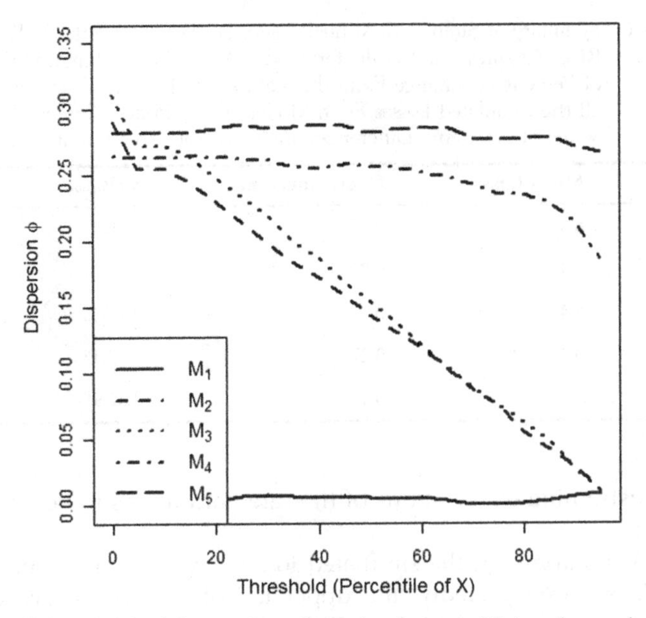

FIGURE 3.3 Overdispersion $\phi(u)$ of the number of simulated events $N(u)$ each year having losses exceeding threshold u. The threshold is expressed in terms of percentiles on the horizontal axis, i.e., $(1-\text{CEP}(u)) \times 100\%$. The static approach (model M1) has no overdispersion, models M2 and M3 have dispersion that decreases with loss intensity, and models M4 and M5 have dispersion that is constructed to stay constant with threshold.

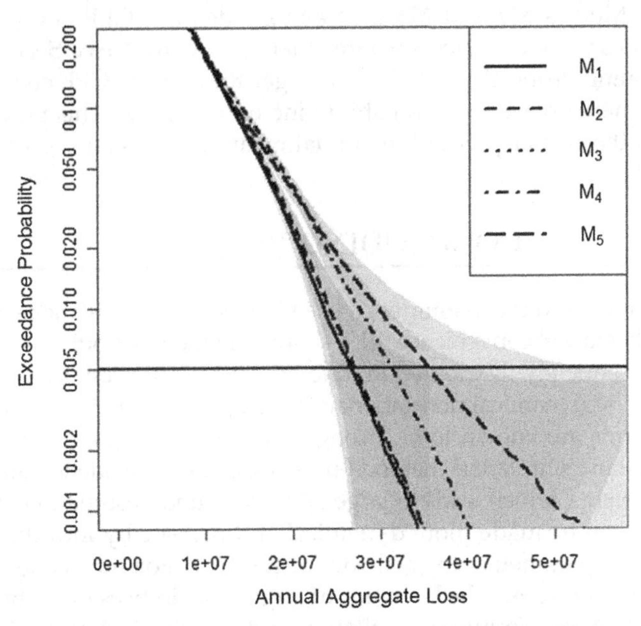

FIGURE 3.4 AEP curves for the loss models. Gray shading indicates the 95% Dvoretzky-Kiefer-Wolfowitz bounds (Hunter, 2014).

TABLE 3.1 Summary of Simulated Annual Aggregate Losses. For the VaR and *ES* Risk Measures, the Results for Models M2–M5 are Reported in Terms of Percentage Change From the Values for M1. The Overdispersion for all the Simulated Losses From Models M2–M5 are Close to the Chosen Value of 0.3—Any Differences are Solely Due to Sampling Variations

Model	Mean Loss	Overdispersion	$VaR_{0.995}$	$ES_{0.995}$
M1	6.4	0	26.8	30.5
M2	6.4	0.30	+1.4%	+3%
M3	6.4	0.32	+1.3%	+3%
M4	6.4	0.27	+17.1%	+19%
M5	6.4	0.29	+32.9%	+50%

consequence of constraining expectations of the rate parameters to be equal to the original rates for each row.

Fig. 3.5 shows AEP curves for the simulated losses conditional on specific ranges of the mixing variable Z. Not unexpectedly, the upper tail of the AEP curve depends strongly upon the value of the mixing variable for all the dynamical models M2–M5. The effect is greatest for the models with constant overdispersion.

Conditional Exceedance Probability

Fig. 3.6 shows CEP curves for the simulated losses conditional on specific ranges of the mixing variable Z. Models M2 and M3 give almost identical CEP curves to that from the M1 static approach as to be expected for rates that scale with Z (see Section Choice of Row Overdispersion: Scaling Properties). This is no longer the case for CEP curves for simulations from M4 and M5 where exceedance probability increases with Z. This positive association is clearly apparent in the scatter plot of the annual mean losses versus Z values simulated by model M5 (Fig. 3.7).

CONCLUDING REMARKS

This chapter has reviewed the assumptions used in stochastic simulations of natural catastrophe losses. For mathematical convenience, ELT simulation methods often assume that losses are identically and independently distributed in different years. Unfortunately, this static approach is inconsistent with the dynamical nonstationary behavior of the natural environment: natural hazards such as storms are known to be strongly influenced by climatic conditions that vary in time. Therefore, static simulation methods underestimate dependency and clustering especially for large loss events, which leads to large systematic underestimation of aggregate losses.

Loss calculations can be made more dynamical and realistic by introducing random variables to modulate the parameters in each row of the ELT. For hurricanes and windstorms, these mixing variables represent the behavior of large-scale indices describing modes of variability such as the El Nino/Southern Oscillation and the North Atlantic Oscillation. Such a dynamical mixture approach is similar to recent approaches used to model portfolio credit

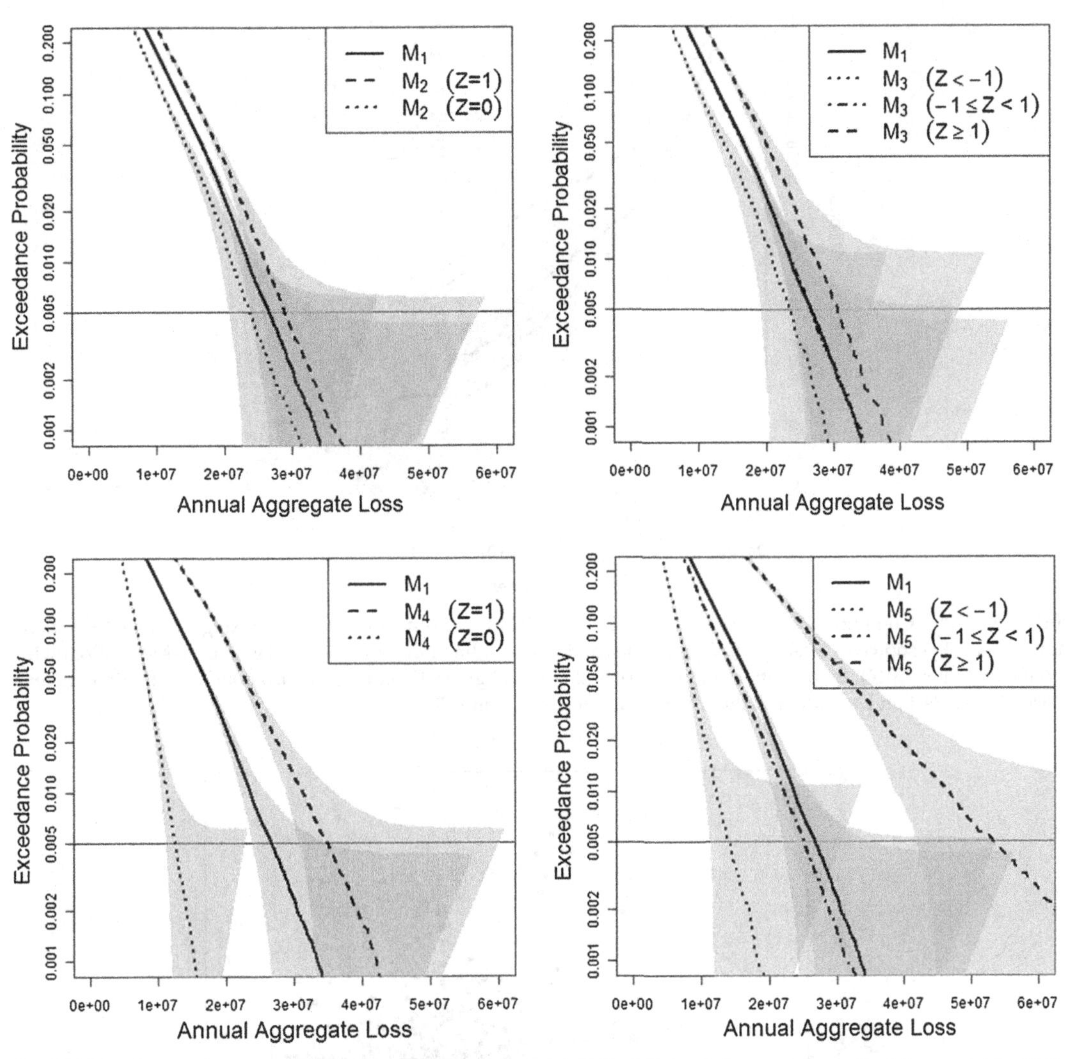

FIGURE 3.5 Conditional AEP curves simulated by models M2–M5 for different ranges of mixing variable. CEP curve for model M1 is shown as reference (*solid black line*) and gray shading indicates the 95% Dvoretzky-Kiefer-Wolfowitz bounds (Hunter, 2014).

risk, e.g., the CreditRisk + method based on binary mixing variables. This chapter has illustrated these approaches by applying four different toy mixing models to an ELT constructed to be typical of that used in practice for European windstorms. All of these models increase the probability of large aggregate losses (i.e., thicken the tail of the AEP curve), especially those models designed to maintain clustering for large loss events.

To make further progress, it is necessary to develop better more reliable methods for quantifying how clustering of hazards depends on intensity; for example, does it stay constant or sometimes even increase with threshold? To properly capture intensity-rate correlations,

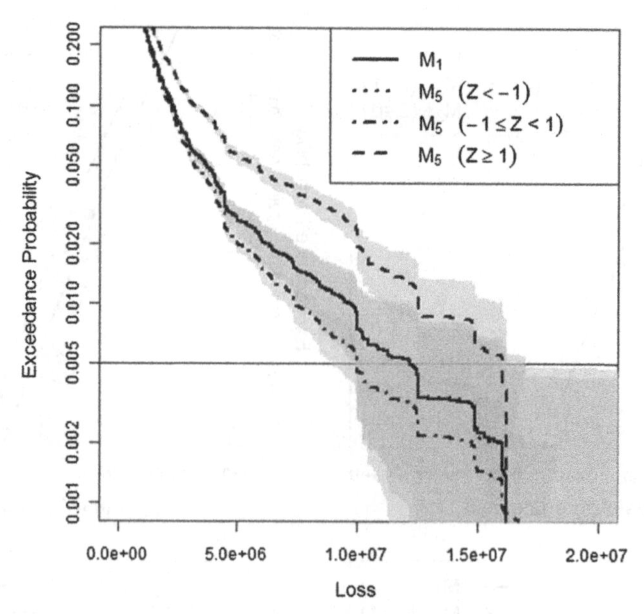

FIGURE 3.6 Conditional CEP curves simulated by model M5 for different ranges of mixing variable. CEP curve for model M1 is shown as reference (*solid black line*) and gray shading indicates the 95% Dvoretzky-Kiefer-Wolfowitz bounds (Hunter, 2014). By definition, models M1, M2, and M3 give CEP curves that do not depend on the mixing variable (see Section Estimation of the Mixing Parameters: $\beta_{0,i}$ and $\beta_{1,i}$).

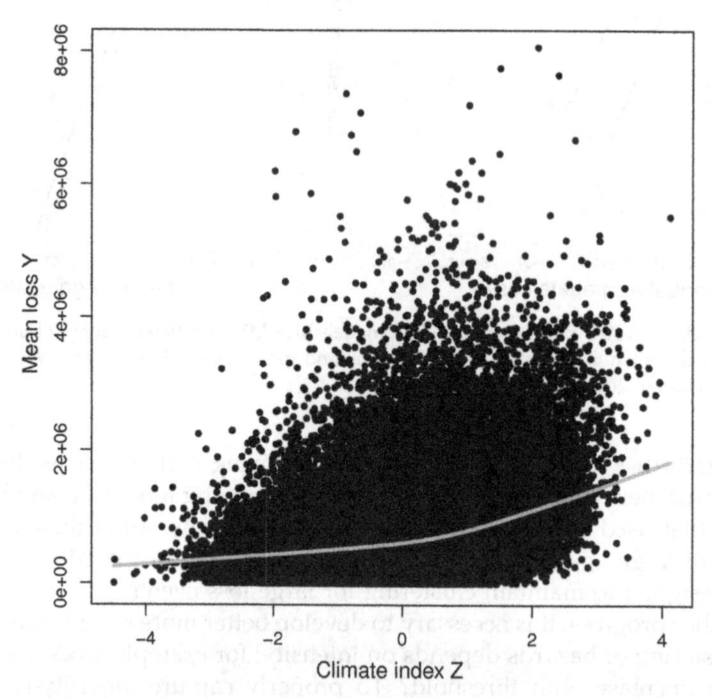

FIGURE 3.7 Scatter plot of mean loss per event (i.e., ratio of aggregate loss and number of events per year) versus mixing variable Z for 100,000 years simulated by model M5. Mean loss shows a strong nonlinear dependence on the mixing variable as indicated by the robust smoothing curve (LOWESS trend; *gray line*).

mixing models need to be developed that can allow the parameters of the loss distribution to be modulated rather than just the rate parameters. More robust estimation methods need to be developed for finding the beta parameters for each row of the ELT, e.g., for the case of increasing clustering profiles in the presence of secondary uncertainty.

Acknowledgments

We wish to thank the Willis Research Network for fellowship funding of Benjamin Youngman and for stimulating discussions especially with Rick Thomas. We are also grateful to Jonty Rougier for providing comments on an earlier draft of the chapter.

References

Crouhy, M., Galai, D., Mark, R., 2000. A comparative analysis of current credit risk models. Journal of Banking and Finance 24 (1), 59–117.

Dowd, K., Blake, D., 2006. After VaR: the theory, estimation, and insurance applications of quantile – based risk measures. Journal of Risk and Insurance 73 (2), 193–229.

Economou, T., Stephenson, D.B., Pinto, J., Shaffrey, L.C., Zappa, G., 2015. Serial clustering of extratropical cyclones in a multi-model ensemble of historical and future simulations. Quarterly Journal of the Royal Meteorological Society 141 (693), 3076–3087.

Embrechts, P., Frey, R., McNeil, A., 2005. Quantitative risk management, vol. 10. Princeton Series in Finance, Princeton.

Frey, R., McNeil, A.J., 2003. Dependent defaults in models of portfolio credit risk. Journal of Risk 6, 59–92.

Friedman, D.G., 1972. Insurance and the natural hazards. The ASTIN Bulletin: International Journal for Actuarial Studies in Non-Life Insurance and Risk Theory 7 (1), 4–58.

Grossi, P., Kunreuther, H., 2005. In: Catastrophe Modeling: A New Approach to Managing Risk, vol. 25. Springer, p. 245.

Hunter, A., 2014. Quantifying and Understanding the Aggregate Risk of Natural Hazards (Ph.D. thesis). University of Exeter, 163 pp. Available from: http://ethos.bl.uk.

Hunter, A., Stephenson, D.B., Economou, T., Holland, M., Cook, I., 2016. New perspectives on the collective risk of extratropical cyclones. Quarterly Journal the of Royal Meteorological Society 142 (694), 243–256.

Khare, S., Bonazzi, A., Mitas, C., Jewson, S., 2015. Modelling clustering of natural hazard phenomena and the effect on re/insurance loss perspectives. Natural Hazards Earth System Sciences 15, 1357–1370.

Kvamstø, N.-G., Song, Y., Seierstad, I., Sorteberg, A., Stephenson, D.B., 2008. Clustering of cyclones in the ARPEGE general circulation model. Tellus A 60 (3), 547–556.

Mailier, P., Stephenson, D., Ferro, C., Hodges, K., 2006. Serial clustering of extratropical cyclones. Monthly Weather Review 134 (8), 2224–2240.

Mumby, P.J., Vitolo, R., Stephenson, D.B., 2011. Temporal clustering of tropical cyclones and its ecosystem implications. Proceedings of the National Academy of Sciences USA 108, 17626–17630.

Pinto, J.G., Bellenbaum, N., Karremann, M.K., Della-Marta, P.M., 2013. Serial clustering of extratropical cyclones over the North Atlantic and Europe under recent and future climate conditions. Journal of Geophysical Research: Atmospheres 118 (22), 12476–12485.

Pinto, J.G., Gómara, I., Masato, G., Dacre, H.F., Woollings, T., Caballero, R., 2014. Large-scale dynamics associated with clustering of extratropical cyclones affecting Western Europe. Journal of Geophysical Research: Atmospheres 119 (24), 13704–13719.

Rolski, T., Schmidli, H., Schmidt, V., Teugels, J., 2009. In: Stochastic Processes for Insurance and Finance, vol. 505. John Wiley & Sons, p. 654.

Villarini, G., Smith, J.A., Vitolo, R., Stephenson, D.B., 2013. On the temporal clustering of US floods and its relationship to climate teleconnection patterns. International Journal of Climatology 33 (3), 629–640.

Vitolo, R., Stephenson, D., Cook, I., Mitchell-Wallace, K., 2009. Serial clustering of intense European storms. Meteorologische Zeitschrift 18 (4), 411–424.

CHAPTER

4

Empirical Fragility and Vulnerability Assessment: Not Just a Regression

Tiziana Rossetto, Ioanna Ioannou

Department of Civil, Environmental & Geomatic Engineering, University College London, London, UK

INTRODUCTION

Most state-of-art catastrophe models are nowadays composed of four main components. The first three characterize the hazard, exposed assets (e.g., people, buildings, infrastructure), and the assets' fragility. In the latter component, fragility functions are commonly used to describe the propensity of physical assets (e.g., buildings) to sustain damage under hazardous events. These fragility functions can be developed empirically, heuristically, but also analytically (i.e., where a numerical model simulates the response of a structure under increasing hazard intensities), or by a combination of these procedures (hybrid). Fig. 4.1A shows an example set of fragility curves relating the probability that the damage state (*DS*) of a building will equal or exceed a number of different damage states (ds_i) given the occurrence of intensity measure (IM) levels of increasing size. The final component of the catastrophe model is the financial module, which converts the estimated asset damage into loss, whether financial or human casualty. In this framework, calibration factors can be applied within the financial module to adjust the socioeconomic loss estimates to account for local policies, maximum payouts, and other factors that may affect the loss evaluation.

According to the Global Earthquake Model (GEM) terminology, vulnerability is defined as the susceptibility of assets (people, infrastructure etc.) exposed to hazardous events to incur losses (e.g., deaths and economic loss). In simple terms, vulnerability functions can replace the fragility and financial components of the catastrophe model. Indeed vulnerability functions can be derived "indirectly" from the combination of a fragility function and a damage-to-loss model, where the latter relates values of loss to thresholds of damage. Approaches for constructing indirect vulnerability functions from fragility functions and loss models are explained in detail in Rossetto et al. (2014). However, vulnerability functions can also be derived "directly" by fitting a statistical model to postevent loss data (empirical)

Risk Modeling for Hazards and Disasters
http://dx.doi.org/10.1016/B978-0-12-804071-3.00004-5

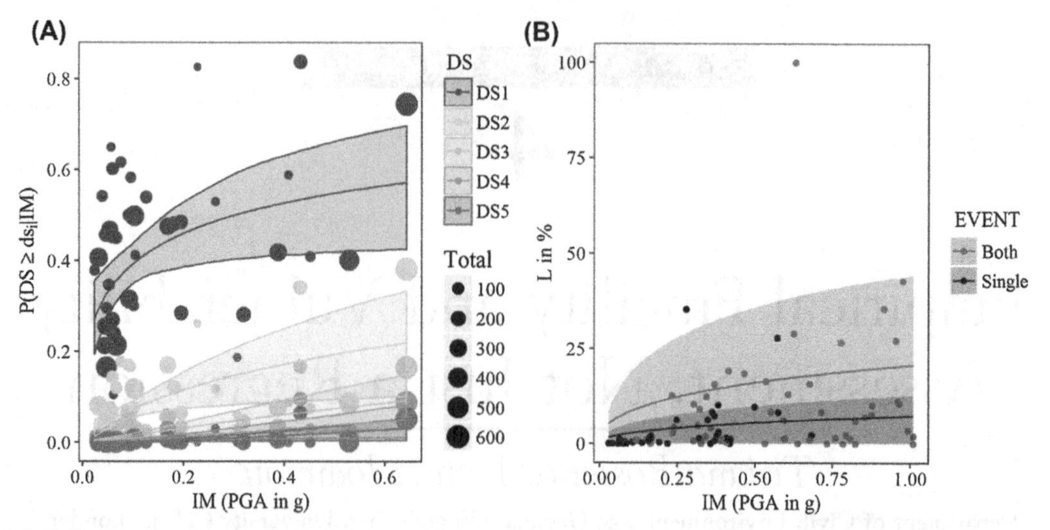

FIGURE 4.1 (A) Empirical fragility curves and their 90% confidence intervals for class C (least vulnerable building class according to Braga et al., 1982) Italian buildings constructed from damage observed after the 1980 Irpinia Earthquake, Italy. Damage is expressed in terms of the six state MSK-76 damage scale, which ranges from no damage to collapse. (B) Vulnerability curves with their associated 90% confidence intervals for Icelandic buildings surveyed in the aftermath of two successive earthquake events on the 17th and 21st June 2000. Vulnerability curves are presented using the loss data for areas affected only by a single event and the cumulative loss from areas affected by both events. *L* expresses the ratio of repair to replacement cost of the buildings as a percentage. *Both sets of curves have been developed by the authors.*

or through the elicitation of expert opinion (heuristic). An example of an empirical vulnerability curve is shown in Fig. 4.1B.

A wealth of fragility and vulnerability models for different exposures exist in the literature for different individual natural hazards. Significant efforts have been made recently to assemble databases of these functions for earthquakes and floods (e.g., Yepes-Estrada et al., 2016; Crowley et al., 2014; Pregnolato et al., 2015). These efforts highlight the lack of both vulnerability and fragility functions for many countries at risk, (particularly those in the developing world; see Fig. 9 in Yepes-Estrada et al., 2016), and a strong predominance of fragility over vulnerability functions and damage-to-loss models. Furthermore, empirical methods are seen to be the dominant approach in the derivation of seismic vulnerability and damage-to-loss functions, with a large proportion of seismic fragility functions also being empirical in nature. The same trends exist for other natural hazards, with an extreme case posed by tsunami where 90% of existing fragility functions are for Japan, and all but two published fragility studies adopt empirical approaches (Petrone et al., 2017).

Empirical approaches to exposure, fragility, and vulnerability are extensively used in the insurance industry, where they are commonly reputed the "gold standard" and to have high credibility because they are based on pastevent data (Crowley and Pinho, 2011). However, the authors have found that in practice there exists no perfect and unbiased event loss dataset from which to construct fragility and vulnerability functions. The reliability of published empirical vulnerability and fragility functions is seen to vary greatly depending on the quality and quantity of the empirical data used, and the procedures adopted to account for data biases and to fit statistical models to the data. Furthermore, despite the significant uncertainties associated with empirical data, it is observed that these are not appropriately communicated in existing vulnerability or fragility studies and that the uncertainty around fragility or vulnerability curves is rarely quantified.

These observations do not undermine the potential value of empirical vulnerability and fragility functions, which currently provide the best means of estimating damage and losses to assets that are complex to model (e.g., people and nonengineered buildings). Instead the authors mean to highlight the fact that the use of past event data presents significant challenges and that many sources of uncertainty in empirical fragility and vulnerability functions are often overlooked or ignored.

This chapter presents an overview of some significant observations made by the authors whilst working on earthquake and tsunami empirical fragility and vulnerability functions over the last 10 years. Common biases in postevent damage and loss datasets are presented, the consequences of ignoring the biases are discussed, and possible ways of dealing with them are suggested. The impact of statistical model fitting assumptions is also described and illustrated with examples drawn from empirical earthquake and tsunami fragility and vulnerability studies. Throughout, areas for further research are highlighted and it is observed that some sources of uncertainty remain largely unexplored.

CHALLENGES WITH POSTEVENT LOSS AND DAMAGE DATABASES

Postevent Loss and Damage Data Sources

Following a natural disaster, several surveys are carried out by different agencies to assess the effects of the event on the population, the built environment, and on the economy. With the diffusion of digital technologies and the internet, event loss and damage data are becoming more easily available. Furthermore, significant efforts have been made in recent years to compile such databases and make them available to the general public, e.g., the GEM Consequence database (So et al., 2012), the Cambridge Earthquake Damage and Casualty Database (Spence et al., 2011), the CATDAT database (Daniell et al., 2011), and EM-DAT International Disaster Database (2012), amongst others. The most common sources of loss and damage data listed in these postevent databases, and that have been used in the literature to derive fragility and vulnerability functions, are summarized in Table 4.1. In this table, the typical size, characteristics, and reliability of the resulting loss data are also described.

In empirical fragility assessment, common issues regarding postevent damage databases that can be observed are the following:

- Lack of detailed data on asset characteristics (e.g., buildings) that allow a distinction to be made between asset classes that will respond differently to the hazard effects. This results in fragility functions that are highly specific to the asset types and composition of the built environment in the location affected, hindering their use in the assessment of other locations.
- Small samples of the affected asset may have been surveyed that are unrepresentative of the overall assets in the area affected by the hazard event. A common example is that

TABLE 4.1 General Characteristics of Postevent Damage and Loss Data Collected Using Different Survey Methods. The Table is a Modified Version of That Presented in Rossetto et al. (2013)

Event Data Type	Survey Method	Typical Sample Sizes	Detail of Asset Data	Reliability of Observations	Typical Issues
Damage	Rapid Surveys	Large	Few details on building type/ characteristics and occupancy	Low	Safety or usability rather than damage evaluations. Need other sources to identify asset characteristics
	Detailed Engineering Surveys	Large to Small	Detailed data on buildings	High	Possibility of unrepresentative samples (i.e., subset of building exposure)
	Surveys by Reconnaissance Teams	Very Small	Detailed data on buildings	High	Unrepresentative samples as only small areas are surveyed (and typically most damaged areas)
	Remotely sensed	Very Large	Few details on building type/ characteristics	Low	Only collapse or very heavy damage states may be reliably identified. Misclassification errors
Economic Loss	Tax assessor data	Very large	Information on general asset class, use, and occupancy	High	Often focus on damaged assets only
	Claims data	Very large	Information on general asset class, use, and occupancy	High	Often focus on damaged and/ or insured assets only
Casualties	Government Surveys	Very large	Overall casualty numbers and their general location	Low	Unlikely association with building damage and causes of injuries
	Surveys by NGOs/hospitals	Varies	Overall casualty numbers and their general location	Med	Possibility of unrepresentative samples. Detailed casualty data
	Detailed Casualty Surveys	Very small	Detailed classes	High	Possibility of unrepresentative samples. Detailed casualty data

only damaged buildings are surveyed, with no information available on undamaged building locations and characteristics. This leads to incomplete or unrepresentative damage databases.

- The same data may not be available for all surveyed assets. If these omissions are not completely at random, they can severely bias the resulting fragility functions.
- Surveys being carried out for building safety evaluations may not reflect the sustained damage. Fig. 4.2 shows the distribution of damage states amongst buildings assigned to

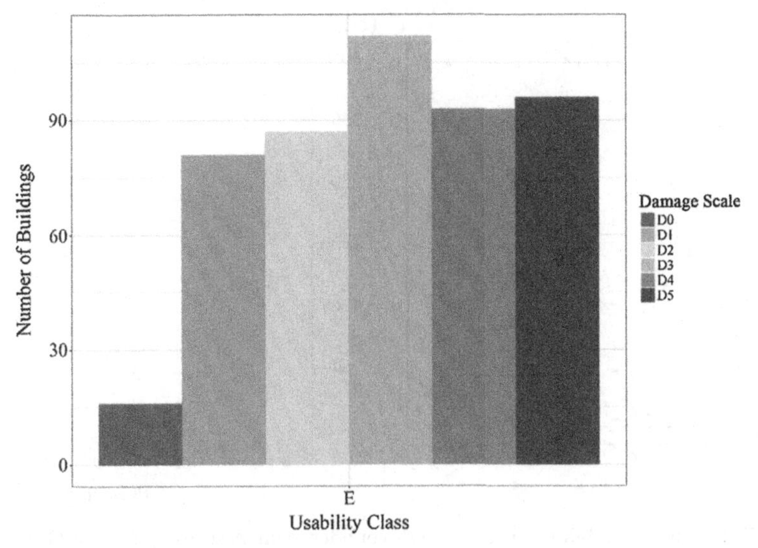

FIGURE 4.2 The number of buildings in damage states D0 (undamaged) to D5 (collapsed) assigned to usability class E (unusable) following the 2009 L'Aquila Earthquake in Italy (Bertelli, 2017).

Class E (unusable) following the L'Aquila, 2009 earthquake in Italy. This shows a spread of building damage between "D0", i.e., undamaged (possibly classed as unusable due to a peril posed by an adjoining structure) and "D5", i.e., collapse. The implication of this is that fragility functions derived using safety classifications should not be converted to equivalent damage state fragilities, and that the common assumption that "unusable" is equivalent to "partial collapse" or similar damage state is invalid.

- Misclassification errors, where either the building type or damage state is wrongly assigned/recorded in the damage database. The influence of misclassification errors are further discussed in the measurement and classification errors section.

A more general issue is that there is a greater abundance of damage data for low hazard intensity values than for large, and consequently more data for low damage states than for heavy damage and collapse. This should be taken into account appropriately in the statistical model fitting and uncertainty evaluation.

Economic loss data sources used in direct vulnerability assessment suffer from similar problems to the damage data sources used in fragility analysis, i.e., lack of detailed data on the asset type, use, and occupancy, and/or incomplete or small datasets. In practice, loss data also commonly comprises local policies for disbursement, maximum payouts, inflation, and other factors that may affect the loss value. This can result in empirical damage-to-loss ratios or vulnerability functions that are highly specific to local conditions and that do not represent the real spread in sustained asset damage. Drawing again on data from the 2009 L'Aquila, Italy, earthquake, Fig. 4.3B shows the vulnerability function derived from payout data for buildings assigned to usability class E (i.e., unusable). It is observed that the vulnerability function is effectively a flat line across all ground motion intensities. This

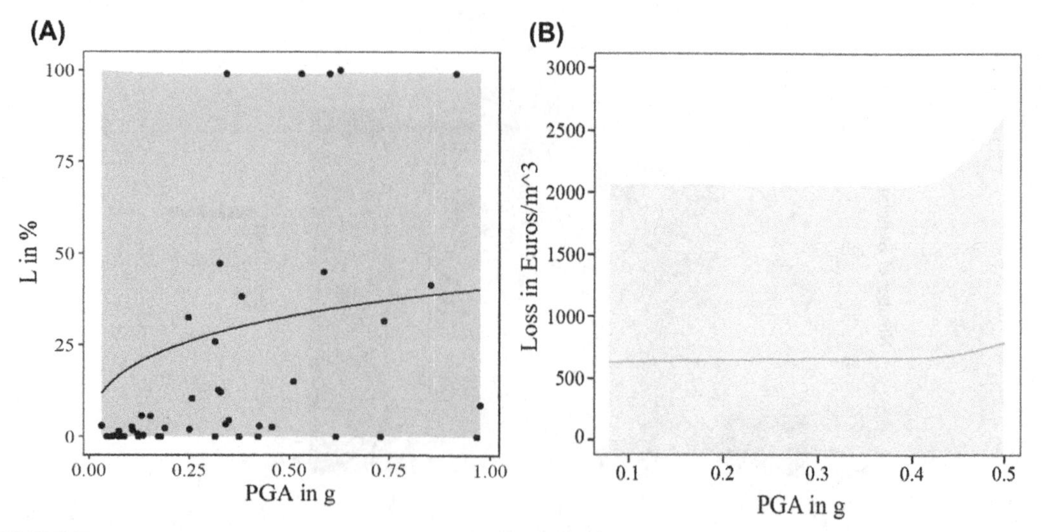

FIGURE 4.3 Empirical vulnerability curves and 90% confidence intervals using (A) the 2002 Icelandic data for pumice buildings and the maximum PGA level for the two successive events developed by the authors and (B) using payout data to buildings affected by the 2009 L'Aquila Earthquake in Italy and assigned to usability class E (unusable). The INGV shakemap has been used to define the peak ground acceleration at the sites of the buildings. *Reproduced with permission from Bertelli, S., 2017. Empirical Vulnerability: Derivation of Empirical Vulnerability and Fragility Functions from the 2009 L'Aquila Earthquake Data. (Thesis for MRes in Urban Sustainability and Resilience). Department of Civil, Environmental and Geomatic Engineering, University College London.*

reflects the local payout policy whereby the maximum amount of money that could be claimed if a building was assigned to usability Class E was approximately 650€/m^2 (Bertelli, 2017). From the spread of damage within Class E seen in Fig. 4.2, one would expect there to be a greater variation, with only some of the buildings being entitled to the maximum payout. But this does not take into account the human factor.

Quoting Jonathon Gascoigne of Willis Towers Watson, it is "more appropriate and potentially powerful to separate out the vulnerability assessment from socio-economic and cultural market factors in current modelling practice, where value-at-risk exposure data capture permits." Indeed, as mentioned in the introduction section, modern catastrophe modeling platforms allow for adjustments to be made to loss estimates that account for the cultural market factors. It is critically important to the portability of derived damage to loss functions or vulnerability functions that only empirical loss data that does not include significant policy or cultural influences be used in their derivation.

Recently, the authors have dealt with a high-quality loss database complied by Iceland Catastrophe Insurance, which includes building-by-building data for all individual residential buildings in Iceland affected by two successive earthquakes on the 17th and 21st June 2002. In this case, the loss dataset records only the cost of repairing the asset as a fraction of the asset replacement cost and does not include significant other influences. Despite this, analysis of the database shows that the ratio of repair to replacement cost of individual buildings is associated with very large levels of uncertainty. This uncertainty is such that it is impossible to depict any trends in the data when this is in a deaggregated form. Therefore this very high-quality data had to be aggregated to construct vulnerability curves. Despite this aggregation, the vulnerability curves for the smallest building class (pumice buildings) are seen to result in vulnerability curves with impractically large prediction intervals, as shown in Fig. 4.3A.

In direct vulnerability functions, casualties are usually characterized by fatality rates, which are commonly defined as the number of deaths as a proportion of the exposed population considered (Jaiswal and Wald, 2010). Similarly to economic loss vulnerability, indirect casualty vulnerability assessment involves combining building damage or collapse estimates from fragility analyses with lethality rates defined for the considered building class. In this case the lethality ratio is defined as the ratio of number of people killed to the number of occupants present in collapsed buildings of a particular class.

A number of sources of casualty data have been used to develop empirical fatality and lethality rates. These include official statistics published by government agencies, hospital admissions data, coroner and medical records, and specific surveys carried out by casualty researchers (e.g., De Bruycker et al., 1985). It is reported in Petal (2011) that casualty data are scarce, are associated with significant quality issues, and are plagued by inconsistencies in definitions and data collection methods. Similar to the case of economic loss data, various factors may affect the casualty data reported. These include the inflation or reduction of official death counts for political reasons (Petal, 2011). When on-the-ground surveys are carried out the resulting casualty data may be of high quality. However, the typical small samples associated with these databases may result in biases that result in estimates of fatality and lethality rates that cannot be projected to the affected population.

Incomplete Data

Incomplete datasets of event loss and damage data arise either due to the survey not having been conducted on all affected assets of interest (incomplete sample) or due to the omission of key information from fields of interest within a portion of the survey forms.

Incomplete samples are common in the insurance sector where data is only available for insured assets. In the case of government-led surveys, incomplete samples typically arise when the owner of the asset needs to request for the survey to be undertaken and results in a database mainly comprising damaged buildings. In the literature incomplete sample bias is addressed in two ways. The first, and preferred approach, involves completing the dataset with data from other sources. This approach commonly relies on the assumption that nonsurveyed assets were undamaged by the event, and a recent census is needed to complete the database (e.g., in Karababa and Pomonis, 2010). The second approach applies to geographically aggregated datasets and involves discarding data from any geographical unit where a minimum proportion of the assets have not been surveyed. However, the value of the latter "threshold of completeness" is seen to vary in the literature between around 60% and 80% of the exposure in earthquake fragility studies (Rossetto et al., 2014). In this approach, the assumption that nonsurveyed assets are not damaged is not made. Instead the distribution of asset damage or loss observed in the incomplete dataset is assumed to represent the complete dataset. Assuming a reasonably high threshold of completeness, this assumption is acceptable on condition that the missing data is missing completely at random. If there is instead a systematic bias in the way the surveys were conducted, the incomplete dataset will not be representative of the losses sustained by the full exposure and the resulting fragility or vulnerability functions will be biased.

For each asset reported within a damage or loss database, the asset location as well as several of its characteristics may be recorded together with the sustained damage/loss. It is commonly found in existing databases that some assets will not have complete entries across all fields, due to errors in the completion of the survey forms or errors in transcription from the forms to the databases. In the fragility and vulnerability literature the most commonly adopted approach in these cases is to conduct a complete-case analysis, i.e., to remove any partial data, such as buildings of unknown material, from their fragility analysis. If the missing data are missing completely at random the complete-case analysis will result in an unbiased fragility or vulnerability function (although the functions lose statistical power due to the reduced number of data used). However, if the missing information is related to the reason that the information is missing (e.g., if data on building design code is missing only for a portion of buildings that are well designed), the information is "Missing Not at Random" (Ware et al., 2012) in which case complete-case analysis would introduce a strong bias. The technique of multiple imputation (MI) is a well-known approach in statistics for estimating missing data from other attributes in the dataset, when the missing data can be shown to be "Missing at Random" (Ware et al., 2012). One of the few studies to have adopted MI in empirical vulnerability studies is Macabuag et al. (2016). They adopt MI to imply the construction material associated with 18.2% of the disaggregated building damage dataset from the 2011 Great East Japan earthquake and tsunami (total 67,125 buildings), to construct building class-—specific fragility functions. The analysis is carried out using four imputations to estimate construction material based on information on the building footprint area, building use, damage state, and inundation depth. They show a significant difference in the mean fragility functions achieved when using MI to complete the data, as compared to removing all data with missing information on construction material (see Fig. 4.4). This observation confirms the potential bias that can arise in vulnerability and fragility functions derived using incomplete data.

Measurement and Classification Errors

Four variables are central to the empirical assessment of asset inventories. These variables include the event intensity at the site where the assets are located, the characteristics of the asset inventory, the damage scale, and the loss. The literature overwhelmingly assumes that all these variables have been accurately measured for each asset. Nonetheless, this is often an unrealistic assumption and recently steps have been taken to address these issues as presented below.

Intensity

Central to the characterization of the empirical fragility or vulnerability of an asset is the choice of IM used and its measurement at sites where the assets are located.

The IM in fragility analysis is supposed to represent the damage potential of the hazard on the asset being assessed. Reviews of the literature (Rossetto et al., 2013; Macabuag et al., 2016) reveal that peak ground acceleration (PGA) and onshore inundation depth are the

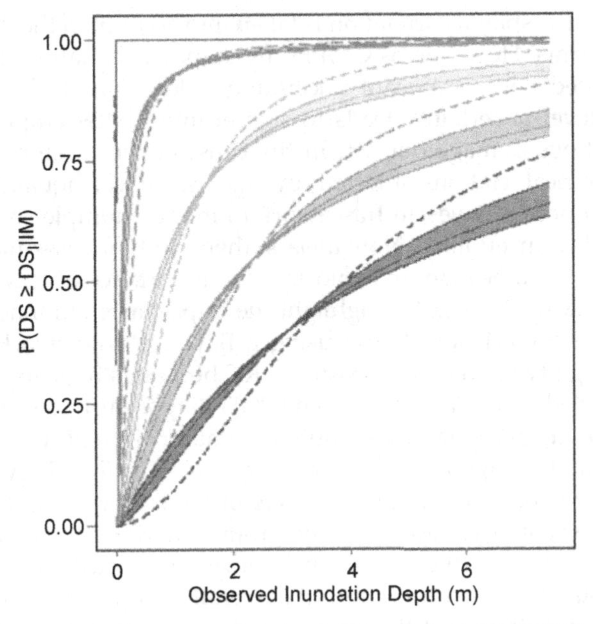

FIGURE 4.4 Empirical fragility curves for engineered structures affected by the 2011 Great East Japan earthquake and tsunami using data from three coastal towns derived by Macabuag et al. (2016). *Dashed lines* show curves formed using complete-case analysis (ignoring missing data). *Solid lines* show the mean curve for the imputed dataset, with the *shaded area* indicating the maximum/minimum values for the mean curves derived separately on each of the four imputations.

predominant IMs used in earthquake and tsunami building fragility studies, respectively. Peak ground velocity and permanent ground deformation are instead used more frequently for seismic fragility studies on underground structures like pipes and cables (e.g., Kongar et al., 2016). Especially in the case of buildings, these choices of IM are based on the ease of evaluation of these measures of intensity rather than on their specific ability to characterize their damage potential on particular assets. This is still true in much of the current fragility and vulnerability literature despite an extensive literature that focuses on proposing more efficient IMs, i.e., measures that reduce the variability in the structural response (e.g., Minas et al., 2015; Macabuag et al., 2016; amongst others).

In fragility and vulnerability studies the IM is assumed to be determined accurately and with complete certainty, as this is fundamental assumption in commonly used statistical model fitting techniques. However, when dealing with postevent observations this is never true.

In the case of most earthquakes, the lack of a dense network of ground motion recording stations means that there may only be a handful of measured ground motion records for each seismic event. These measurements cannot generally be directly extrapolated to determine the IM level at each asset location due to the asset's distance from the recording station or differences in soil characteristics between recording station and asset sites. Instead, Ground Motion Prediction Equations (GMPE) are used to complement data from recording stations and build a more detailed picture of the range of intensities produced by the earthquake.

In state-of-the-art studies, spatial correlation is taken into account in the development of these ground motion footprints. However, as GMPEs are empirical equations themselves, their IM estimates are associated with significant uncertainties. Ioannou et al. (2015) show that the presence of the IM level uncertainty leads to significantly flatter empirical fragility curves in the case of a fictitious damage dataset. In the construction of empirical fragility curves for elements of electrical stations affected by successive earthquakes, Straub and Der Kiureghian (2008) propose a Bayesian framework to model multiple sources of uncertainty, which also includes IM uncertainty. Nonetheless, their study focuses more on the development of the framework rather than on exploring the importance of this particular source of uncertainty. Yazgan (2012, 2013; 2015) highlight the importance of modeling the spatial correlation in accounting for IM uncertainty using a Bayesian framework capable to update existing analytical fragility curves for four-storey RC buildings by using data from 516 individual buildings affected by the 1999 Düzce and 2003 Bingöl earthquakes. More recently, the authors examined the impact of IM uncertainty on empirical fragility curves using large databases, though with data aggregated over geographical units. They propose a flexible Bayesian framework to construct fragility models of increasing complexity, accounting for uncertainty in IM, the spatial variability of the ground motion, as well as for the reduction of uncertainty in locations for which the intensity levels are measured. We show that by accounting for these sources of IM uncertainty the overall uncertainty in the fragility curves is reduced, but they demonstrate that the main source of uncertainty remains the overdispersion in the damage data. Fig. 4.5 depicts the reduction in the 90% credible intervals for the fragility curves corresponding to the collapse of the least vulnerable (class C) buildings in 1980 Irpinia, Italy, earthquake.

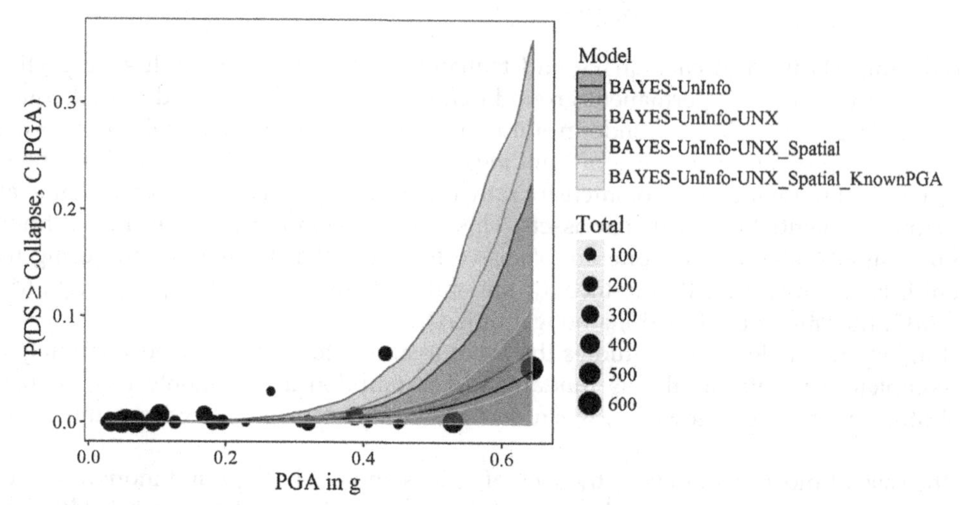

FIGURE 4.5 Mean fragility curves corresponding to the collapse state and associated 90% credible intervals for four models fit to the 1980 Irpinia earthquake data. PGA levels at the centroid of each municipality are estimated from Bindi et al. (2011). 'BAYES-UnInfo' ignores the uncertainty in the GMPE. 'BAYES-UnInfo_UNX' accounts for intra- and interevent uncertainty ignoring spatial correlation. The model termed 'BAYES-UnInfo_UNX_Spatial' accounts for the spatial correlation of the intraevent component of the uncertainty. Finally the model 'BAYES-UnInfo_UNX_Spatial_knownPGA' takes into account recorded ground motions. The size of the data points (i.e., number of buildings) used is shown by dots of different size, as per the legend.

In the empirical earthquake fragility literature there is one source of IM uncertainty that is still ignored. In empirical fragility studies it is a common practice to aggregate damage observations over a geographic area to produce a data point for the fragility function that is characterized by a single IM value, often evaluated at the centroid of the geographic area. However, there will be a natural variation of IM across the geographic area, and hence uncertainty is introduced when a single IM value is adopted. This source of uncertainty is ignored and is particularly worrisome as in single fragility studies vastly different sizes of geographic area can be associated with different datapoints, especially when the analyst tries to achieve a similar number of observations in each data point. Research is still needed to assess the overall importance of this source of uncertainty.

Tsunami empirical fragility studies published to date have not considered uncertainty in IM. In these studies, when inundation height at the asset site is used as the IM, this is either determined directly from field observation (e.g., Suppasri et al., 2013) or through the simulation of the tsunami onshore inundation using a number of different computational fluid dynamics codes (e.g., in Koshimura et al., 2009). Both these approaches to determining the tsunami inundation depth are prone to error. In the former case, mudlines or similar in buildings are used to determine the tsunami inundation height but it is not known if the mudlines are actually representative of the maximum inundation depth or are formed when the water level drops or when the water flows back offshore. In the latter case, numerical tsunami inundation models are currently unable to account for the presence of buildings explicitly when they are run over large urban areas, due to the very large computational expense that would be involved. The interaction between the onshore flow and the urban environment is commonly accounted for through the inclusion of a roughness parameter in the calculations. Although this approach is useful in providing an estimate of the inundation depth, it cannot be considered as accurate. Indeed Macabuag et al. (2016) observe that there are instances where the difference between the numerically predicted inundation depth and that observed in the filed differ by 4–6 m (i.e., the equivalent of 1–2 stories of a building). The determination of tsunami onshore velocity is even more error prone in the case of numerical modeling, and there are almost no observations against which the velocities calculated can be compared. Hence, fragility and vulnerability functions developed on the basis of onshore velocity or IMs determined from velocity (e.g., forces) are potentially prone to large errors. Despite this, uncertainty in tsunami IMs remains largely unexplored.

Asset Characteristics

Ideally the asset (or exposure) classification system used in the development of empirical fragility and vulnerability functions should allow the grouping of postevent observations for assets that will be affected in a similar manner by the hazard being studied. For example, in the case of buildings affected by earthquakes, this means defining classes using those building characteristics that determine its dynamic properties for fragility estimation and, additionally, its nonstructural components, use, and occupancy in vulnerability estimation. GEM and PAGER have developed such asset classification systems for use in earthquake risk assessment, which could be considered to largely represent consensus in the earthquake risk community. These allow different levels of detail to be included in the asset classification, in appreciation that the definition of asset classes, (especially in the case of empirical studies), is determined by more practical considerations. For example, to gather sufficient postearthquake observations from which to construct a fragility or vulnerability function it may be necessary to adopt an asset classification system that is very coarse (e.g., by construction material only). No consensus exists in the asset classification systems for tsunami. Instead, it is observed that published tsunami studies adopt differing asset classification systems, often ignoring important features of the asset that make it vulnerable to tsunami (e.g., openings).

The implication on fragility and vulnerability functions of using either a coarse asset classification system or one that excludes important considerations of the asset's vulnerability to the hazard is to make the empirical functions highly location and event specific. This is because inherent in the empirical data are the other asset features not captured in the asset classification system that however determine the asset performance.

It is well known that asset misclassification errors are also commonly seen in postevent damage and loss data (Rossetto et al., 2013). Such errors can be found both in datasets collected on-the-ground (e.g., due to the inexperience of the surveyor or partial visibility of the asset characteristics) as well as from airborne surveys or remote sensing. The latter are particularly prone to such errors due to the angle of view and inability to look at either the detailed characteristics of assets or their internal damage. Misclassification errors of asset type can only be corrected through further surveys or using auxiliary data sources, but it is difficult to remove them completely from a dataset. These must be acknowledged but little can be done to avoid them.

Damage Scale

Damage scales consist of a finite number of discrete damage states, which for use in postevent damage estimation should be associated with clear descriptions of likely observable damage in different types of structure (see Hill and Rossetto, 2008 for further discussion). The classification of damage by often-inexperienced groups of engineers in speedy surveys can suffer from errors. This misclassification error in damage is particularly significant if the damage observations are based on assessments made using remote sensing techniques (Booth et al., 2011). Ioannou and Rossetto (2013) propose a Bayesian framework that could incorporate this error in the construction of empirical fragility curves and Fig. 4.6 shows the substantial impact this error has on empirical fragility curves constructed from data from the 1978 Thessaloniki earthquake, Greece.

Loss

Postevent loss is typically measured in terms of direct economic loss (i.e., the cost of repairing or replacing assets), casualties, or sometimes as downtime (i.e., time to restoration of normal operations). As mentioned in postevent loss and damage data sources section, the values reported for loss can be influenced by socioeconomic factors that may bias the database.

Measurement errors in loss are difficult to assess after the data collection has occurred, unless new data arise or the affected assets are resurveyed.

Multiple Successive Hazard Events

The recorded damage and loss sustained by an asset in the aftermath of a natural disaster does not always reflect the impact of a single event. It may be the case that the damage or loss

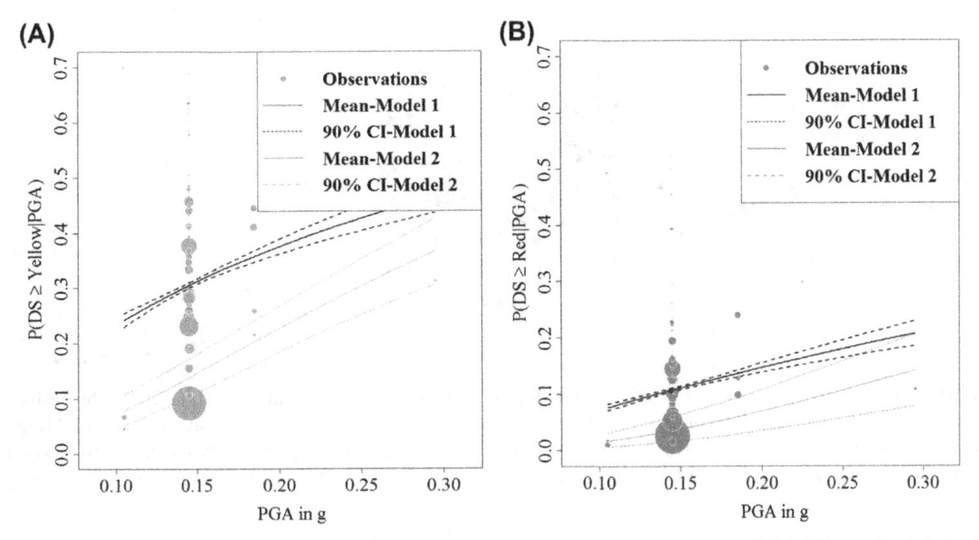

FIGURE 4.6 Mean fragility curves and their 90% posterior uncertainty intervals (90% CI) for Model 1, which ignores the misclassification error, and Model 2, which accounts for the misclassification error, corresponding to (A) *yellow* (*gray in print versions*) and (B) *red* (*dark gray in print versions*) safety levels.

data have been collected in the aftermath of multiple successive events. In the tsunami empirical fragility literature, researchers commonly acknowledge that coastal infrastructure is likely to have been affected by both the tsunami and the preceding earthquake, if situated close to the fault. However, existing studies do nothing to separate out the influence of the two hazards, there usually being an implied assumption that the earthquake damage is minimal and that losses are driven exclusively by the tsunami. In the case of the 2011 Great East Japan event (for which the vast majority of empirical tsunami fragility functions have been developed), the geographic areas affected by strong ground shaking and tsunami are the same, and hence it is not possible to separate out the hazard-affected sites.

In the case of earthquakes, where successive events have affected different and nonoverlapping geographical areas, the earthquakes are treated as separate events. However, where there is an area of overlap between the sites affected by the events, identification of the area that sustained cumulative damage or loss is important for the use of the fragility or vulnerability curves to determine future damages or losses. The authors have constructed empirical fragility curves using damage data collected by the authorities in Emilia-Romagna (Italy) in the aftermath of two successive damaging earthquakes on the 20th and 29th May 2012. Due to the incomplete nature of the database used, the damage data are aggregated at municipality level. This aggregation leads to significant loss of information making the identification of an area where buildings have suffered cumulative damage from both events not feasible. In this case, the IM values used are considered as the maximum values for each municipality accounting for the impact of both events. We contrast this with the case of a high quality loss database that was collected by Iceland Catastrophe Insurance in the aftermath of two successive damaging earthquake events that affected Iceland on the 17th and 21st June 2002. This database includes building-by-building loss data for every single residential

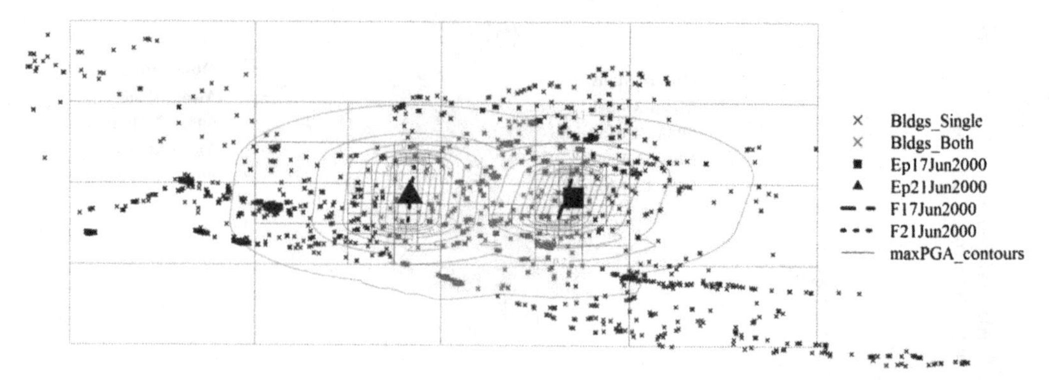

FIGURE 4.7 Map of the area affected by two Icelandic events on 17th and 21st June 2002, the individual buildings, as well as the contours of the maximum value of the PGA levels for the two events. The buildings are aggregated with the use of an adaptive grid and in each box the standard deviation of the PGA is 0.10 g. The map identifies two areas: the area affected by a single event (*black*) and the area affected by both events (*red* (*gray in print versions*)).

building in the affected area, allowing for the identification of an area of cumulative damage by using a novel statistical methodology. The latter aggregates the buildings using an adaptive grid, which ensures that the variance of the intensity level in each square is known and is low. From this analysis it is concluded that the area where the minimum of the two IM levels is equal to 0.10 g includes buildings affected by both events, as depicted in Fig. 4.7. This allows the authors to separately include data concerning each single event in the derivation of vulnerability functions.

An approach to dealing with cumulative damage or loss from an empirical perspective has as yet not been dealt with by the literature (in any significant way), but is likely going to have to be informed by numerical analyses of the response of buildings under successive events.

STATISTICAL MODELS – WHICH MODEL FITS THE DATA BEST?

Introduction

The characteristics of the postevent damage or loss databases vary considerably according to the natural hazard event. In particular, seismic damage appears to be concentrated in the lower IM levels, which are associated with small percentages of buildings with extreme damage states. The fitted fragility curves are often flat and cover a small range of IM levels and rarely cover the total range of the probabilities of exceedance (i.e., 0–1). By contrast, tsunami damage data are observed to be distributed across a wide range of tsunami intensity levels and mostly include collapsed buildings. Tsunami fragility curves are steeper than their seismic counterparts and the data cover the total range of exceedance probabilities (from 0 to 1).

After the empirical data are collected, treated, and separated by asset class, a statistical model is fitted to the data, which determines the shape of the fragility or vulnerability curve as well as the confidence intervals. Practically, a statistical model consists of two components: random and systematic. The random component expresses the probability distribution of the response variable (i.e., the counts of buildings reaching or exceeding a damage state or the economic loss) given the explanatory variables (i.e., IM and building characteristics). The systematic component expresses the mean response as a function of the explanatory variables. This latter component represents the fragility or vulnerability curves based on damage or loss data, respectively.

The GEM methodology (Rossetto et al., 2014) is the first serious attempt to draw upon the state-of-art methodologies for empirical fragility and vulnerability assessment in order to develop a methodology that aims to identify which model fits the available data best. The proposed framework encourages the use of a range of statistical models, which include parametric as well as semiparametric and nonparametric models, and highlights the need of assessing the goodness of fit of each model. Parametric models represent family of distributions, which can be determined by a finite number of parameters. In general, these models are based on strong assumptions and when they are met, they lead to powerful analyses and the ability to extrapolate outside the IM range for which damage or loss data are available. The use of parametric models is based on the underlying assumption that the damage or loss strictly monotonically increases with the increase of the IM intensity. This underlying assumption is relaxed if nonparametric models are adopted, which do not make any prior assumptions regarding the probability distribution of the variables. Semiparametric models represent an intermediate type of models, which have both parametric and nonparametric components and are capable to identify multiple trends in the data.

The parametric models are preferred in the empirical fragility and vulnerability literature, as they are straightforward to interpret. The GEM methodology concentrates on the importance of goodness-of-fit tests, which are typically ignored in the literature, and the need to construct the confidence intervals around the fragility or vulnerability curves. In particular, the empirical fragility literature favors the expression of the fragility curves in terms of a lognormal cumulative distribution function. A range of questionable assumptions regarding the random component are instead observed in the literature. The significance of these assumptions is explored in this section by briefly discussing the three most commonly used statistical models and fitting them to the 1980 Irpinia earthquake damage database.

The use of semiparametric or nonparametric models in the empirical fragility or vulnerability curves is limited. Gaussian kernel smoothers (i.e., a type of nonparametric model) have been used as part of an exploratory analysis of the 2012 Emilia data by the authors. These models have also been fitted to the 1980 Irpinia earthquake data in Rossetto et al. (2014). However, the use of nonparametric models to detect trends in a small numbers of data points that are mostly concentrated in the low IM range is problematic. By contrast, a more rigorous use of the semiparametric models has been noted in Macabuag et al. (2016), which aimed to identify the most efficient tsunami IM measures.

Parametric Statistical Models

The lognormal cumulative distribution function has been fit to postevent damage data to construct empirical fragility curves using three models, namely: nonlinear, linear, and generalized linear models (GLMs) (see Seber and Wild (1989) and Chatterjee and Hadi (2006) for further explanation of these approaches). In the following section a brief discussion of the advantages and disadvantages of these three models and the optimization procedures adopted to fit them to the postevent data are discussed.

Nonlinear Models

Past empirical fragility assessment studies (e.g., Rossetto and Elnashai, 2003; Amiri et al., 2007; Rota et al., 2008) have constructed nonlinear models (NLM) by expressing the response variable, P_i, in terms of the probability of a damage state, ds_i, being reached or exceeded. The explanatory variable, *IM*, is expressed in terms of an *IM* type (e.g., *PGA* for earthquakes). Fragility functions are then constructed based on the assumption that the response variable for each IM level is normally distributed:

$$P_i|im \sim normal\left(\mu_i, \sigma^2\right) \tag{4.1}$$

This distribution is characterized by the mean response, μ_i, which is considered to be equal to a nonlinear parametric function of the *IM* with constant variance, σ^2, for each level of *IM*. However, as P_i can only have values in the range between [0,1] the normality assumption is violated near these value extremes. In addition, the assumption of constant variance is also likely to be violated given that the uncertainty in seismic performance for large ground motion or tsunami inundation intensities is considered lower than for the intermediate IM values.

In the literature, the systematic component, which represents the fragility curve, has been expressed either in terms of lognormal cumulative functions or in terms of exponential functions:

$$\mu_i = \Phi(\theta_{1i}\ln(im) + \theta_{0i}) \tag{4.2}$$

where $\boldsymbol{\theta}_\mathbf{i} = [\theta_{0i}, \theta_{1i}]$ is the vector of the parameters describing the parametric function. It should be noted that Eq. (4.2) is the most commonly used expression of empirical fragility curves. Its popularity can be attributed to its three main properties. Firstly, it is constrained in the y-axis between (0, 1), which is ideal for fitting data points expressing aggregated probabilities. Secondly, values in the x-axis are constrained in (0, +∞) and approach zero for very small levels of *IM*. The former agrees with the range of almost all *IMs*, and the latter is a realistic expectation given that for very small *IM* levels the probability of damage is negligible. Finally, this distribution appears to be skewed to the left, and thus it can, theoretically at least, provide a better estimate for the smaller intensities, where the majority of the data typically lies. Instead, the exponential function is unconstrained in both x- and y-axis. The use of a non-probability distribution function to express the fragility curves may have implications in the overall risk assessment, which requires its coupling with a hazard curve to produce the annual probability of reaching or exceeding a damage state.

Having constructed the model, its unknown parameters, $\boldsymbol{\theta}$, can be estimated numerically by minimizing the sum of the weighted squares of the errors (i.e., observed − predicted values), in the following form:

$$\boldsymbol{\theta}_i^{opt} = \arg\min\left[\sum_{j=1}^{m_{bins}} w_j \varepsilon_{ij}^2\right] = \arg\min\left[\sum_{j=1}^{m_{bins}}\left[w_j\left(p_{ij} - \mu_{ij}\right)^2\right]\right] \tag{4.3}$$

where ε_{ij} is the error between the observed and estimated levels of the response variable, P_i, which is considered independently and normally distributed for each j with mean zero and constant variance, σ^2; w_j is the weight assigned to the error in bin j; p_{ij} is the proportion of buildings, corresponding to bin j, whose observed level of damage reaches or exceeds damage state i:

$$p_{ij} = \frac{n_{ij}}{n_j} = \frac{\sum_{i=i}^{m} y_{ij}}{\sum_{i=0}^{m} y_{ij}} \tag{4.4}$$

where n_{ij} is the count of surveyed buildings in bin j, whose observed level of damage reaches or exceeds a given ds_i; and n_j is the total number of buildings in bin j. Most existing studies fit NLMs by assuming that the errors are equally weighted:

$$w_j = 1.0 \tag{4.5}$$

This consideration, however, suggests that the error based on observations made on 20 assets is as reliable as one based on 1000 assets. This limitation can be addressed by using a weighting scheme, which considers bins containing a larger number of asset observations to be more reliable than bins with smaller sample sizes:

$$w_j = \sum_{i=0}^{m} y_{ij} \tag{4.6}$$

The assumptions made and issues raised here pose severe limitations to the adoption of NLMs in the construction of fragility and vulnerability curves and raise questions about the reliability of past fragility relationships derived in this way.

Linear Models

The drawbacks of the nonlinear statistical models regarding the random component can be addressed by the use of linear statistical models (LM). These models have often been used in the empirical fragility assessment literature to fit mainly lognormal cumulative distribution functions (e.g., Miyakoshi et al., 1997 and Liel and Lynch, 2017, amongst others), and in few cases, normal cumulative distribution functions to the postearthquake damage data. Linear models (LMs) have been obtained by linearizing the systematic component, as

$$\Phi^{-1}(P_i) = \theta_{1i}\ln(im) + \theta_{0i} \tag{4.7}$$

In the case of LM, the response variable is expressed in terms of the inverse cumulative standard normal distribution of the probability that a damage state is being reached or exceeded, $\Phi^{-1}(P_i)$. The response is considered to follow a normal distribution at any given IM value (im) with conditional mean equal to the linear expression of the explanatory variable in the right hand side of Eq. (4.7) and constant variance, σ^2:

$$\Phi^{-1}(P_i)\big|im \sim normal(\mu_i, \sigma^2), \quad \text{where } \mu_i = \theta_{1i}\ln(im) + \theta_{0i} \tag{4.8}$$

Similar to NLMs, the optimum parameter values for these models are also calculated by minimizing the sum of the weighted square errors, through a closed form solution:

$$\boldsymbol{\theta}_i^{opt} = \arg\min\left[\sum_{j=1}^{m_{bins}} w_j\varepsilon_{ij}^2\right] = \arg\min\left[\sum_{j=1}^{m_{bins}}\left[w_j\left(\Phi^{-1}(p_{ij}) - (\theta_{1i}\ln(im_j) + \theta_{0i})\right)^2\right]\right] \tag{4.9}$$

Most studies have not used a weighting scheme in the optimization procedure described by Eq. (4.9). Nonetheless, in some studies this may not pose a significant problem given that the bins have similar sample sizes of buildings.

In the case of LM, the response variable does not suffer from the variable constraints used for the construction of NLMs. This means that the assumption that the response follows a normal distribution appears to be reasonable. However, the expression of the systematic component in terms of a cumulative normal distribution may violate the physical constraints of the selected *IM* type. This may occur given that this function is unconstrained in the x-axis, which means that the *IM* can vary from $(-\infty, +\infty)$. This suggests that the fragility curve may not approach zero for very low *IM* levels, which is physically unrealistic.

Nonetheless, the main limitation of this model is considered to be the transformation of the damage data, which is necessary for the estimation of the parameters of the LM (Eq. 4.6). This limitation is more evident when p_{ij} is equal to 0 and 1, which leads to response variable values equal to infinity. To date, problematic data points have been addressed by removing the corresponding bins of observations altogether, leading to a, perhaps substantially, biased database. For $p_{ij} = 0$, Porter and Kennedy (2007) state that one could assume that one asset has sustained damage equal to or exceeding the damage state of interest. However, this assumption seems questionable as it introduces a bias in the damage data by not accounting for the bins where no buildings have suffered damage equal to or exceeding the damage state of interest (Baker, 2011).

Generalized Linear Models

Empirical fragility curves, corresponding to a given damage state, can be constructed by fitting a GLM to the data. This model is based on the underlying assumption that the seismic response of each buildings is a Bernoulli trial, i.e., which is equal to 1 if the sustained damage is $DS \geq ds_i$ and 0 otherwise. Given that the damage data are typically aggregated in bins of similar IM, the response is considered to follow a binomial distribution for every level of IM (*im*):

$$N_i|im \sim \binom{n}{n_i}\mu_i^{n_i}[1-\mu_i]^{n-n_i} \tag{4.10}$$

where N_i is the number of assets that have been damaged to a level equal or exceeding a predetermined damage state, $DS \geq ds_i$; n is the total number of assets in a bin associated with IM level im. This distribution is characterized by the conditional mean response, μ_i, as

$$g(\mu_i) = \theta_{1i}\ln(im) + \theta_{0i} \tag{4.11}$$

where $g(.)$ is termed the link function and relates the mean response to the explanatory variable expressed either in terms of the intensity or as the natural logarithm of the intensity. The link function can take three forms: probit, logit, and complementary log-log. The "probit" link function has been used to relate the mean response with the explanatory variable expressed in terms of the natural logarithm of intensity (which is essentially the inverse cumulative lognormal distribution) in the construction of empirical seismic fragility curves for buildings by Ioannou et al. (2012) and bridges by Shinozuka et al. (2000). The "logit" function has also been used to relate the mean response with the intensity for the construction of empirical fragility curves for bridges (Basoz et al., 1999) and steel tanks (O'Roorke and So, 2000). The "complementary log-log" link function is an asymmetrical function, which has not been adopted to date in the empirical construction of fragility curves.

The parameters of the GLMs are then estimated by maximizing the likelihood function through an iterative least squares algorithm, as

$$\boldsymbol{\theta}_i^{opt} = \max\arg[L(\boldsymbol{\theta}_i)] = \max\arg\left[\prod_{j=1}^{m_{bins}}\binom{n_j}{n_{ij}}\mu_{ij}^{n_{ij}}\left[1-\mu_{ij}\right]^{n_j-n_{ij}}\right] \tag{4.12}$$

Contrary to the NLMs and LMs, GLMs provide a more realistic (in theory at least) representation of the postearthquake data given that (1) they recognize that the fragility curves are bounded in [0,1], (2) they successfully relax the assumption of constant variance of residuals by accommodating for smaller uncertainty in the tails of the fragility curve, and higher in the middle, (3) they do not require the transformation of the observations necessary for the fitting of the LM in the data, and (4) they take into account the fact that some data points have a larger overall number of assets associated with them than others, without the need of a weighting system.

Case Study: How Important Is It to Use a Suitable Parametric Statistical Model?

The Irpinia earthquake took place in Italy on November 23, 1980 with magnitude $M_w = 6.9$. The map of the affected area in Campania-Basilicata is depicted in Fig. 4.8 and highlights the 41 municipalities for which there are postearthquake damage data. It also shows 364 adjacent municipalities, for which no data is available but which are used in the present analysis. The total number of buildings in the 41 municipalities was surveyed in 1982. The damage sustained by the surveyed buildings is classified in six discrete damage states according to the MSK-76 scale, which ranged from zero damage (ds_0) to collapse (ds_5).

This case study is used here to investigate the importance of modeling the uncertainty in *IM* levels, using only the damage data for the least vulnerable, class "C" buildings reported in the survey (these buildings would be deemed highly vulnerable according to current standards). These buildings include masonry buildings (i.e., field stone, hewn stone, and brick masonry) built with reinforced concrete (RC) floors, brick masonry buildings with wooden or steel floors, and RC buildings built with low or no seismic design code. *PGA* is used here as the *IM*, as it is the most widely adopted *IM* type in empirical fragility assessment studies, and the authors would like to see what the effect of statistical model choice might

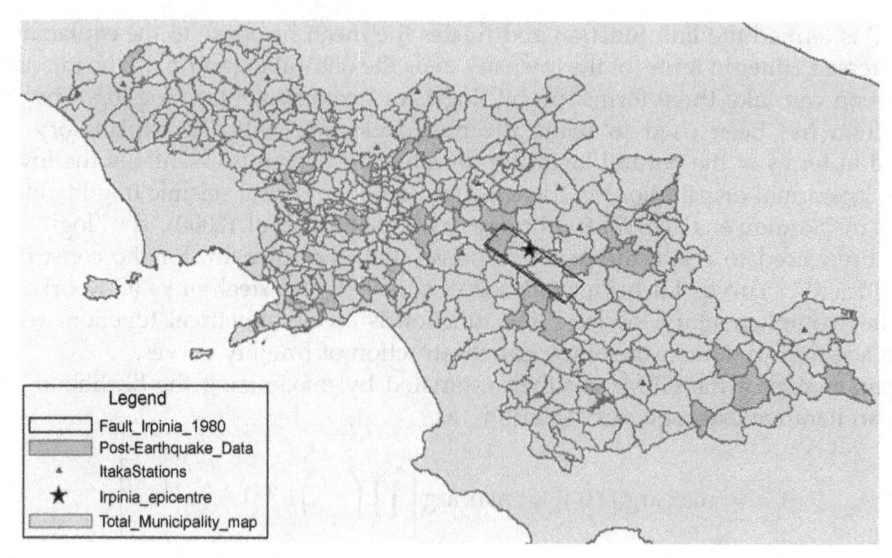

FIGURE 4.8 Map of the area affected by the 1982 Irpinia earthquake in Italy. Municipalities for which damage data are available are *shaded red* (*gray in print versions*). The location of ground motion recording stations is shown by *red dots* (*dark gray in print versions*).

have on determining the reliability of currently published studies. The level of *PGA* is estimated at the administrative center of each municipality and is assumed to remain constant throughout the municipality, in line with most existing studies. For each municipality, the *IM* level is provided from the recent GMPE proposed by Bindi et al. (2011), which accounts explicitly for the interevent and intraevent uncertainty. The soil class at the center of each municipality is also determined by the Italian geological map with scale 1:50,000.

The percentage of class C buildings which sustained damage state, ds_i, or above is plotted in Fig. 4.9 against the median PGA level at the center of the 41 municipalities for $i = 1-5$. It can be observed that although the PGA levels range from 0.025 to 0.64 g, most data points are clustered in the lower *PGA* levels, which is a common observation made in most post-earthquake damage databases. Overall, the buildings appear to have performed well during the earthquake, as most of these buildings sustained either no damage or slight ds_1 damage. In Fig. 4.9, the scatter in the percentage of buildings that suffered damage $\geq ds_1$ appears to be substantial and it reduces with the increase in the damage severity. It should be noted that only 53 class C buildings were seen to collapse ($DS = ds_5$) in the 1980 Irpinia earthquake.

Fragility curves corresponding to the 5 damage states are constructed by fitting GLM as well as the LM and are depicted in Fig. 4.9 together with their associated 90% confidence intervals constructed using bootstrap analysis. The LM is fit assuming that all data points (irrespective of the total number of class C buildings in each municipality) have the same weighting. The LM is then also fit assuming that the contribution of each data point to the construction of the fragility curve depends on the number of buildings it represents (LM_w). For ds_1, the large overdispersion and the absence of municipalities where no class C buildings

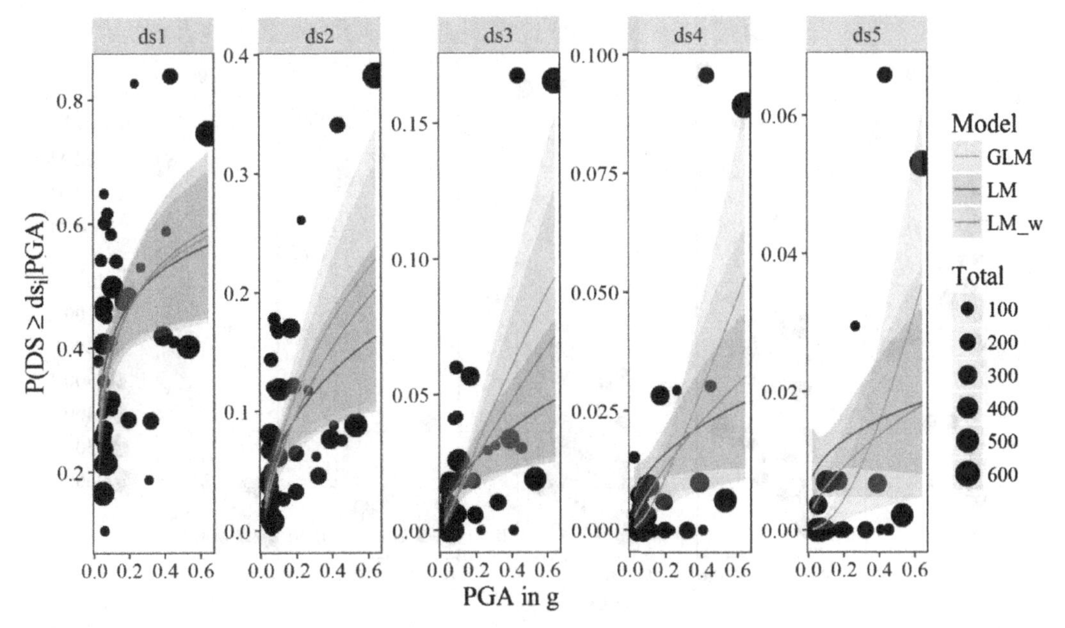

FIGURE 4.9 Fragility curves for the 1980 Irpinia database of damage to Class "C" buildings developed using GLM, LM, and weighted linear (LM_w models). The empirical fragility curves and their 90% confidence intervals are shown for five damage states ranging from slight damage (ds_1) to collapse (ds_5).

were damaged is seen to result in all three models having very similar fits. Instead, the differences in the fitted models becomes more obvious for fragility curves corresponding to higher damage states (ds_2–ds_5), which are associated with a reduced overdispersion as well as an increase in data points with zero frequency of damage reaching or exceeding a given damage state. For these four damage states, the use of a GLM model results in fragility curves that are systematically steeper than when a LM is used. In addition, the 90% confidence intervals appear to be significantly wider if a GLM model is used. Comparing between the LMs used, the fit is observed to be improved when a weighting system based on number of buildings in each municipality is used, i.e., model LM_w. This model results in steeper fragility curves and wider confidence intervals. Despite the improvement in the fit, LM_w yields flatter fragility curves with smaller confidence intervals than its GLM counterpart.

Fig. 4.10 depicts the fragility curves and their 90% confidence intervals using a GLM as well as NLM and weighted nonlinear models (NLM_w). Overall, the differences in the three models appear to depend on the data and vary for each damage state. With regard to the NLM, the use of a weighting system that accounts for the number of buildings in each municipality instead of equal weights appears to have a notable effect (i.e., steeper fragility curves and wider confidence intervals) only for ds_1 and ds_2, and no impact for the remaining three damage states. For ds_1 and ds_2, the GLM fit appears to lead to steeper fragility curves and wider confidence intervals than the NLM. The NLM fit seems to minimize these differences with GLM for ds_1. For ds_2, the GLM fit leads to wider confidence intervals than NLM and more narrow than NLM_w.

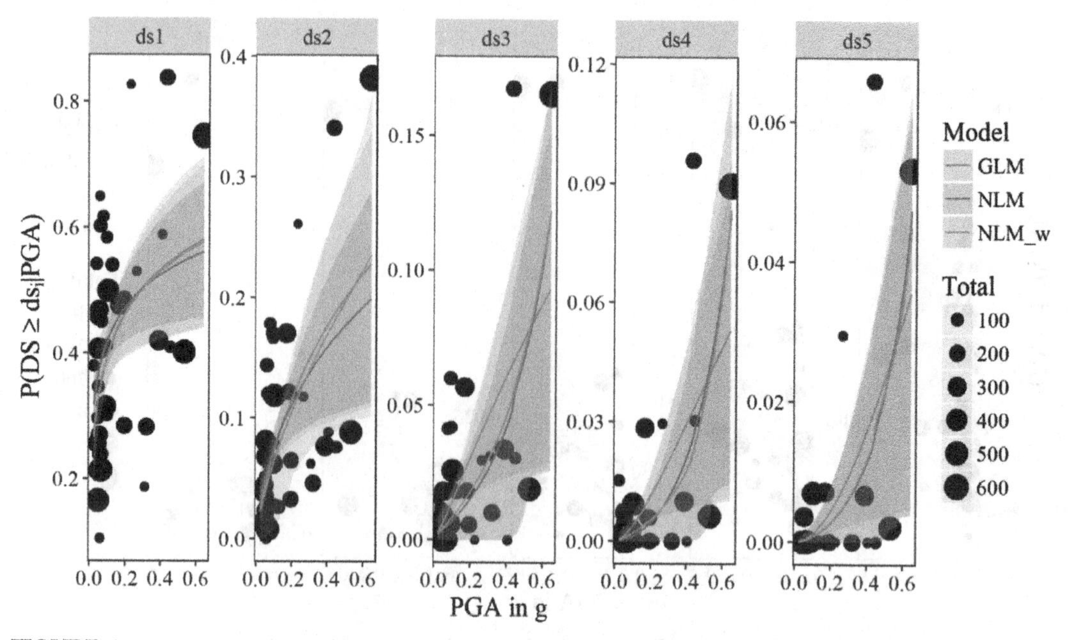

FIGURE 4.10 Fragility curves for the 1980 Irpinia database of damage to Class "C" buildings developed using GLM, NLM, and NLM_w model. The empirical fragility curves and their 90% confidence intervals are shown for five damage states ranging from slight damage (ds_1) to collapse (ds_5).

Overall, the differences in the parametric models depend on the data and can vary from very little to significant. Nonetheless, given that the damage data are overwhelmingly associated with low *IM* levels and low percentages of moderate or extreme damage, significant differences are seen in the fragility functions associated with higher damage states if different statistical models are adopted. In the recent applications of the GEM methodology more complex parametric models, called cumulative linear models, have been favored as they can fit the fragility curves for all damage states and building classes at the same time using the total database instead of subsets of data.

Identifying the Best Model Fit

Contrary to the overwhelming majority of the empirical fragility or vulnerability studies, the GEM methodology (Rossetto et al., 2014) advocate for the fitting of a range of parametric as well as nonparametric models and the trial of different types of *IM*. How can the best model be selected? Goodness-of-fit tests and the study of the residuals are recommended to choose between model fits (e.g., Maqsood et al., 2016). However, the tests used for this depend on whether different statistical models are being tested (e.g., in Charvet et al., 2014 the Akaike Information Criterion is used to select amongst statistical models) or whether *IM* measures are being compared for the same statistical model (e.g., Macabuag et al., 2016 uses the 10-fold cross-validation approach to identify which tsunami measure fitted the damage best).

CONCLUSIONS

Postevent damage and loss observations present a valuable source of data for the construction of fragility and vulnerability functions, respectively. They reflect reality and the natural variation in the hazard and its effects on the natural and built environment. However, due to the typical survey approaches used to collect the observations, empirical data can suffer from strong biases and quite often may include the influence of multiple or cascading hazard events. These present significant challenges when using empirical data to develop fragility and vulnerability functions, and it is noted that not every source of uncertainty is modeled in existing studies. Moreover, existing studies commonly do not appropriately communicate the overall uncertainty in the vulnerability or fragility functions.

As an understanding of the reliability of existing fragility functions is crucial in determining whether they should be adopted in risk assessments, this chapter presents an overview of some significant observations made by the authors whilst working on earthquake and tsunami empirical fragility and vulnerability functions over the last 10 years.

It is shown that the treatment of data bias is of great importance and has strong influences on the resulting functions. This is especially the case with incomplete data, which if ignored or not properly accounted for can result in drastically different fragility or vulnerability functions. It is emphasized that where biases cannot be dealt with in the databases, then at least these should be acknowledged and their potential influence on the reliability of the resulting fragility functions be discussed. It is also noted that few vulnerability functions exist in the published literature and that loss data, irrespective of the quality of the database used, appear to have very large uncertainties associated with them. This is an area where further research is necessary.

Efforts by Rossetto et al. (2014) have highlighted the need for rigorous statistical methods to be used in the development of fragility and vulnerability functions and the need to use more complex models to account for multiple sources of uncertainty. Such uncertainties, in particular those associated with the evaluation of IM values, have been systematically ignored by the empirical fragility and vulnerability literature. New approaches are being proposed to account for these but this is an area where further research is still needed.

It is hoped that the reader will consider some of the issues discussed in this text either when choosing to adopt an existing empirical function or when deriving their own empirical fragility or vulnerability functions.

References

Amiri, G.G., Jalalian, M., Amrei, S.A.R., 2007. Derivation of vulnerability functions based on observational data for Iran. In: Proceedings of International Symposium on Innovation & Sustainability of Structures in Civil Engineering. Tongji University, China.

Baker, J., 2011. Fitting Fragility Functions to Structural Analysis Data Using Maximum Likelihood Estimation Working Paper-PEER.

Basoz, N.I., Kiremidjian, A.S., King, S.A., Law, K.H., 1999. Statistical analysis of bridge damage data from the 1994 Northridge, CA, earthquake. Earthquake Spectra 15 (1), 25–54.

Bertelli, S., 2017. Empirical Vulnerability: Derivation of Empirical Vulnerability and Fragility Functions from the 2009 L'Aquila Earthquake Data (Thesis for MRes in Urban Sustainability and Resilience). Department of Civil, Environmental and Geomatic Engineering, University College London.

Bindi, D., Pacor, F., Luzi, L., Puglia, R., Massa, M., Ameri, G., Paolucci, R., 2011. Ground motion prediction equations derived from the Italian strong motion database. Bulletin of Earthquake Engineering 8 (5), 1209–1230.

Booth, E., Saito, K., Spence, R., Madabhushi, G., Eguchi, R.T., 2011. Validating assessments of seismic damage made from remote sensing. Earthquake Spectra 27 (S1), S157–S177.

Braga, F., Dolce, M., Liberatore, D., 1982. Southern Italy November 23, 1980 Earthquake: A Statistical Study of Damaged Buildings and an Ensuing Review of the M.S.K.-76 Scale. Report, Rome, Italy.

Charvet, I., Ioannou, I., Rossetto, T., Suppasri, A., Imamura, F., 2014. Empirical fragility assessment of buildings affected by the 2011 Great East Japan tsunami using improved statistical models. Natural Hazards 73 (2), 951–973.

Chatterjee, S., Hadi, A.S., 2006. Regression Analysis by Example-4th Edition. Wiley & Sons, Hoboken, New Jersey, USA.

Crowley, H., Pinho, R., 2011. Global earthquake model: community-based seismic risk assessment. In: Dolšek, M. (Ed.), Protection of Built Environment Against Earthquakes. Springer, Netherlands, Dordrecht.

Crowley, H., Colombi, M., Silva, V., 2014. Epistemic uncertainty in fragility functions for European RC buildings. In: Pitilakis, Crowley, Kaynia (Eds.), SYNER-G: Typology Definition and Fragility Functions for Physical Elements at Seismic Risk: Buildings, Lifelines, Transportation Networks and Critical Facilities. Springer, Netherlands.

Daniell, J.E., Khazai, B., Wenzel, F., Vervaeck, A., 2011. The CATDAT damaging earthquakes database. Natural Hazards and Earth System Sciences 11 (8), 2235.

De Bruycker, M., Greco, D., Lechat, M.F., Annino, I., De Ruggiero, N., Triassi, M., 1985. The 1980 earthquake in Southern Italy—morbidity and mortality. International Journal of Epidemiology 14 (1), 113–117.

EM-DAT, Centre for Research on the Epidemiology of Disasters, 2012. The International Disaster Database.

Hill, M., Rossetto, T., 2008. Comparison of building damage scales and damage descriptions for use in earthquake loss modelling in Europe. Bulletin of Earthquake Engineering 6 (2), 335–365.

Ioannou, I., Rossetto, T., 2013. Sensitivity of empirical fragility assessment of buildings to the misclassification error of damage. In: Proceedings of the 11th International Conference on Structural Safety & Reliability. New York, USA.

Ioannou, I., Douglas, J., Rossetto, T., 2015. Assessing the impact of ground-motion variability and uncertainty on empirical fragility curves. Soil Dynamics and Earthquake Engineering 69, 83–92.

Ioannou, I., Rossetto, T., Grant, D.N., 2012. Use of regression analysis for the construction of empirical fragility curves. In: Proceedings of 15th World Conference on Earthquake Engineering, Lisbon, Portugal.

Jaiswal, K., Wald, D., 2010. An empirical model for global earthquake fatality estimation. Earthquake Spectra 26 (4), 1017–1037.

Karababa, F.S., Pomonis, A., 2010. Damage data analysis and vulnerability estimation following the August 14, 2003 Lefkada Island, Greece, Earthquake. Bulletin of Earthquake Engineering. Online.

Kongar, I., Giovinazzi, S., Rossetto, T., 2016. Seismic performance of buried electrical cables: evidence-based repair rates and fragility functions. Bulletin of Earthquake Engineering. Online First. http://link.springer.com/article/10.1007/s10518-016-0077-3.

Koshimura, S., Oie, T., Yanagisawa, H., Imamura, F., 2009. Developing fragility functions for tsunami damage estimation using numerical model and post-tsunami data from Banda Aceh, Indonesia. Coastal Engineering Journal 51 (03), 243–273.

Liel, A.B., Lynch, K.P., 2017. Vulnerability of reinforced concrete frame buildings and their occupants in the 2009 L'Aquila. Natural Hazards Review (In press).

Macabuag, J., Rossetto, T., Ioannou, I., Suppasri, A., Sugawara, D., Adriano, B., Imamura, F., Eames, I., Koshimura, S., 2016. A proposed methodology for deriving tsunami fragility functions for buildings using optimum intensity measures. Natural Hazards 84 (2), 1257–1285.

Maqsood, T., Edwards, M., Ioannou, I., Kosmidis, I., Rossetto, T., Corby, N., 2016. Seismic vulnerability functions for Australian buildings by using GEM empirical vulnerability assessment guidelines. Natural Hazards 80 (3), 1625–1650.

Minas, S., Galasso, C., Rossetto, T., 2015. Spectral shape proxies and simplified fragility analysis of mid- rise reinforced concrete buildings. In: Proceedings of the 12th International Conference on Applications of Statistics and Probability in Civil Engineering (ICASP12), Vancouver, Canada, July 12–15, 2015.

Miyakoshi, J., Hayashi, Y., Tamura, K., Fukuwa, N., 1997. Damage ratio functions of buildings using damage data of the 1995 Hyogo-Ken Nambu earthquake. In: Proceedings of 7th International Conference on Structural Safety and Reliability, Kyoto, Japan.

O'Roorke, M.J., So, P., 2000. Seismic fragility curves for on-grade steel tanks. Earthquake Spectra 16 (4), 801–815.

Petal, M., 2011. Earthquake casualties research and public education. In: Spence, R., So, E., Scawthorn, C. (Eds.), Human Casualties in Earthquakes: Progress in Modelling and Mitigation. Springer, Netherlands, Dordrecht.

Petrone, C., Rossetto, T., Goda, K., Eames, I., 2017. Tsunami analysis of structures: comparison among different approaches. In: Proceedings of 16th World Conference on Earthquake Engineering, Santiago, Chile.

Porter, K., Kennedy, R., 2007. Creating fragility functions for performance-based earthquake engineering. Earthquake Spectra 32 (2), 471–489.

Pregnolato, M., Galasso, C., Parisi, F., 2015. A compendium of existing vulnerability and fragility relationships for flood: preliminary results. In: Proceedings of 12th International Conference on Applications of Statistics and Probability in Civil Engineering, ICASP12, Vancouver, Canada.

Rossetto, T., Elnashai, A., 2003. Derivation of vulnerbaility functions for European-type RC structures based on observational data. Engineering Structures 25 (10), 1241–1263.

Rossetto, T., Ioannou, I., Grant, D.N., 2013. Existing Empirical Fragility and Vulnerability Relationships: Compendium and Guide for Selection. Report. GEM Foundation, Pavia, Italy.

Rossetto, T., Ioannou, I., Grant, D.N., Maqsood, T., 2014. Guidelines for Empirical Vulnerability Assessment Report. GEM Foundation, Pavia, Italy.

Rota, M., Penna, A., Strobbia, C.L., 2008. Processing Italian damage data to derive typological fragility curves. Soil Dynamics and Earthquake Engineering 28 (10–11), 933–947.

Seber, G.A.F., Wild, C.J., 1989. Nonlinear Regression. John Wiley & Sons, New York, USA.

Shinozuka, M., Feng, M.Q., Lee, J., Naganuma, T., 2000. Statistical analysis of fragility curves. Journal of Engineering Mechanics 126 (12), 1459–1467.

So, E.K.M., Pomonis, A., Below, R., Cardona, O., King, A., Zulfikar, C., Koyama, M., Scawthorn, C., Ruffle, S., Garcia, D., 2012. An introduction to the global earthquake consequences database (GEMECD). In: The Proceedings of the 15th World Conference on Earthquake Engineering 2012, Lisbon, 10 pp.

Spence, R., So, E., Jenkins, S., Coburn, A., Ruffle, S., 2011. A global earthquake building damage and casualty database. In: Human Casualties in Earthquakes. Springer, Netherlands, pp. 65–79.

Straub, D., Der Kiureghian, A., 2008. Improved seismic fragility modeling from empirical data. Structural Safety 30 (4), 320–336.

Suppasri, A., Mas, E., Charvet, I., Gunasekera, R., Imai, K., Fukutani, Y., Abe, Y., Imamura, F., 2013. Building damage characteristics based on surveyed data and fragility curves of the 2011 Great East Japan tsunami. Natural Hazards 66 (2), 319–341.

Ware, J.H., Harrington, D., Hunter, D.J., D'Agostino, R.B., October 4, 2012. Missing data. New England Journal of Medicine 367, 14.

Yazgan, U., 2015. Empirical seismic fragility assessment with explicit modeling of spatial ground motion variability. Engineering Structures 100, 479–489.

Yazgan, U., 2012. A new approach to building stock vulnerability modeling based on Bayesian analysi. In: Proceedings of 15th World Conference on Earthquake Engineering, Lisbon, Portugal.

Yazgan, U., 2013. Empirical vulnerability modeling considering geospatial ground motion variability. In: Proceedings of ICOSSAR, New York, USA.

Yepes-Estrada, C., Silva, V., Rossetto, T., D'Ayala, D., Ioannou, I., Meslem, A., Crowley, H., 2016. The global earthquake model physical vulnerability database. Earthquake Spectra 32 (4), 2567–2585.

SECTION II

MODEL CREATION, SPECIFIC PERILS AND DATA

CHAPTER

5

The Use of Historic Loss Data for Insurance and Total Loss Modeling

James E. Daniell, Friedemann Wenzel, Andreas M. Schaefer

Geophysical Institute & Center for Disaster Management and Risk Reduction Technology (CEDIM), Karlsruhe Institute of Technology (KIT), Karlsruhe, Germany

INTRODUCTION

Insurance modeling of natural risks requires three key components of information: (1) a catalog of all possible events that can occur within a certain time period, (2) an exposure database comprising the built parameters of insured structures or the population exposed to the hazard (usually in the form of a portfolio), and (3) the susceptibility of the structures or population at risk to the hazard.

Creating a catastrophe model for insurance purposes has a few key data needs for each of these components. This chapter uses CATDAT data (Daniell et al., 2011) to provide trends in insured as well as noninsured market losses. In addition, statistics of the historic losses helps understanding trends as well as offers implications for future losses. In addition, we discuss normalization of losses and how historic (empirical) loss data are best applied for model calibrations. Modeling earthquake risk for Australia provides a specific case study.

EMPIRICAL LOSS STATISTICS

We are aware of several key sources for losses from historic events. The most well-known databases for insurance data come from MunichRe's NATCAT data and from SwissRe's Sigma services. Additional databases include CATDAT (Daniell et al., 2011) discussed herein. These additional databases are not necessarily insurance specific but provide additional events in time as well as regionally. Best modeling results are achieved by combining information from as many as possible databases.

The CATDAT databases include the key exposure components needed for the estimation of total economic losses. Changes in building standards, risk mitigation features, as well as

Risk Modeling for Hazards and Disasters
http://dx.doi.org/10.1016/B978-0-12-804071-3.00005-7

TABLE 5.1 Characterization of Wanted Variables in Global Socioeconomic Loss Consequence Databases for Insurance Purposes of Structural Components

Information Theme	Variables in Database
Intensity parameters	Location, magnitude, speed, flow, velocity, depth of hypocenter, variability, intensity etc.
Date information	When did the disaster occur? Time of day?
Country data	In what country? What location? What region?
Socioeconomic event indicators	Population; socioeconomic data; indicators relating to wealth, social status, health etc.; exchange rates; construction cost indices; normalization parameters
Infrastructure and other loss parameters	Building damage classes, infrastructure damage classes, other nonstructural damage, losses per associated building class
Secondary effect parameters	Socioeconomic losses due to tsunami, liquefaction, landslide, fault rupture, fire, NaTech (Natural Disasters causing Technological Disasters etc.)
Economic loss parameters	Direct, indirect, aid, sectoral, capital stock vs. production losses
Economic cost parameters	Reconstruction as well as replacement cost methods not taking into account building exposure
Exposure information	No. of buildings, typologies (i.e., PAGER classes), building code use, vulnerability details
Age information	Losses per building age class, building groups

economic effects including price changes need to be considered in addition to the historical loss information.

A clear definition of what is considered in the given loss calculation is vital. Earthquakes have in general a high impact on infrastructure, whereas very little effect on the agricultural sector; this is opposed to extratropical storms, floods, or hurricanes that often creates the largest losses in the agriculture and production sectors (Table 5.1).

Most existing databases contain dates and times of the event and the country in which the event occurred, along with which countries were affected by the event. In addition, most datasets contain the hazard information of the disaster, i.e., include intensity, and location of the events and (especially for more recent events) may include the approximate latitude and longitude of the highest intensity, epicenter, or e.g., the location of the six hourly storm track. Most databases do however not include socioeconomic parameters for the country at the time of the event, thus giving little context to the environment of the losses. Most databases collect death toll statistics but many do not look at the number of homeless and/or the overall number of people affected by the events. In terms of economic losses, very few databases compare losses to capital stock and gross domestic product (GDP) at the time of the event. This provides a measure for the actual net worth relevant at the time of the earthquake. Direct and indirect losses are often not separated and definitions tend to be used inconsistently. For example, the MunichRe's NatCat Service uses the term economic loss as the overall loss, whereas EM-DAT defines the economic loss for the estimated damage. We have worked on rectifying these definitions through various DATA projects (see IRDR, 2014).

Secondary effects of earthquakes, including social and indirect economic losses, are often not addressed, either. This includes separating flood and wind losses from historical hurricane/cyclone/typhoon events or separating shake and tsunami, or liquefaction, mass movement, and other second-order losses from earthquakes, or on-flood-plain versus off-flood-plain losses for major floods. Insured losses change significantly depending on how claims are paid and/or what policy conditions, sublimits etc. are valid at the time of the event. These conditions may change over time.

General Statistics on Losses From Historic Events Over the Last 116 Years

Normalized losses are calculated via loss statistics over a certain number of years using normalization techniques based on country consumer price index (CPI) and wage data. This allows approximating average annual losses (AAL) for a country and by disaster type. Although the CATDAT database extends back beyond 1950, only events from 1950 onwards are used given that data are deemed reasonably complete since then (although some minor countries are missing altogether before the 1970s for some smaller disaster types).

The focus of the database is on earthquakes, floods, storms (tropical cyclones along with extratropical and some larger convective systems), and volcano eruptions. The modified database with larger damaging events includes c.25,000 natural disaster events from 1900 to 2015 (8000 + earthquake and tsunami events, 6000 + storm events, 7000 + flood events [including annual assessments], 1000 + volcano events, 1500 + drought/temperature events, and 750 + bushfire events).

Storm losses have been divided into 75% storm losses and 25% flood losses in the absence of more detailed information and based on averages for 20 large events for which the split was available. In a study by NWS (National Weather Service of the USA) of specific large hurricane losses, flood losses amounted to approximately half of the overall losses showing a ratio of 47% of flood as opposed to 53% wind-, hail-, and lightning-induced losses. Japanese typhoon losses have been divided into "wet" and "dry" events with some of the largest losses in history having been dominated by heavy rainfall and widespread flooding (s. the Isewan event in 1959; https://en.wikipedia.org/wiki/Typhoon_Vera). The database is deemed complete for earthquakes and volcanoes since 1900, for floods since 1950, and for storms since 1950. To compare results from different perils, we have concentrated on the period between c.1950 and 2015 herein. Care was taken to include the full record of Chinese floods from 1949 onwards and typhoon losses in Japan from 1950 (Fig. 5.1).

Converting Historic Losses Into Today's Dollar Terms

To compare risk and to be able to use data for insurance purposes, historical losses need to be brought to a currency reference.

CPI is commonly used to unlevel losses. The problem with CPI normalization is however that it may underestimate the true impact of historical events given that CPI is not directly related to building typologies (CPI unleveling might however lead to overestimate insured losses in case tariffs have changed and/or in case the considered portfolio does not follow a common trend). A hybrid index closer to construction cost indices is used below. To do this, country-based wages and CPI series are used, to convert the economic value based on the original local currency rather than on the US CPI series. This makes a significant

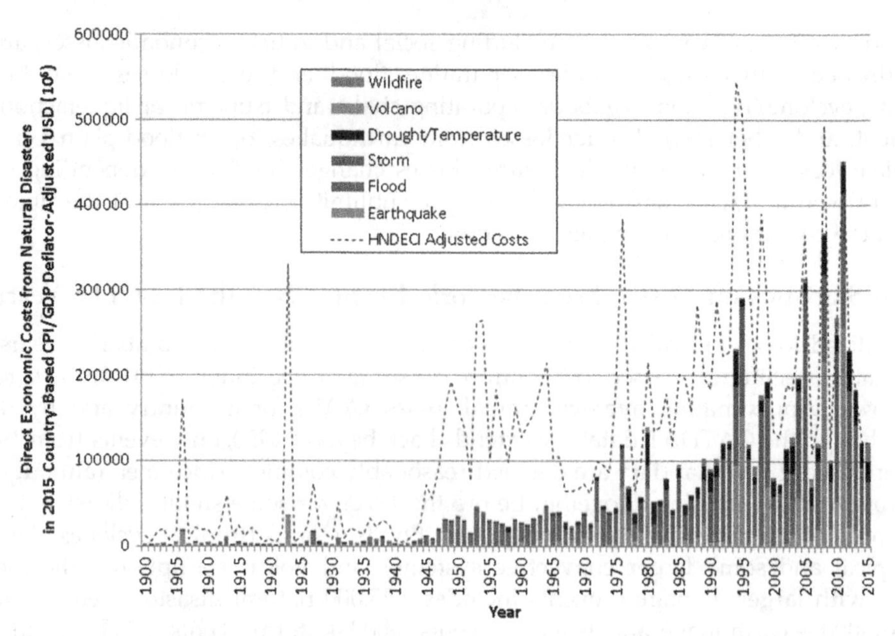

FIGURE 5.1 Global direct economic costs of natural disasters from 1900 to 2015 in 2015 USD (Country-based CPI adjusted) in the CATDAT database.

difference for countries such as Japan. A loss of $334 billion in 2011 of around 27 trillion JPY (i.e., the approximate Tohoku earthquake losses) may act as an example for this. If the US CPI or wage series converted this loss, this would amount to $365 billion in 2015 dollars. However, a value of $210 billion would result from using the Japan CPI and wage series given the large difference in exchange rates and quality of living since the event.

Without taking into account local conditions losses may be massively overestimated (or underestimated) in today's terms. This has been discussed in greater detail in Daniell et al. (2012b).

Local construction cost indices providing unleveled results are close to unskilled wage indices due to the fact that labor costs make up the majority of the construction cost. Using the GDP deflator alone, however may lead to underestimating the cost of a disaster as the GDP deflator allows for changing consumption and investment changes over time, but not necessarily taking into account the change in the cost of production or new materials. There is no perfect metric however as seen in the case of New Zealand for inflation metrics (Fig. 5.2). The HNDECI method prefers a construction cost index style approach with a large unskilled wage index component. The HNDECI method uses a spatiotemporal combination of country-specific wages, construction costs, workers' production, GDP, CPI, and other tools, to better account for the historical cost of events in today's terms (see Daniell et al., 2012a). By retaining the scale of the original economy, this might lead to inflating the costs of historical events beyond to what they would be today. Within developing countries such as China, marked increases in values compared to the rather static US and European economies have occurred over the past several years.

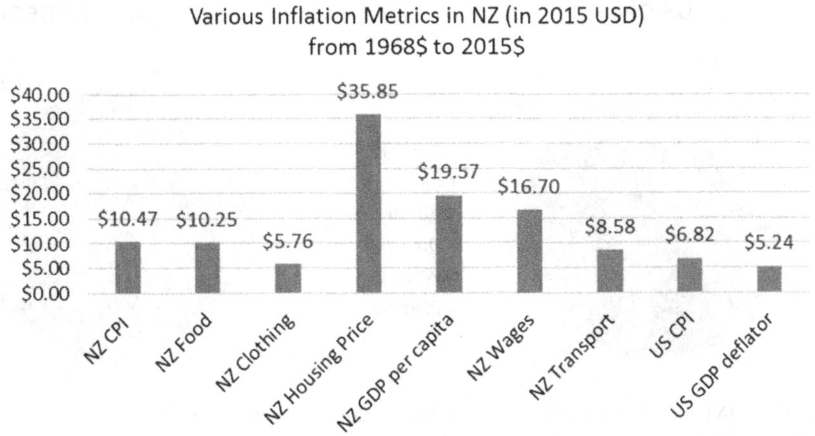

Various Inflation Metrics in NZ (in 2015 USD)
from 1968$ to 2015$

$40.00
$35.00
$30.00
$25.00
$20.00
$15.00
$10.00
$5.00
$0.00

$10.47 $10.25 $5.76 $35.85 $19.57 $16.70 $8.58 $6.82 $5.24

NZ CPI
NZ Food
NZ Clothing
NZ Housing Price
NZ GDP per capita
NZ Wages
NZ Transport
US CPI
US GDP deflator

FIGURE 5.2 The difference of inflation metrics in NZ from 1968 through to 2015 (1968 = Inangahua earthquake).

As mentioned above, the CPI is not always a useful indicator, as prices of goods do not accurately represent dollar value losses of historic events in today's dollars. China had a USD value of GDP per capita, three times less than the Ivory Coast in 1990 (preadjustment of exchange rate), yet GDP per capita values were five times larger for China than the Ivory Coast in 2015. Due to price inflation, the GDP deflator however shows a very different value when compared to wage or housing indices. An earthquake like Tangshan in 1976 with a direct loss of around 6 billion USD at the time of the event (c.11 billion CNY) would be the equivalent of c.27 billion USD using the US Construction Cost Indices, yet reaches c.100 billion USD for 2015 using the China Construction Cost Indices. Similar trends are seen in Taiwan with large differences between consumer price and construction cost indices.

The disaster data are often taken as representing complete losses. Insurance losses however depend on the insurance policies (with limits and deductibles), resulting in payout patterns that can deviate significantly from the underlying actual loss. In addition, the book value of the loss may be quoted, whereas the replacement cost is what insurers pay after an event. The reconstruction cost of the disaster may in addition not be covered fully given that buildings may only be reconstructed to the old rather than new standards with the latter not being covered by insurance policies. A fluke that often happens for global loss databases is such that Japanese losses (which are reducing over time with deflation) cancel out the underestimation of the Chinese losses (which increase with inflation), which means that globe average values look reasonable.

An overall economic loss of $12.4 trillion (in 2015 dollars) is registered using the country-based HNDECI method for the period between 1950 and 2015 (c.$200 billion per year i.e., c.0.1% of World Capital Stock) as opposed to $6.8 trillion in 2015 dollars from 1950 to 2015 using US CPI conversion (c.$105 billion/year or 0.05% of World Capital Stock) (Fig. 5.3). The real value is deemed somewhere around the HNDECI number when converting dollars for construction cost. The difference can be seen via the relative influence of the Japanese earthquake losses (higher in US CPI given current economic downturn) versus Chinese flood losses (much higher in country-based HNDECI as based on Chinese wage data). Fig. 5.4 shows the Chinese Construction Cost Index (using actual exchange rates) versus US CPI. These figures highlight the potential difference in the loss multiplication factors for both methods.

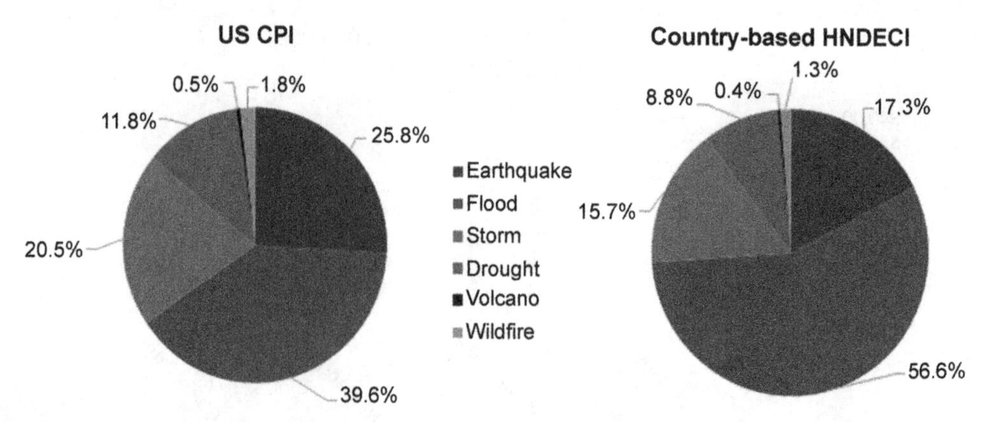

FIGURE 5.3 CATDAT natural disaster losses from 1950 to 2015 in terms of US CPI adjusted (left) and country-based Hybrid Natural Disaster Economic Conversion Index (HNDECI) (right) estimates of losses in 2015 dollars.

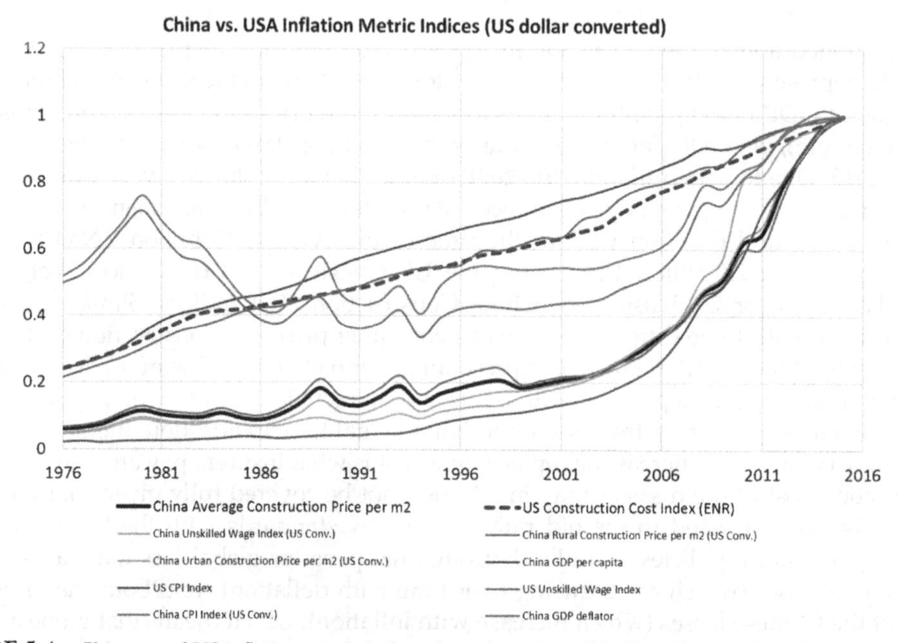

FIGURE 5.4 Chinese and US inflation metric indices from the point of the Tangshan earthquake in 1976 to today. Major differences are seen between wage, construction cost, and consumer price indices due to the rapid rise in development as well as the exchange rate differences.

As a percentage of total losses, insured losses have however significantly increased over the last 15 years due to increasing insurance penetration. Insurance premium is also growing for nonlife insurance (see SwissRe, 1999–2015 data from their Sigma database). However, comparing total economic natural disaster losses to insured losses suggests that global capital stock is growing faster than premium (especially in the developing and emerging markets),

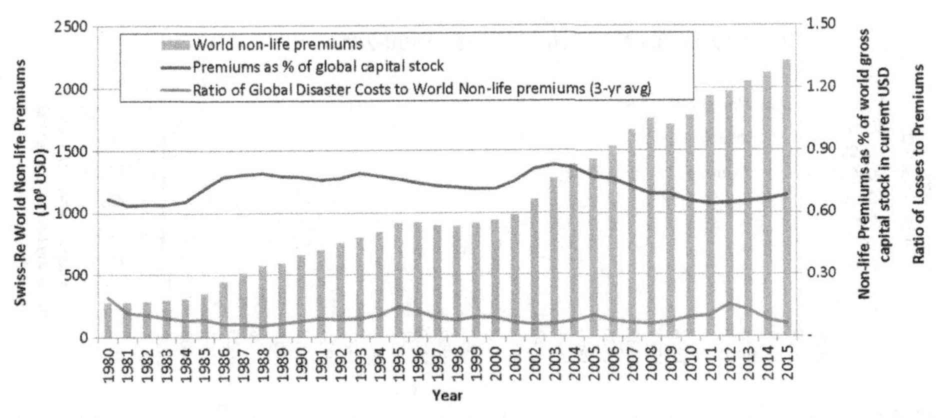

FIGURE 5.5 Insurance premiums globally versus global capital stock and natural disaster economic losses.

suggesting that the net global insurance penetration has been decreasing (relatively) rather than increasing over the last 15 years (see Fig. 5.5, green line) and/or that insurance rates for natural catastrophes have been decreasing. Comparing economic costs versus premiums for 2011 and 2012 suggests that the large recent peak in costs (3-year average) has not been compensated by an adequate increase in premium. Part of this might be explained by additional capital influx and reduced rates after 2008 when the capital market had started to look for alternative investments (as asset returns vanished otherwise). Investment opportunities were detected in the insurance industry leading to deemed lower than adequate price increases after the 2011 losses (Michel oral communication). Higher take-up rates (more complete insurance coverage) also leads (at least globally) to more efficient insurance coverage, which allows reducing costs (lower volatility). Further decrease in the ratio of cost through premium (Fig. 5.5) after 2012 is due to the lack of severe hurricane losses that make up a significant portion of both natural catastrophe insurance premium and losses. Over the longer term insurance penetration versus disaster costs have however remained reasonably constant. Absolute insured losses are of course increasing with greater takeout.

Fig. 5.6 depicts insurance penetration and losses for hurricanes in the US derived using the method of Collins and Lowe (2001). The figure includes event data from 1900 to 1999 (complemented by additional analytics in this study). In total, 299 hurricane and tropical storm event losses are recorded from 1900 to 2015. The figure shows that the insurance penetration depends on where the hurricane hits. Overall insured losses are uncertain as not all companies report losses. Overall estimates hence differ depending on the reporting entity (i.e., Sigma SwissRe, MunichRe NATCAT Service, government insurance agencies, or brokers such as GC, Willis, or AON). PCI (Property Casualty Insurers Association of America) the deemed most accurate source for insured losses in the US also does not have losses from all companies. Their loss estimates provide however one official source for industry loss warranties (ILWs) that allow trading with industry losses in the US. This product has a significant market penetration due to its ease of use and its reasonable accuracy. ILWs are considered by various investors as well as insurance and reinsurance companies that aim at covering peaks in their tail risk without considering long-term needs or large broker costs.

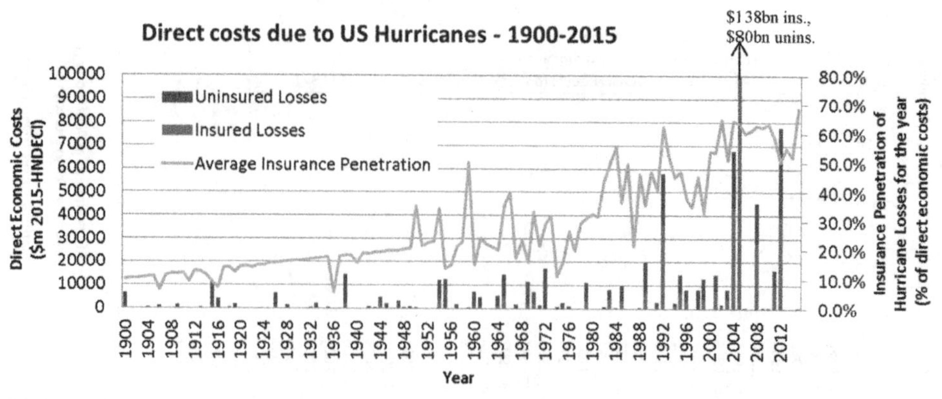

FIGURE 5.6 Direct costs due to US Hurricanes and tropical storms from 1900 to 2015 in CATDAT, compared to the insurance penetration and insured losses. This shows the increase in insurance penetration, yet also the marked increase in losses due to hurricane and tropical storm damage (including flooding, surge, and wind).

Economic losses reported by reinsurers are often derived from the insured loss divided by the insurance penetration. In addition, the DALA (Damage and Loss Assessment) methodology from ECLAC has been used extensively throughout the world to evaluate postdisaster losses. The DALA method uses an agreed methodology of counting economic losses for various sectors. The Chinese and the Japanese governments use their own methods of counting economic losses postdisaster. GB/T18208.4-2005 is one of such standards for the assessment of direct loss for China. An example of these DALA sectors is shown below in the earthquake example for loss breakdowns.

Both insurance loss databases and economic total loss databases have their place in insurance loss modeling. The advantage of global databases in comparison to insurance-based loss databases is that disaster losses are covered in greater detail especially in countries where insurance penetration is small (Asia outside Japan, Africa, or Latin America other than, e.g., Chile). This hence allows for better coverage of emerging and developing markets. One disadvantage however is that loss numbers do not include demand surge (Olsen and Porter, 2008) and other postearthquake effects. So-called "Post event loss amplification" (insurance lingo) due to changes in wages following higher than average demand is a function of the overall economy, infrastructure, and more in addition to the size and direct impact of the event. These numbers are rarely included in the replacement cost assessment at the time of the event. Demand surge along with business interruption as well as costs due to additional living expenses (hotel etc. costs due to houses being uninhabitable) are thence time and market dependent. Modeling these costs using the size of the event only may be misleading.

HISTORIC AAL PER LOCATION PER PERIL

Using the historical losses from 1950 to 2015 and dividing these losses by the total capital stock of a country can help assessing peril-specific AALs. This is done here using flat integration over the time period of 65 years.

North Korea (1995 floods), Montenegro (1979 earthquake), and Montserrat (1989 storm) have been the hardest hit countries in the past 65 years. Small island states and small countries are often highly affected by a direct tropical cyclone or earthquake. This is due to the fact that a small country may be affected entirely by the same event, i.e., TC Winston in Fiji in 2016 or the 1979 Montenegro earthquake. This is opposed to larger nations that experience much smaller relative losses unless risk is highly concentrated along certain populated areas of the coast or to the capital (see, e.g., the Maule, Chile, earthquake in 2010 for which the loss was 5% of the overall capital stock of the country, a relatively large number due to the earthquake hitting close to the main exposure concentration of the country). AALs for small countries may hence be at the same order of magnitudes as PMLs (=value at risk 1 in 100) for some larger countries as a percentage of capital stock (Table 5.2).

Normalization of Disaster Losses

Different normalization methodologies have been undertaken historically. Normalization of losses can help assessing potential trends in vulnerabilities to different disaster types as well as help forecasting losses in case vulnerabilities have not changed significantly.

Normalization strategies have included multiplying historical losses by an economic conversion index herewith bringing losses to today's costs. One avenue is to use the difference in current population versus historical population in an area to create normalized losses. Bouwer (2011) reviewed 22 papers describing methodologies used for trending natural disaster losses. The only existing earthquake-only normalization study is from Vranes and

TABLE 5.2 AAL as a Percentage of Total Capital Stock From Historical Disasters Since 1950

Rank	Flood Rank	FL AAL (%)	Earthquake Rank	EQ AAL (%)	Storm Rank	ST AAL (%)
1	North Korea	0.448	Montenegro	1.084	Montserrat	1.239
2	China[a]	0.224	Macedonia	0.879	Belize	1.035
3	Pakistan	0.198	Armenia	0.342	Saint lucia	0.887
4	Mozambique	0.163	Turkmenistan[b]	0.264	Tokelau	0.852
5	Cambodia	0.120	Chile	0.174	Guam	0.748
6	Bangladesh	0.107	Haiti	0.164	Dominica	0.726
7	Japan	0.082	Nicaragua	0.136	Vanuatu	0.668
8	Yemen	0.074	Guatemala	0.110	Tuvalu	0.649
9	Nepal	0.055	Wallis and Futuna	0.106	Virgin Islands, US	0.632
10	Guyana	0.048	Nepal	0.097	Cayman Islands	0.543

[a]*China floods created large changes in loss costs over different periods. Losses ranged from 0.665% for the time period 1950–2015 to 0.444% for 1960–2015 and 0.224% for 1970–2015. These changes are due to improvements in flood control standards after the 1954 and 1956 floods rather than any significant changes in hazard.*

[b]*We also included Turkmenistan from the database from 1948 to 2015, given the large nature of the 1948 Ashgabat earthquake event dominating the overall losses.*

Pielke (2009) for US earthquakes from 1900 to 2005. This study uses wealth (GDP), inflation, mitigation, and population to describe changes in overall values over time. Pielke (2005) have used a wealth and inflation factor using construction cost index to determine US hurricane losses from 1900 to 2005. Due to a lack of socioeconomic data, Miller et al. (2008) used only GDP per capita and population changes for their worldwide analysis of storms from 1950 to 2005. Neumayer and Barthel (2011) provide an in-depth study, using the MunichRe NatCat database from 1980 to 2009 with a similar methodology of normalizing the GDP deflator, a wealth factor, and changes in population. In this study, an alternative method is proposed that uses only the difference in wealth per capita over time. The authors propose a worldwide methodology based on capital stock. The method has however not been applied uniformly across the globe due to lack of relevant data. Instead Neumayer and Barthel (2011) use the G-ECON dataset, which takes a geocell product of purchasing power parity (PPP; adjusting to the prices of a country for like goods)–adjusted GDP mostly based on 1995 GDP data, from 180 nations. One downside of this work is (besides the low quality of the spatial economic datasets utilized) that only 29 years of data were used, due to the short nature of the NatCat database. The shorter the time series of annual loss data available is, the greater is the standard error of the estimated results.

Barredo (2010) used an inflation, population, wealth, and PPP approach to normalize data. A similar approach to that of Crompton and McAneney (2008) for meteorological hazards has been utilized to normalize historic Australian earthquakes. This study uses the value of dwellings. Daniell and Love (2010) undertook a normalization approach for Australian earthquakes from 1788 to 2010 using (1) the HNDEC indexed (see above) values, (2) the current population versus historical population in the area, as well as (3) an index of the level of vulnerability of the building stock compared to the historical building stock. In their 2012 paper on normalizing global earthquake losses, Daniell et al. (2012b) use two methods: the original method of the earlier paper and a second method that takes into account the change in capital stock herewith making inflation part of the change in capital stock over time. Changes in vulnerability are paramount in places where major change in construction or any other change in standards, mitigation etc. has occurred in past years. These changes include flood control structures such as defenses, dykes, among others. Defenses have been erected in most countries especially after large recent losses.

The following equations denote the methodology of the capital stock method used to adjust losses to today's values:

$$\mathrm{NormD}_{2012,\mathrm{loc}} = \mathrm{D}_{\mathrm{y,loc}} \times \mathrm{CSF}_{\mathrm{y,loc}}$$

This methodology simply multiplies loss at the time of the event with the infrastructure wealth factor change for the area considered from the time of the event to today. This is applied here using the multilevel CATDAT capital stock database from 1900 to 2015.

$$\mathrm{CSF}_{\mathrm{y,loc}} = \text{Capital Stock Value}_{2015,\text{location}} / \text{Capital Stock Value}_{\text{year of event,location}}.$$

We normalize the losses from the CATDAT Natural Disaster databases including c.25,000 natural disaster events from 1900 to 2015 (8000 + earthquake and tsunami events, 6000 + storm events, 7000 + flood events [including annual assessments], c.1000 + volcano events, 1500 + drought/temperature events, 750 + bushfire events).

Using this method, the Great Kanto, Japan earthquake of 1923 would result in a loss, similar to that of the 1976 Tangshan, China earthquake and the 1906 San Francisco, USA, and Valparaiso, Chile events if the vulnerabilities were the same today as they were historically (which is unlikely as building standards are much better today), and in case the todays exposure would have been there when the event occurred.

Flood control structure improvement is crucial to any normalization methodology as flood losses would otherwise result in unrealistically high numbers for previous large events (especially from the early 1950s) compared to today's better protected stock. Most of the high (unprotected) losses in the 1950s and 1960s are from Chinese floods. These changes of vulnerability from historic events is brought in often through an age-based vulnerability or fragility function, or separate functions for better building standards and/or other improvements. Similarly for flood control structures and flood modeling, these components are critical when modeling flood potential, as much of the uncertainty of flood losses is dependent on up-to-date modeling of flood control structures. National systems for flood modeling that include up-to-date defense and vulnerability estimates (such as HWGK Flood Hazard Maps in Germany and the DFIRMs in USA) are hence likely to have much greater accuracy and relevance than global flood modeling procedures that do not consider these idiosyncrasies. The above results and challenges in accurately accounting for changes in vulnerabilities also hint to why it is unlikely that we may be able to detect small trends in losses, e.g., due to climate change (LSE "Munich Re Programme", 2015).

Building Toward a Loss Model: A Concentration on Earthquake Loss Data

Fig. 5.7 shows earthquake losses from 1900 to 2015 from the CATDAT Damaging Earthquakes Database. This database includes the total economic costs associated with previous earthquakes along with the relative percentage of losses from second-order effects such as infrastructure losses. Over 50% of earthquake fatalities have occurred due to failures of masonry buildings. This is despite the fact that property damage from the collapse of masonry structures has produced much lower economic losses than other typologies. The reason for this is that masonry structures are fairly cheap to rebuild. Usually and as a rule of thumb, around 45%–50% of losses of an earthquake are from damage to housing with the rest due to damage of industrial and commercial facilities (see Daniell et al., 2012b for more information).

Damage breakdowns by occupation are often available for large events (several USD bn events). This is opposed to smaller events for which insurance loss data are often aggregated with some percentage estimates for broad occupancy types. Detailed level data are available for only a limited number of cases from independent studies undertaken by governments or universities. Dividing loss figures into residential, commercial, industrial, infrastructure etc. typologies are necessary to assign losses to particular building classes as well as to calibrate loss models.

Earthquake losses examined in the CATDAT database distinguish between two key components: (1) the capital stock loss (building and infrastructure losses) and (2) the GDP loss (as the production losses). The cost of an earthquake includes these two components

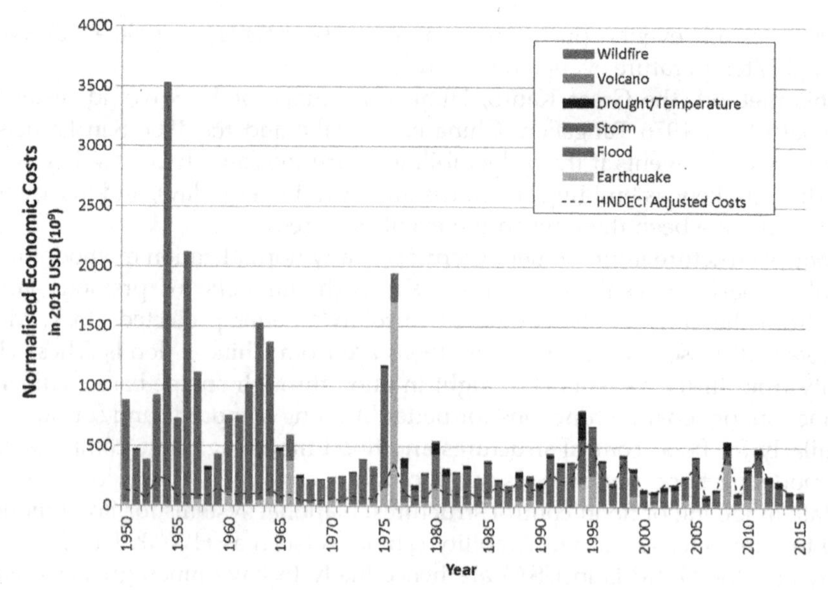

FIGURE 5.7 CATDAT natural disaster losses from 1950 to 2015 in terms of normalized losses due to each peril.

along with a third component: the capital improvement component, i.e., the increase in value an area undergoes as a result of the reconstruction from net (depreciated) to gross (new/replaced) capital stock. Improvement components suggest that buildings become safer (i.e., preserve lives). Safer building typologies are however more expensive and may result in greater reconstruction costs if damage occurs to those buildings. Safer and hence more expensive buildings are however more likely to be insured. This difference in value is seen as a key reason why insurance penetration does also need to increase even if overall insurance penetration might not rise after an event, as new more expensive stock is built in place of old dilapidated stock after a disaster (Fig. 5.8).

The disaggregation of losses into the subcomponent hazards of earthquakes show that only 62% of total losses are directly attributed to shaking, whereas 38% come from other forms of secondary effects such as tsunami (2004 Indian Ocean; 2011 Tohoku), liquefaction (Christchurch 2011), landslides (2005 Kashmir, 2008 Sichuan), and fire (1906 San Francisco, 1923 Great Kanto).

USING EMPIRICAL STATISTICS IN INSURANCE MODELING

Examples of Empirical Functions for Disasters

Using the historical loss data from disasters is only possible for building insurance models if hazard information for the events is correlated with the losses. Relationships of hazard metrics to the economic losses allow for the production of vulnerability functions. Vulnerability functions depict the damageability of certain building or infrastructure typologies given certain hazard intensity. Wind speeds, flow and rainfall data, or ground shaking can often

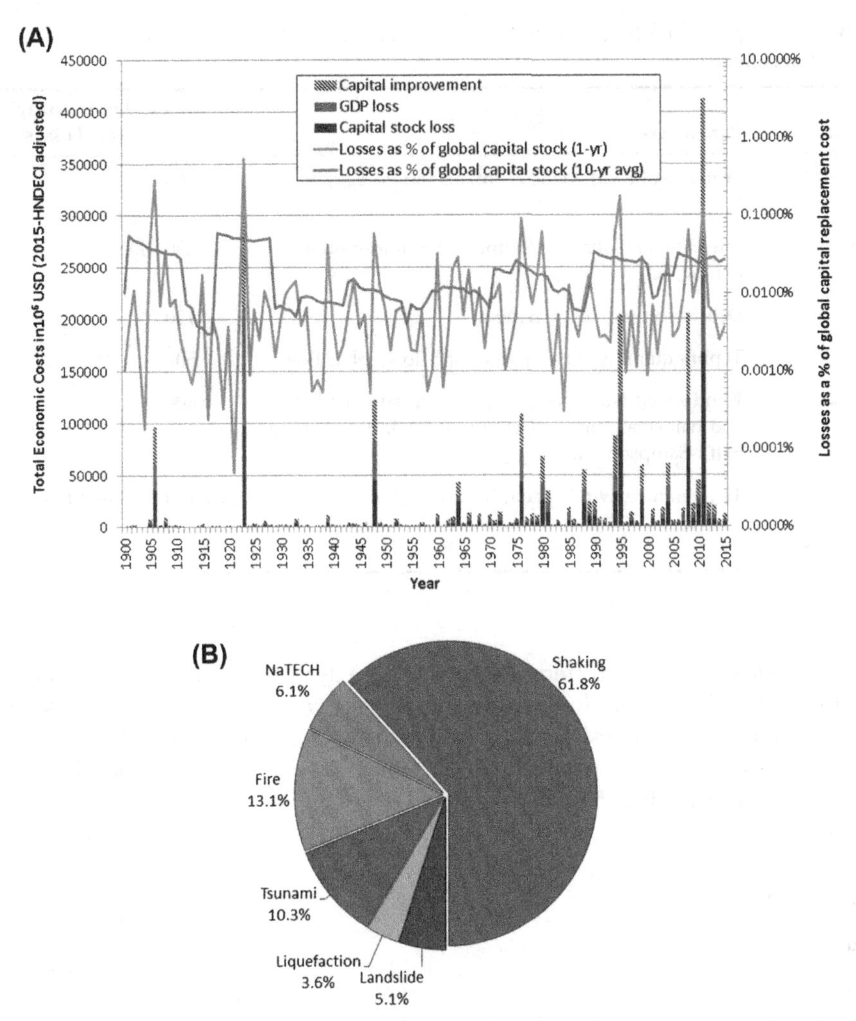

FIGURE 5.8 (A) Economic losses and costs from earthquakes occurring 1900–2015, as well as the relative cost versus the global gross capital stock. A reduction over time can be seen as a percentage of gross capital stock or GDP. (B) The disaggregated total economic costs due to secondary effects of earthquakes in addition to shaking cumulated from 8200 + damaging earthquakes (right). *Lq,* liquefaction; *NaTECH,* natural hazard triggering a technological disaster. Dollar amount in right-hand figure was adjusted to 2015 dollars using the HNDECI. *Calculations based on data in CATDAT. (A) Based on unskilled wage/combination series.*

be inferred from measured data over the past few decades. The further studies go back in history, the higher the uncertainties in these parameters. Combinations of the reported damage (such as macroseismic intensity assignments) help to infer a historical magnitude of an event (subject to the underlying uncertainties).

Different intensity parameters have been developed for different disaster typologies. For example, water depth and the water velocity (in m/s) are common parameters for floods (Table 5.3).

TABLE 5.3 The Type of Disasters, Hazard Parameters and the Number of Damaging Events Per Year Seen Globally

Disaster Type	Parameters	No. of Major Damaging Events Per Year (Ec. Losses > $0.1 m)
Flood, storm surge, tsunami	Water depth (m), velocity (m/s), and energy, flow	c.600+
Earthquake	Intensity, and shaking footprint; Ground motion (Sa, Sv, Sd)	c.250–300
Landslide	Debris volume, displacement	c.150
Volcano	Tephra quantity (kPa), pyroclastic flow, lahar flow	c.50
Wind/Typhoon/Hail	Wind speed (sustained, or gust), pressure, hail track and hail size (mm), reflectivity (dBz), kinetic Energy, Saffir-Sampson scale	c.800+
Extreme temperature, bushfire, drought	Temperature, wind speed, heat output, energy	c.100 (where measured)

Differences in loss ratios for different events can be derived by examining some direct and simple empirical relationships. Modifications in losses, even for similar intensities, result from (1) different building typologies, (2) changing nonstructural and content losses, (3) natural variability in event characteristics, and (4) varying base exposure values in different (or even the same) countries (Fig. 5.9).

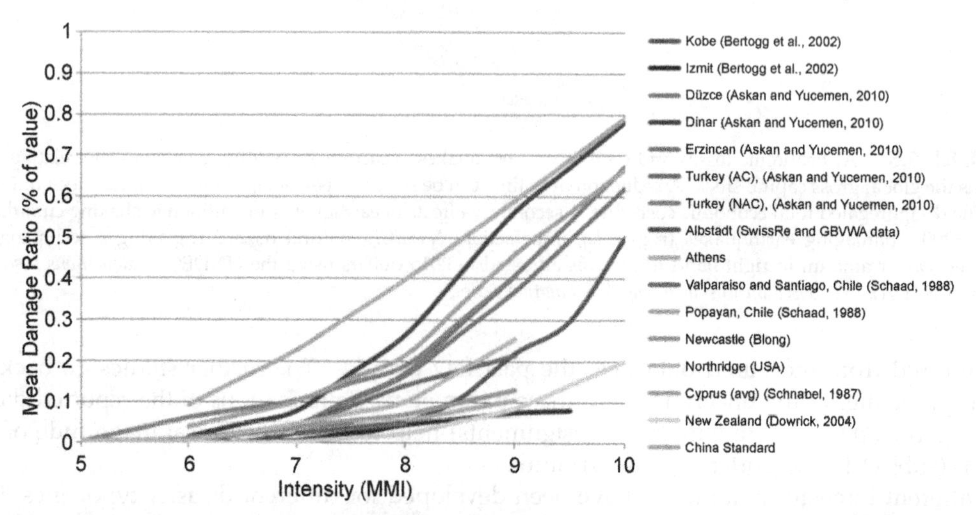

FIGURE 5.9 Various empirical loss ratios given intensity of ground shaking for earthquakes (combined building loss from earthquake events).

The Age of Infrastructure and the Impact of Building Standards Over Time

Loss ratios often fluctuate markedly with building standards, quality of buildings, and the total amount of exposure. Losses from the Kobe earthquake (as well as many other and more recent events) show large differences in losses based on the age of buildings. The total losses of the building stock based on book value are shown to be markedly different if compared to the total replacement cost.

In numerous cases such as for the 1978 Albstadt, the 1983 Popayan earthquakes investigated by Schaad (1988) and Cochrane and Schaad (1992), or for the 1999 Athens earthquake examined by Eleftheriadou and Karabinis (2012), the loss ratios of older structures have been four to six times those of younger structures. The total number and value of these structures also contribute to the differences seen in the amount of net capital stock. For the Athens 1999 earthquake, 85% of the building stock was built before 1985 incurring an average 7.8% median damage compared to 3.1% for the 1986–95 stocks and 1.7% for the buildings erected after 1995. In these cases, the net to gross capital stock effect can approach a loss difference value of 0.3–0.4 from the actual losses to the traditional replacement cost.

For the Sichuan earthquake, a greater loss than the initial net capital stock thus occurred, with the final actual loss of the capital stock being 47.8% of the initial 101.3 billion USD assessment, as the older depreciated stock is replaced (48.4 billion USD). Therefore, approximately 52.9 billion USD can be estimated as the improvement or replacement cost that occurs in the region, whereas the actual loss to capital stock is 48.4 billion USD. The production losses in many cases will not be affected by this concept of net versus gross capital stock. However, much work is required in accurately characterizing these losses and the overlap between indirect and direct production loss.

The 1995 Kobe earthquake losses show similar results. In this earthquake, the direct cost estimate at the time of the event was based on replacement cost for new values. Loss statistics are from the Kobe city reconstruction and damage documents, with losses being estimated using the work of Toyoda (2008). The final numbers were derived by combining multiple Japanese estimates. The direct replacement cost was estimated at around 133 billion USD.

For the building and infrastructure components the losses totaled approximately USD 94 billion. These were calculated using a pricing system of replacement cost per m^2 for various typologies, which is comparable to using gross capital stock. Out of the USD 94 billion of overall capital stock costs, about 58 billion USD were representing building losses of various typologies as shown in Fig. 5.10. Commercial and industrial buildings generally incur less damage than residential buildings; however, as only the total net loss is attempted to be calculated, the relative difference of survival life and damage ratio is assumed to be constant.

As seen in Fig. 5.11A, 99% of the buildings destroyed or demolished in the Kobe earthquake were built before 1980 (as shown in red) however, only 64% of buildings from the total building stock were constructed before 1980, thus showing that the 36% of total buildings built after 1980 only had 1% of the total destroyed buildings in the event.

Using the gross to net capital stock ratio of 1.685 in 1995 and comparing it to today's ratios, the actual loss is equal to USD 66.5 billion as opposed to USD 112 billion replacement cost/repair cost quoted postdisaster; given the net capital stock at the time. However, in earthquakes, the loss ratios are generally much higher for the older stock than the newer stock (Tiedemann, 1990). In Kobe, the average net capital stock construction age (for the value of all buildings) was from 1976 (i.e., 19 years old), which can be seen in Fig. 5.11.

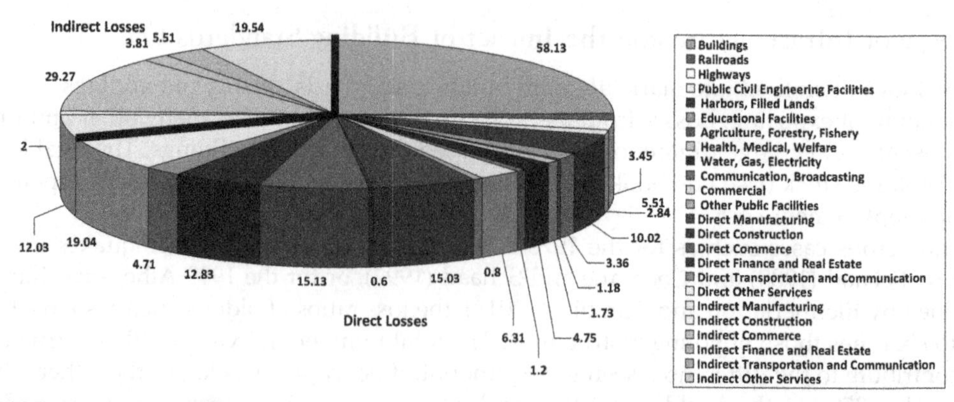

FIGURE 5.10 Sectorial distribution of losses in the 1995 Kobe Earthquake (Daniell and Wenzel, 2014).

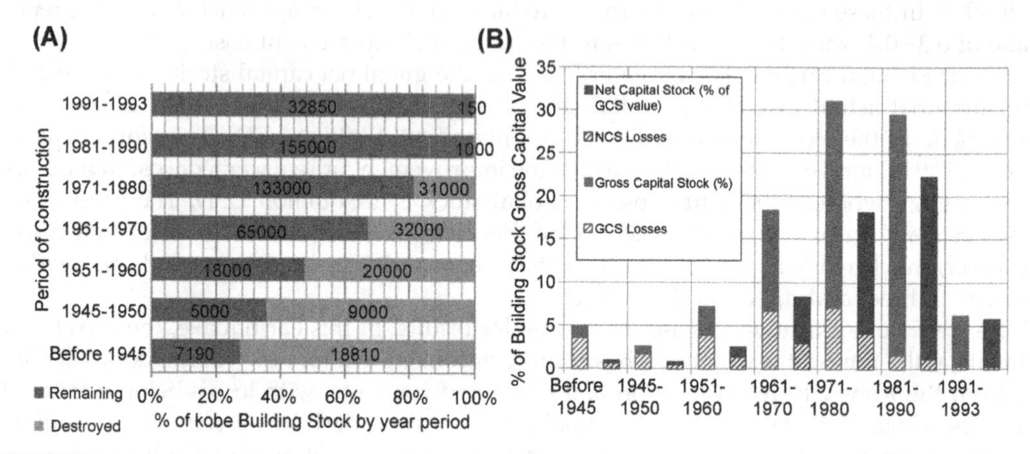

FIGURE 5.11 (A) The losses of the city of Kobe looking at the number of destroyed and remaining buildings after the event. (B) Calculation of the net capital and gross capital stock estimates for the dwelling portion of the losses/costs incurred in the 1995 Kobe earthquake. *(A) Adapted from the Japanese statistics provided by the Kobe municipal government.*

Calculating average repair costs of 5% for all other than the reported destroyed buildings, and considering destroyed buildings at costs of 100%, the current loss value is about 2.2 times less than what the replacement costs were. Considering the year of construction and using the weighted losses for each building, the average year of loss (as weighted by their dollar values) was 1966; thus, there was a 10-year difference between the average year of construction for net capital stock versus what was lost in the earthquake. This results in a book value loss of USD 51.1 billion instead of USD 66.5 billion based on simply stock ratios, thus showing the large difference versus the replacement costs for the event. In smaller earthquakes, or for earthquakes for which losses occur equally for old and new buildings, this effect approaches zero. To achieve smaller replacement costs, higher losses must be seen in the newer houses relative to the average net capital stock house. Additional capital influx in an area increases both its value and its need for insurance (Table 5.4).

TABLE 5.4 The Capital Stock Losses Adjusted for the Three Definitions in 1995 Kobe Compared to 2008 Sichuan

Earthquake (Event Year Billion USD)	1995 Kobe EQ[a]	2008 Sichuan EQ[b]
Actual loss	51.1	48.4
Replacement cost	112.0	101.3
Reconstruction cost/total spent on new capital	133.2	133.7[b]
Additional capital impact (improvement of capital)	+82.1	+85.3

[a]*The reconstruction cost at Kobe was reported at 159 billion USD for all components—19% increase on the replacement cost.*
[b]*Not all of the reconstruction cost in Sichuan translates into capital change—a lot was GDP based—on the final reconstruction estimate of the government. The value was taken as 32% of the 51% additional loss as first put forward via the gross capital stock estimate.*

Replacement cost is again based on gross capital stock and defined by a price per m^2 for various building types. This does not take into account service life and depreciation. The additional capital impact is calculated as the reconstruction cost minus the actual loss (direct damage at book value). The offset impact is then described as the additional capital impact minus the actual loss. Retrofitting older buildings is often less costly than waiting for a loss as the difference of the repair cost to the final loss value makes the repair money better spent.

Damage Functions Using Cost Data and Exposed Value

Exposed value to be replaced can be determined in a number of ways including (1) a straight replacement cost either by direct multiplication of floor area and value or (2) indexing an original value using construction or a "new value index" as determined by previous insurance claims. Correctly indexing building values is important as different parameters can lead to very different (if not contradicting) results. Construction costs in disasters however are often difficult to assess as the final paid value may not be representative of a simplified damage state (as shown previously).

Most models use damage ratios for particular damage states based on claims data and damage surveys despite many uncertainties prevalent in these values. Data are often missing for the higher intensities where most damage occurs. This is due to the fact that most of the damage occurs at long return periods that was often not observed in recent history. In addition, exact intensities might not be available at the location of the loss due to coarse measurement networks. Losses at lower return periods may in addition not be driven by mean values but rather by some transients, which makes correlation between loss and intensities misleading. The use of empirical loss functions is hence limited and often may need to be combined with other methodologies.

The mean damage ratio (repair cost divided by replacement cost) is only one expression of the damage distribution. Another one is business interruption and downtime that is a function of the distribution of damaged and destroyed buildings along with occupancy and (risk) management processes. Empirical loss data are usually represented as a lognormal, beta or binomial distribution around a mean damage ratio. Loss states might be sampled from ranges of derived clouds of data. Even for buildings of the same typology, such as a brick house in Australia, there is significant scatter in mean damage functions. This requires using methods such as a logic tree approach among others to select the preferred functions in the absence of the one unique best-fitting distribution. Analytical solutions may be matched to existing empirical data to avoid over- or underestimation of result. An example for this is the EQRM curve shown below.

The data depicted in Fig. 5.12 comes from various events (over several years) as well as different building stock typologies. Difference in age, adherence (or not) to code, as well as varying construction or occupancy types and/or subtle differences in cladding, rooks, adjacent sheds, interactions of buildings, or inefficiencies/idiosyncrasies in construction material (in addition to "random" differences in site conditions) etc. can lead to changes in damage ratios. Increasing adherence to building codes, as well as global improvements in engineering quality has increased the quality of buildings, meaning that fewer losses are occurring now than in the past. However, increasing exposure can offset this effect as mentioned earlier.

Number of stories and other parameters as shown in the electronic appendix are often needed to supplement basic empirical loss data. Too many parameters however may increase rather than decrease uncertainty meaning that a model for which results are not additive is required to ensure convergence of the model. A too large number of parameters used to create a model hence decreases rather than increases accuracy (otherwise known as overfitting—like the example of fitting a sixth order polynomial to four points of data), a fact that might be detected using rigorous testing only, although the regressions of 10 parameters may show a better fit, the model when used in a stochastic modeling environment.

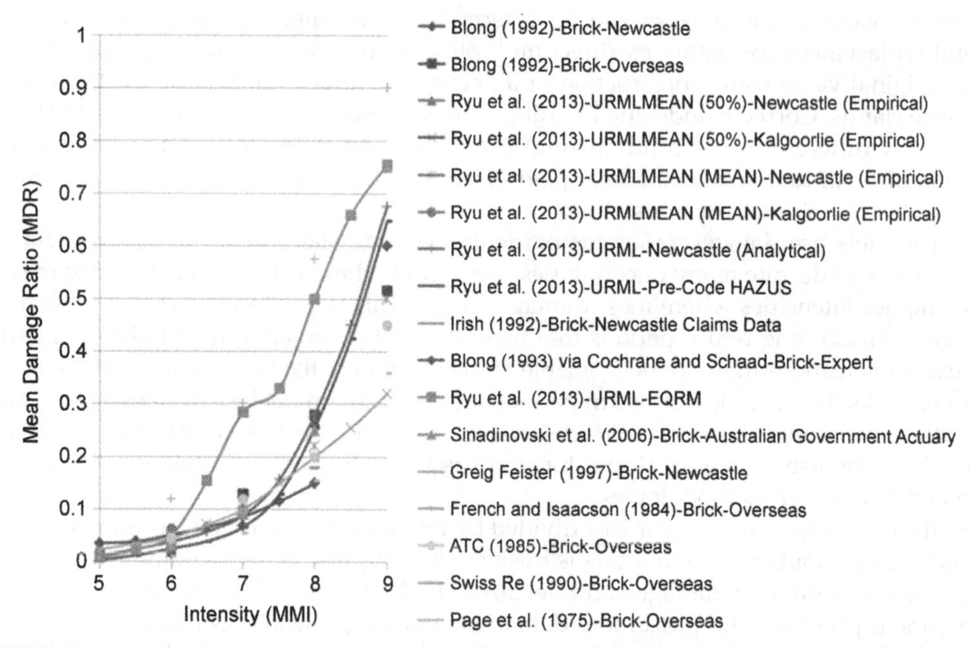

FIGURE 5.12 Brick/URM buildings showing the damage ratio versus intensity from mainly Australian functions.

Once exposure, hazard, and vulnerability components are derived, all components are present for the production of the risk results. Larger events alone might not suffice to understand spatial distribution of losses and return periods of events and a significant number of smaller events may be needed to calibrate losses. An example model for Australian earthquake is described below.

CASE STUDY: AUSTRALIAN EARTHQUAKE MODEL

The below Australian earthquake model is an attempt to build a rapid, but robust metric model herewith exploring the available depth of risk data along with providing a tool for discussions into usable risk metrics and assumptions describing Australian earthquake risk.

Scarcity of events as well as analytical rather than empirical loss functions can make creating an Australian earthquake model challenging. In the absence of data for one country, data from other countries around the globe or inferences about the shape of damage function (beta, tanhyp, lognormal) may assist and have allowed creating reasonably accurate models for various territories.

Many attempts have been undertaken to identify the seismic risk in Australia. Key hazard, exposure, and/or vulnerability models were derived by EQRM from Geoscience Australia (Edwards et al., 2004) as well as through Risk Frontiers and their QuakeAUS model (https://www.riskfrontiers.com/quakeaus.html).

Damage data exist from only a limited number of earthquakes given the fact that Australia is an area with only moderate to low earthquake activity and the fact that Australia has only a limited history of around 400 years; data from other areas and analytical procedures are hence needed to fill in missing earthquake damage records. To create the Australia earthquake model, risk calculations resulting in the number of fatalities and economic costs of earthquakes needed to be included to complement the hazard work of Schäfer and Daniell (2014). Fulford et al. (2002) provided an early view of risk outputs from the Geoscience Australia EQRM model (Robinson et al., 2005), incorporating user workshop data on vulnerability functions. This has helped to alter a version of HAZUS for the use for Australia.

The Christchurch (NZ) earthquakes have taught us various lessons as to the potential impacts of large damaging earthquakes for unreinforced masonry (URM) buildings (Moon et al., 2014; Ingham and Griffith, 2011) and for low-code RC buildings such as those built in Australia. The sister city of Christchurch is Adelaide, which has an extremely high level of URM and light timber Brick Veneer building typology percentage as defined by NEXIS (National Exposure Information System).

Hazard Modeling

Stochastic modeling in general starts with revisiting historical earthquake catalogues including magnitudes, locations, and intensities of events. Frequency-magnitude distributions are then calculated as a basis for creating a stochastic set of events for a particular temporal period and area.

Below is a short summary of the parameters used in the hazard modeling for Australia. A fuzzy source zonation was linked to a smooth seismicity model. This approach uses historic and instrumental earthquake data since 1800 to develop a stochastic earthquake catalogue. The resulting events were then used to compute the ground motion for specified locations. Earthquake data for Australia is both sparse and inconsistent. This also holds for both space and time. However earthquake observations have been recorded and retrieved from historic records and some events date back to the years of the earliest colonial settlements in the late 18th century. For this study, historic observations from 1800 onwards, based on the work of McCue (2013a,b,c, 2014), have been combined in the Geoscience Australia earthquake database. Considering a minimum magnitude of 2.0 since 1800, the catalogue consists of 24,034 earthquakes. These earthquakes were declustered using the methodology of Reasenberg (1985), which reduces the total number of events to 11,838.

The source parameters were stored at 10 km resolution grid. The Gutenberg-Richter a-value of a certain pixel is hereby representing the weighted average of the normalized smooth seismicity of the catalogue and the assigned a-value from the distance correlation with the area sources. A bilinearly truncated Gutenberg-Richter distribution was used, which doubled the b-value for magnitudes larger than six to account for the very long return periods common of a stable crust. The total resolution of the model is then run at 1 km. For the computation of ground motion, four ground motion prediction equations have been selected. These include both Atkinson and Boore (2006) and Lin and Lee (2008) representing a globally calibrated equation for shallow earthquakes as well as Allen (2012) and the noncraton equation of Somerville et al. (2009). The two latter equations have been developed using Australian earthquake data. Differentiating between cratonic and noncratonic crust has been neglected for the sake of a uniform ground motion prediction scheme for the whole continent and to keep the hazard curves comparable.

For further details, see Schäfer et al., 2015. Below is a list of the various components used in the earthquake model described herein (Table 5.5; Fig. 5.13).

Exposure Modeling

Exposure modeling requires collecting census data along with values of the buildings (or building types) for which the vulnerability is calculated. The NEXIS database (Nadimpalli et al., 2007) provides a census of data on wall materials, roof materials, and storey heights along with information of associated cost. This was used from SA1 level data (Australian Bureau of Statistics smallest statistical area unit [Level 1], depicting around 55,000 units across Australia). In addition, mesh block data (a unit smaller than SA1) for population information and for the number of dwellings are found on a mesh block level (c.300,000 units in Australia). Downscaling techniques, i.e., splitting the mesh blocks into an individual number of buildings, are then used to evaluate the losses at each point. This is done by randomly distributing losses based on street density. Unfortunately, building footprint data through Open Street Map is not available for all relevant cities, thus meaning that downscaling data to building levels is only possible for certain major cities.

Population adjustments were made using growth trends to calculate the June 2015 population for each of the locations modeled based on 2011 data. A weighted analysis, by the

TABLE 5.5 Hazard Components Used in the Model

Historical data used	Geoscience Australia, McCue (2013a,b,c, 2014)
Completeness periods	Automated per seismic source and pixel
No. of events (declustered & not)	24034 (total), 11838 (declustered)
No. of random years calculated	500,000
Seismic source zonation method	Fuzzy domains
Site effects	USGS vs 30, 1/3 weighting within GMPE selection
GMPEs used	Atkinson and Boore (2006), Lin and Lee (2008), Allen (2012), Somerville et al. (2009)
Uncertainty accounted for	Spatial uncertainty/seismic migration (large smoothing kernel), incomplete data record (deterministic scenarios), seismic source mechanism/b-value (fuzzy logic)
PGA-MMI relationships	Atkinson and Kaka (2007) with checks of Bilal (2013), Tselentis and Danciu (2008) via Greenhalgh et al. (1989) as per Daniell (2014)

GMPE, ground motion prediction equation; *MMI*, modified mercalli intensity; *PGA*, peak ground acceleration; *USGS*, United States Geological Survey.

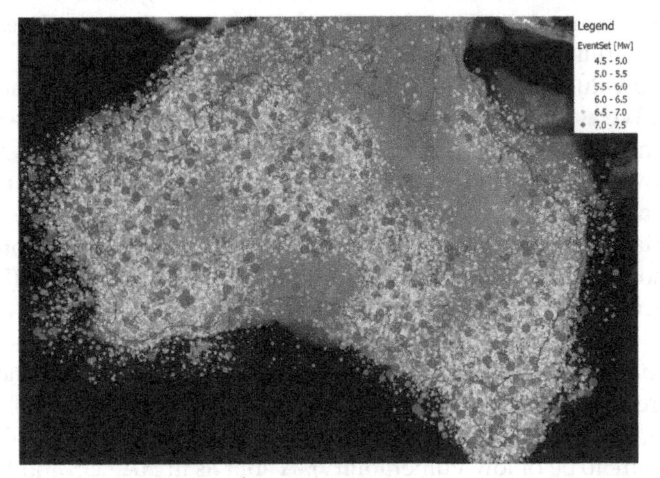

FIGURE 5.13 Stochastic earthquake catalogue for 100,000 years from the 500,000 years of events created in our model.

number of dwellings, was made in addition to the population adjustments to check the difference this makes for the results.

Costs of dwellings versus nondwelling construction was examined using the methodology described for LAC (Latin-American and Caribbean) risk profiles of the World Bank (Gunasekera et al., 2015) where gross capital stock (replacement costs) is calculated top-down from investment to value along with using a bottom-up replacement costs approach

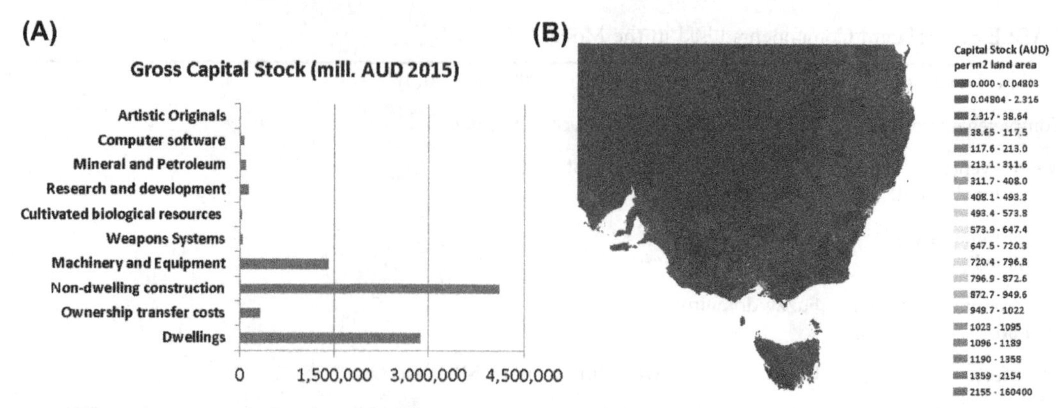

FIGURE 5.14 (A) Gross capital stock estimates as calculated using ABS (Australian Bureau of Statistics) (2014) and Daniell (2014). (B) Gross Capital stock per m^2 land area in SE Australia.

(per square meter and multiplied by the area of the built stock) (see the work of NEXIS). In this case a difference of more than 70% was revealed when comparing results from the two methodologies. The net capital stock (actual value of all dwellings, including contents) is calculated at c.$1.7 trillion AUD (Australian dollars). When taking the Perpetual Inventory Method into account, a value of c.$2.9 trillion AUD, including fixed contents, is calculated. Taking into account demand surge and nonfixed contents and service lives, results reached values of up to $3.6 trillion AUD (Fig. 5.14).

By comparison, a value of over $4.5 trillion is shown in NEXIS for a combination of dwellings and contents. We assume that part of this offset might be due to differences in the classification of buildings as dwellings versus commercial buildings. We suggest that the NEXIS results would be lower by around 18% if the World Bank methodology would be applied. For nondwelling construction, a reduction of 3% is made for commercial structures and no change for industrial structures when making the same adjustments. The total values of gross capital stock for nondwelling construction types include (1) road values, (2) port infrastructures, and (3) agricultural risks along with several other components that are not accounted for. Capital stock values of $6.2 trillion are calculated for all other assets for Australia. Of these, $5.1 trillion of the building stock are deemed subject to modeled earthquake risk (buildings, infrastructure etc.), leaving the nonmodeled exposure components (i.e., removing weapons, research stocks etc.) at $1.1 trillion (we consider earthquake vulnerability for the nonmodeled exposure to be of low vulnerability). A split as in HAZUS and EQRM as to structural and nonstructural components were (acceleration sensitive and drift sensitive) undertaken.

Vulnerability Modeling and Calibration With History

Much effort has been commenced in the past in terms of vulnerability modeling for Australian (Edwards et al., 2004). For this study, we reviewed papers from the AEES (Australian Earthquake Engineering Society) over the last 25 years, in addition to other external papers (e.g., Edwards et al., 2004; Blong, 1992; Sinadinovski et al., 2005; Fulford et al., 2002). Over 45 vulnerability functions were found for various typologies derived from loss data from different historical events. Results for some of the URM functions are shown above for residential structures. Splits of commercial, residential, and industrial buildings have been made by the local government Geoscience agency, Geoscience Australia.

The nonstructural acceleration-sensitive part of Australian building typologies is subject to significant damage, as seen in recent earthquakes such as the Moe and Kalgoorlie events. This might require that the damage ratios around and before yielding of the structure are taken into account while building a model. We did consider this in our calibration work (contact first author for more details). The original EQRM functions were high and provided an overestimation of loss at high intensities when compared to historical events such as for the 1989 Newcastle earthquake. The insured loss value in 1989 from the event was around $862 million at the time of the event, with estimates of insurance penetration between 30% and 70% of the final total losses. A full historical reanalysis like the above was undertaken in Daniell and Love (2010) for Australian earthquakes, as well as subsequently by McCue (2014, 2013a,b,c), providing the necessary historical and empirical data.

The problem in low seismicity countries is that there is generally a lack of damage data above intensity VII or of ground motions above peak ground acceleration (PGA) of 0.1 g. This means that information from other countries is required to examine the potential loss effects for stronger events. Upper bound measures as collected through Christchurch and similar events reaching above intensity VIII provide invaluable insights and data for potential future events for Australia. In modeling risk for insurance purposes, proxies are often used from deemed comparable typologies in neighboring countries.

We used the capacity spectrum method to reproduce functions in terms of precode (assuming the buildings are not built to earthquake resistant code) and low-code (assuming the buildings are built to the current code standard in Australia). A conversion from PGA to modified mercalli intensity was then applied for the calibration of vulnerabilities. We used an adjustment based on Atkinson and Kaka (2007), and a set of calibrated functions from the European Macroseismic method (Giovinazzi, 2005). This was done via equating the mechanical outputs to intensity modification in terms of the loss functions. This allowed further comparison of results to empirical data. Uncertainties include the spectral response used, uncertainty in the damage data around a mean value (beta distribution used) as well as extrapolation of empirical losses above existing loss data for the area. In addition, the seismic quality of the building stock was only allowed via a factor approach based on empirical data from the Newcastle 1989 and Moe 2012 earthquake with a basic change based on the NEXIS age blocks available (pre-1980 vs. post-1980 stock). For the mechanical method we calibrated calibration data further to check differences in "Precode" and "Low-code" results (Fig. 5.15).

Existing EQRM functions revealed much higher estimates than the adjusted functions later by Ryu et al., 2013 in their reassessment of European and US (HAZUS) typologies adapted to fit the Australian model data (see also Ryu et al., 2013). For unknown vulnerability functions (for construction types with no data), estimated parameters of code influence, ductility, and additional system response were kept similar to the fitted system of adapting HAZUS CSM typologies. This resulted in a reasonable set of vulnerability functions for Australia.

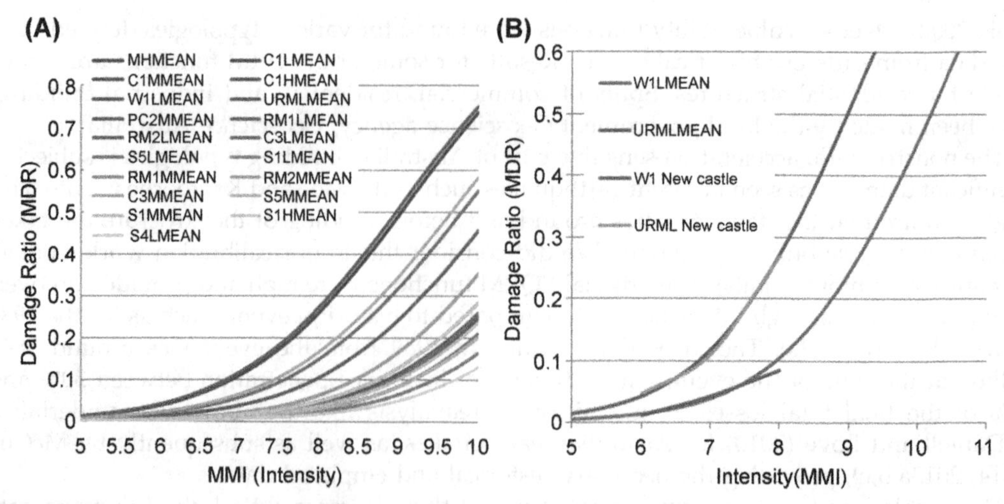

FIGURE 5.15 (A) MDR of the vulnerability functions used in the study for the 37 typologies. (B) A comparison of URM and timber building loss vulnerability functions used in the study, with the empirical loss data from Newcastle (the most damaging recorded earthquake in Australia).

Risk Modeling

Australia earthquakes locations are well distributed across the country (more so than in many other countries of a similar size and distribution of population). In a recent study of Central American nations, PML250 (Probable Maximum Loss event at a 250-year return period) values of around 12% of capital stock are quite common (Gunasekera et al., 2015). In Australia, however, lower seismicity (despite high vulnerability) along with the distributed nature of earthquakes give a PML250 around 0.25% (for structural losses). However, given the AUD 5.1 trillion of Australian capital stock (total) that can be affected by an earthquake, 1 in 250 economic costs might reach losses as high as $11 billion (AUD) (Fig. 5.16).

Around 75% of the AAL comes from events below PML1000 across Australia. Fulford et al. (2002) suggested that for the Newcastle area, 82% of AAL comes from below PML1000. With acceleration sensitive components being particularly at risk, nonstructural as opposed to structural damage is likely to dominate losses for smaller events.

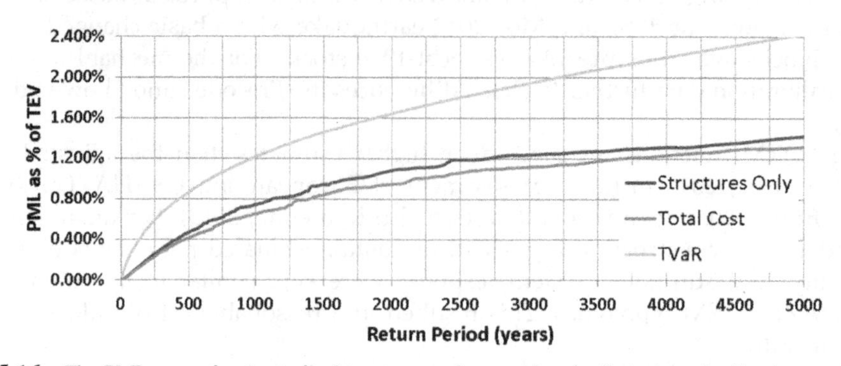

FIGURE 5.16 The PML curve for Australia (structures only considered; all stock and tail value at risk).

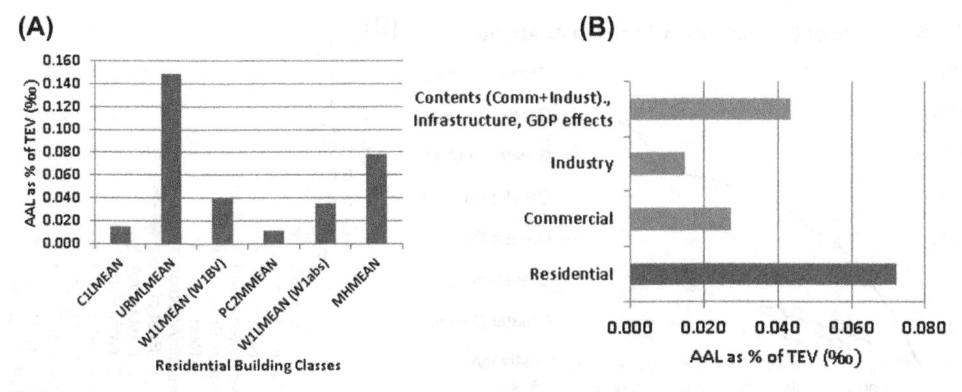

FIGURE 5.17 (A) Residential building class AALs (‰). (B) AAL (‰) in each use class/economic class.

Tail Value at Risk (TVaR) represents the average (economic) loss given that an event occurs above a certain return period. This is a useful value for making decisions as to the tail of the risk distribution. TVAR has several advantages to VAR (Value at Risk or "PML") including the fact that TVAR produces sets of "smooth" values as numbers are integrated over the tail of the distribution rather than representing as single value. $TVAR_{100}$ often approaches values close to VAR_{200} to VAR_{250}. The $TVaR_{250}$ was calculated to be 0.587%, as compared to 0.226% for the VAR_{250} (=PML_{250}).

Studying the AAL (integral over the entire risk curve) for the various typologies is a good way to start examining the most vulnerable features. URM buildings (damage seen in the 2011 Christchurch earthquake) are among the most vulnerable building typologies, with a total AAL of around 0.14‰ (Fig. 5.17). This is three times higher than that of the brick veneer-/timber-based buildings. These results are comparable to the results of Ryu et al. (2013) who showed that URM damage from a Mw = 5.35 event was 2.5 times, and from a Mw = 6.5 event 3 times, higher than that of brick veneer-/timber-based building. Buildings of different ages depicted moderate differences in loss behavior across various suburbs from major cities, with older buildings showing slightly higher loss ratios. Different storey heights also showed some minor difference, with mid-rise buildings often having slightly higher loss ratios than high- or low-rise buildings depending on the spectral period adjustments used.

The AAL derived for structural (only) losses totals around AUD 429 million. This equals a damage ratio (as a percentage of exposure) of 0.047‰. This is close to the AAL of around 0.04‰ calculated through the historic 1950–2015 AALs.

The calculated life AAL is AUD 378 million, leading to interesting implications for accident and health or casualty and life insurance due to earthquake.

For an overall exposure value of $2.5 trillion, Walker (2003) calculates an AAL of around AUD 210 million (0.08‰) hence suggesting AAL to be roughly twice the numbers estimated herein (0.047‰). Uncertainty is large but part of the difference might be the recent decease in risk due to newer buildings along (not considered in the Walker, 2003 study) and the lower vulnerabilities of the commercial and industrial buildings included herein (this study). Risk results considered herein are significantly higher than those of GAR (2015).

PML curves aggregated over the greater city regions (urban areas, only) approximate Australia's city risk. Depending on the stochastic catalogue used, Perth and Adelaide interchange as the cities showing the highest risk. Canberra has the third highest earthquake

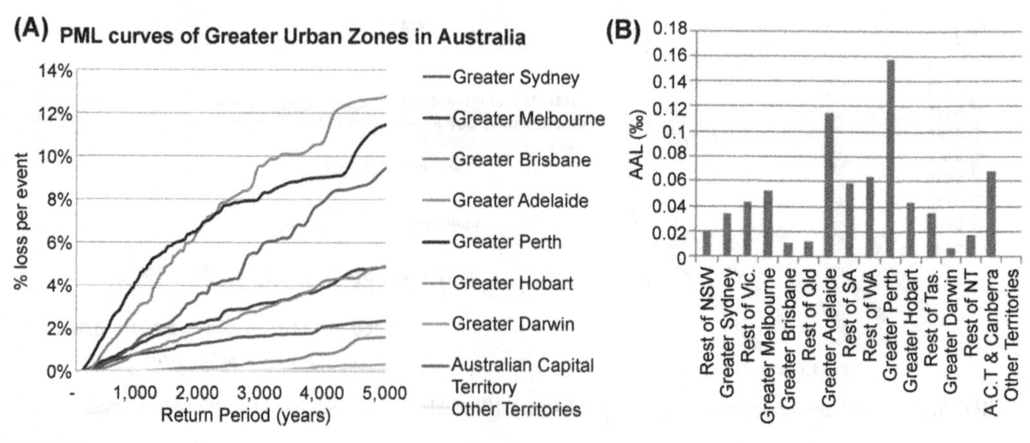

FIGURE 5.18 (A) PML curves for capital cities. (B) AAL (‰) for each larger zone.

risk at both long periods (i.e., for the tail of the distribution) as well as in terms of AAL (measured in percentage of exposure) (Figs. 5.18 and 5.19).

Insurance premiums are in general based on a combination of perils and may (more recently) be based on modeled AALs derived by integrating over the risk curves for areas considered for technical pricing. Insured values used in these calculations are in general

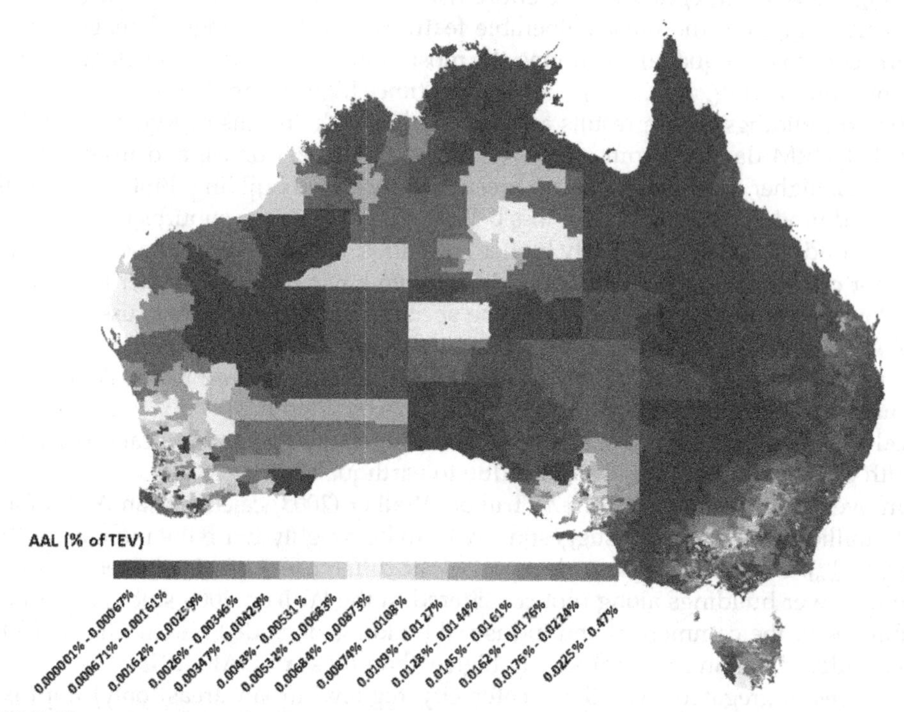

FIGURE 5.19 AAL aggregated to SA1 statistical units. It shows higher AALs across SW Australia, Adelaide, Tennant Creek, and parts of SE Australia.

based on reconstruction costs (see above for discussion). As reconstruction costs are not unique, undervaluation might occur. Insurers might adjust undervaluation when new evidence demands it. Insurance is however understood to be "crowdsourcing" of risk, which in turn requires insurers to deviate from a very strict risk-based pricing concept. In other words, although models might be used as rough guides for rating insurance business, various levels of "smoothing" is taking place before the actual/final prices (especially for homeowners insurance) are determined. This means that those in low-risk areas are likely to pay for those in higher risk areas as well. Models have their fair value however for insurance portfolio management, capital allocation, and hedging (buying reinsurance).

CONCLUSION

This chapter acts as a guide to incorporating empirical loss data into stochastic risk modeling for insurance and other purposes. One focus has been on introducing the reader to the value of CATDAT, the Natural Disaster and Socioeconomic Loss Database. Understanding and calibration of modeled historical losses (such as flood) is paramount given the recent changes in risk due to mitigation including better flood protection, among others. The evaluation of CATDAT reveals that flood losses have historically had the highest financial impact from any natural disasters. This is opposed to the last 10 years during which earthquakes and hurricane losses have dominated the global records.

Gross insured losses are increasing with greater exposure and greater insurance penetration. Global capital stock is increasing at around the same rate as insurance premiums for developed countries as well as in emerging markets, meaning that net insurance penetration might stall with absolute amounts of uninsured losses increasing rather than decreasing (as shown in the Kobe 1995 and Sichuan 2008 earthquake examples). This might not necessarily be taken into account in published numbers.

Prudent collection of historical event data and using correctly adjusted historic losses at today's terms allow calibrating catastrophe models with loss data and loss ratios. To do this, we need to (1) account for changes in vulnerabilities over time; (2) separate out secondary effects (and their changes over time); and (3) consider changes in building typologies (residential, commercial, industrial etc.), age classes, among others. Inflating historical losses by premium or changes in capital stock alone, may however fail to produce appropriate results. Due to scarcity of historical loss information, neighboring country data or analytical modeling might be used to replace historical evidence. This was shown for the earthquake model created for Australia. This highlights the fact that global model experience might be needed to create reasonably accurate regional or local models. "The more the better" might not necessarily be true for model creation and in doubt, it is better to fall back to a lesser rather than larger number of variables well backed by historical or analytical parameters to create a natural catastrophe loss model. More complete information, remodeling of historical events and losses, as well as the more we understand changes inventory regionally as well as over time and around the globe will undoubtedly help improving our understanding of risk.

References

ABS, 2014. 5204.0-Australian System of National Accounts, 2013–14. Australian Bureau of Statistics.

Allen, T., 2012. Stochastic Ground-Motion Prediction Equations for Southeastern Australian Earthquakes Using Updated Source and Attenuation Parameters. Geoscience Australia Record, 2012/69.

Atkinson, G., Boore, D., 2006. Earthquake ground-motion prediction equations for Eastern North America. Bulletin of the Seismological Society of America 96 (6), 2181–2205.

Atkinson, G., Kaka, S., 2007. Relationships between felt intensity and instrumental ground motions for earthquakes in the central United States and California. Bulletin of the Seismological Society of America 97, 497–510.

Barredo, J.I., 2010. No upward trend in normalised windstorm losses in Europe: 1970–2008. Natural Hazards and Earth System Sciences 10 (1), 97–104.

Bilal, M., 2013. Relationships Between Felt Intensity and Recorded Ground Motion Parameters for Turkey (Doctoral thesis). Middle East Technical University.

Blong, R.J., 1992. Domestic insured losses and the Newcastle earthquake. Earthquake resistant design & insurance in Australia. In: AEES and Specialist Group on Solid-Earth Geophysics Conference, Sydney.

Bouwer, L.M., January 2011. Have disaster losses increased due to anthropogenic climate change. American Meteorological Society, BAMS 39–46.

Cochrane, H.C., Schaad, W.H., 1992. Assessment of earthquake vulnerability of buildings. In: Proceedings of the 10th World Conference on Earthquake Engineering. Taylor & Francis, Madrid, Spain (I).

Collins, D.J., Lowe, S.P., 2001. A macro validation dataset for U.S. hurricane models. Casualty actuarial society. Winter Forum 217–252.

Crompton, R.P., McAneney, K.J., 2008. Normalised Australian insured losses from meteorological hazards: 1967–2006. Environmental Science and Policy 11, 371–378.

Daniell, J.E., Love, D., 2010. The socio-economic impact of historic Australian earthquakes. In: AEES 2010 Conference, Perth, Australia, vol. 21. Paper No. 8.

Daniell, J.E., Khazai, B., Wenzel, F., Vervaeck, A., 2011. The CATDAT damaging earthquakes database. Natural Hazards and Earth System Science 11 (8), 2235–2251.

Daniell, J.E., Khazai, B., Wenzel, F., Vervaeck, A., 2012a. The worldwide economic impact of historic earthquakes. In: 15th WCEE.

Daniell, J.E., Wenzel, F., Vervaeck, A., 2012b. The Normalisation of socio-economic losses from historic worldwide earthquakes from 1900 to 2012. In: 15th WCEE, Lisbon, Portugal, Paper No. 2027.

Daniell, J.E., 2014. The Development of Socio-economic Fragility Functions for Use in Worldwide Rapid Earthquake Loss Estimation Procedures (Doctoral thesis) (Karlsruhe, Germany).

Daniell, J.E., Wenzel, F., 2014. The Economics of Earthquakes: A reanalysis of 1900–2013 historical losses and a new concept of capital loss vs. cost using the CATDAT Damaging Earthquakes Database, Paper No. 1505, 15th ECEE (European Conference of Earthquake Engineering). Istanbul, Turkey.

Edwards, M.R., Robinson, D., McAneney, K.J., Schneider, J., 2004. Vulnerability of residential structures in Australia. In: In 13th World Conference on Earthquake Engineering, Vancouver. Paper (No. 2985), August.

Eleftheriadou, A.K., Karabinis, A.I., 2012. Seismic vulnerability assessment of buildings based on the near field Athens (7-9-1999) earthquake damage data. International Journal of Earthquakes and Structures 3 (2), 117–140.

Fulford, G., Jones, T., Stehle, J., Corby, N., Robinson, D., Schneider, J., Dhu, T., 2002. Earthquake Risk in Newcastle and Lake Macquarie. Geoscience Australia Record 2002/15. Geoscience Australia, Canberra, pp. 103–122.

GB/T18208.4-2005, 2005. Post-earthquake Field Works Part 4: Assessment of Direct Loss (in Chinese).

Giovinazzi, S., 2005. The Vulnerability Assessment and the Damage Scenario in Seismic Risk Analysis (Ph.D. thesis). Technical University Carolo-Wilhelmina, Braunschweig, Germany.

Greenhalgh, S.A., Denham, D., McDougall, R., Rynn, J., 1989. Intensity relations for Australian earthquakes. Tectonophysics 166, 255–267.

Gunasekera, R., Ishizawa, O., Aubrecht, C., Blankespoor, B., Murray, S., Pomonis, A., Daniell, J.E., 2015. Developing an adaptive global exposure model to support the generation of country disaster risk profiles. Earth-Science Reviews 150, 594–608.

Ingham, J., Griffith, M., 2011. Damage to unreinforced masonry structures by seismic activity. The Institution of Structural Engineers The Structural Engineer 89 (3), 14–15.

IRDR, 2014. Peril Classification and Hazard Glossary. http://www.irdrinternational.org/wp-content/uploads/2014/04/IRDR_DATA-Project-Report-No.-1.pdf.

Lin, P.-S., Lee, C.-T., 2008. Ground-motion attenuation relationships for subduction-zone earthquakes in northeastern Taiwan. Bulletin of the Seismological Society of America 98 (1), 220–240.

McCue, K., 2013a. Darwin Northern territory – an earthquake hazard. In: Australian Earthquake Engineering Society 2013 Conference, Nov 15–17. Hobart, Tasmania.

McCue, K., 2013b. Historical earthquakes in the Northern territory. Australian Earthquake Engineering Society.

McCue, K., 2013c. Some historical earthquakes in Tasmania with implications for seismic hazard assessment. In: Australian Earthquake Engineering Society 2013 Conference, Nov 15–17. Hobart, Tasmania.

McCue, K., 2014. Balancing the earthquake budget in NSW. In: Australian Earthquake Engineering Society 2014 Conference, Nov. 21-23, p. 43 (Lorne, Victoria).

Miller, S., Muir-Wood, R., Boissonade, A., 2008. An exploration of trends in normalized weather-related catastrophe losses. In: Diaz, H.F., Murnane, R.J. (Eds.), Climate Extremes and Society. Cambridge University Press, New York.

Moon, L., Dizhur, D., Senaldi, I., Derakhshan, H., Griffith, M., Magenes, G., Ingham, J., 2014. The demise of the URM building stock in Christchurch during the 2010–2011 Canterbury earthquake sequence. Earthquake Spectra 30 (1), 253–276.

Nadimpalli, K., Edwards, M., Mullaly, D., December 2007. National exposure information system (NEXIS) for Australia: risk assessment opportunities. In: MODSIM 2007 International Congress on Modelling and Simulation, Modelling and Simulation Society of Australia and New Zealand, pp. 1674–1680.

Neumayer, E., Barthel, F., 2011. Normalizing economic loss from natural disasters: a global analysis. Global Environmental Change 21, 13–24.

Olsen, A., Porter, K., 2008. A Review of Demand Surge Knowledge and Modeling Practice. White Paper. Willis Research Network. Available from: http://www.willisresearchnetwork.com.

Pielke Jr., R.A., 2005. Are there trends in hurricane destruction? Nature 438, E11. http://dx.doi.org/10.1038/nature04426.

Robinson, D., Fulford, G., Dhu, T., 2005. EQRM: Geoscience Australia's Earthquake Risk Model: Technicalmanual: Version 3.0. Geoscience Australia Record 2005/01. Geoscience Australia, Canberra.

Ryu, H., Wehner, M., Maqsood, T., Edwards, M., 2013. An enhancement of earthquake vulnerability models for Australian residential buildings using historical building damage. In: Australian Earthquake Engineering Society 2013 Conference, Hobart, Tasmania, 15–17 November.

Schaad, W.H., 1988. Earthquake loss analyses from the insurer's standpoint. In: Proceedings of the 9th World Conference on Earthquake Engineering, Tokyo, Japan.

Schäfer, A.M., Daniell, J.E., 2014. A stochastic hazard assessment for Australia based on an external perspective. In: Australian Earthquake Engineering Society 2014 Conference, Lorne, Vic., 21–23 November.

Schäfer, A.M., Daniell, J.E., Wenzel, F., 2015. The seismic hazard of Australia – a venture into an uncertain future. In: Proceedings of the Tenth Pacific Conference on Earthquake Engineering, Building an Earthquake-Resilient Pacific, Sydney, Australia, 6–8 November.

Sinadinovski, C., Edwards, M., Corby, N., Milne, M., Dale, K., Dhu, T., Jones, A., McPherson, A., Jones, T., Gray, D., Robinson, D., White, J., 2005. Natural Hazard Risk in Perth, Western Australia. Geoscience Australia, Canberra, pp. 143–208.

Somerville, P., Graves, R., Collins, N., Song, S.G., Ni, S., Cummins, P., 2009. Source and ground motion models for Australian earthquakes. In: The Australian Earthquake Engineering Society Conference. The Australian Earthquake Engineering Society, Newcastle.

SwissRe, 1999–2015. Sigma, Economic Research and Consulting. Swiss Reinsurance Company Ltd.

Tiedemann, H., 1990. Newcastle: The Writing on the Wall. Swiss Reinsurance Company.

Tselentis, G.A., Danciu, L., 2008. Empirical relationships between modified Mercalli intensity and engineering ground-motion parameters in Greece. Bulletin of the Seismological Society of America 98 (4), 1863–1875.

Vranes, K., Pielke Jr., R.A., 2009. Normalized earthquake damage and fatalities in the United States: 1900–2005. Natural Hazards Review 10, 84–101.

Walker, G., 2003. Insurance of Earthquake Risk in Australia. www.aees.org.au/wp-content/uploads/2013/11/25-Walker.pdf.

Further Reading

Allen, T., Wald, D., 2009. On the use of high-resolution topographic data as a proxy for seismic site conditions (VS30). Bulletin of the Seismological Society of America 99 (2A), 935–943.

Applied Technology Council (ATC), 1985. Earthquake Damage Evaluation Data for California (ATC-13). Redwood City, California, USA.

Askan, A., Yucemen, M.S., 2010. Probabilistic methods for the estimation of potential seismic damage: application to reinforced concrete buildings in Turkey. Structural Safety 32 (4), 262–271.

Bertogg, M., Hitz, L., Schmid, E., 2002. Vulnerability functions derived from loss data for insurance risk modelling: findings from recent earthquakes. In: Proceeding of the 12th European Conference on Earthquake Engineering, London, Great Britain.

Blong, R.J., 1997. Earthquake PML: Household Buildings-Sydney II. Greig Fester.

Blong, R.J., 2004. Residential building damage and natural perils: Australian examples and issues. Building Research and Information 32 (5), 379–390.

Center for Research on The Epidemiology of Disasters (CRED), 2009. EM-DAT: The International Disaster Database. Retrieved from: http://www.emdat.be.

China Statistics Bureau, 2015. China Statistical Yearbook. China Statistics Press, Beijing.

Clark, D., Leonard, M., 2014. Regional variations in neotectonic fault behaviour in Australia, as they pertain to the seismic hazard in capital cities. In: Australian Earthquake Engineering Society 2014 Conference, Nov. 21–23. Lorne, Vic.

Crompton, R.P., McAneney, K.J., Leigh, R., 2005. Indexing the Insurance Council of Australia Natural Disaster Event List (Report prepared for the Insurance Council of Australia, Risk Frontiers).

Daniell, J.E., Vervaeck, A., Wenzel, F., 2011. A timeline of the socio-economic effects of the 2011 Tohoku Earthquake with emphasis on the development of a new worldwide rapid earthquake loss estimation procedure. In: Australian Earthquake Engineering Society 2011 Conference.

Daniell, J.E., Wenzel, F., Vervaeck, A., Khazai, B., 2012. Worldwide CATDAT damaging earthquakes database in conjunction with earthquake-report.com – presenting past and present socio-economic earthquake data. In: 15th WCEE.

Daniell, J.E., Wenzel, F., Khazai, B., Santiago, J.G., Schaefer, A., 2014. A worldwide seismic code index, country-by-country global building practice factor and socioeconomic vulnerability indices for use in earthquake loss estimation. In: Paper No. 1400, 15th ECEE, Istanbul, Turkey.

Daniell, J.E., Schäfer, A., Wenzel, F., 2015. A tale of eight cities: earthquake scenario risk assessment for major Australian cities. In: Australian Earthquake Engineering Society Conference, Sydney, Australia.

Daniell, J.E., 2003-2016. The CATDAT Damaging Natural Disasters Databases. Searchable Integrated Historical Global Catastrophe Database, Digital Database, Updates v1.0 to Latest Update v7.0.

Dowrick, D.J., Cousins, J., Rhoades, D.A., 2004. Earthquake risk reduction in New Zealand. In: Proceedings of the 13th World Conference on Earthquake Engineering. Vancouver, BC, Canada.

Edwards, M., Griffith, M., Wehner, M., Nelson, L., Corby, N., Jakab, M., Habili, N., 2010. The Kalgoorlie earthquake of the 20th April 2010: preliminary damage survey outcomes. In: Proceedings of the Australian Earthquake Engineering Society 2010 Annual Conference.

French, S.P., Isaacson, M.S., 1984. Applying earthquake risk analysis techniques to land use planning. Journal of the American Planning Association 50 (4), 509–522.

G-Econ, 2010. Geographically Based Economic Data. http://gecon.yale.edu/.

Geoscience Australia, 2015. Earthquake Database (Online). Available at: http://www.ga.gov.au/earthquakes/searchQuake.do.

Irish, J.L., 1992. Earthquake damage functions for Australian houses and the probable maximum loss for an insurance portfolio. In: Earthquake Resistant Design & Insurance in Australia, AEES and Specialist Group on Solid-earth Geophysics Conference, Sydney.

Jaiswal, K.S., Wald, D.J., 2008. Creating a Global Building Inventory for Earthquake Loss Assessment and Risk Management. U.S. Geological Survey Open-File Report 2008–1160.

Jones, T., Middelmann, M., Corby, N., 2004. Natural Hazards in Perth, Western Australia. Comprehensive report. Geoscience Australia.

Marin, S., Avouac, J.P., Nicolas, M., Schlupp, A., 2004. A probabilistic approach to seismic hazard in metropolitan France. Bulletin of the Seismological Society of America 94 (6), 2137–2163.

Maskrey, A., Safaie, S., April 2015. GAR global risk assessment. In: EGU General Assembly Conference Abstracts, vol. 17, p. 6494.

Munich Re, 2009b. Globe of Natural Disasters, MRNATHAN DVD and NATCATService. Munich Reinsurance Company.

National Institute of Building Sciences (NIBS), 2003. HAZUS-MH Technical Manual. Federal Emergency Management Agency, Washington, D.C., USA.

Page, R.A., Joyner, W.B., Blume, J.A., 1975. Earthquake shaking and damage to buildings; recent evidence for severe ground shaking raises questions about the earthquake resistance of structures. Science 189 (4203), 601–608.

Risk Frontiers, 2015. QuakeAUS Model. https://www.riskfrontiers.com/quakeaus.html.

Rossetto, T., Ioannou, I., Grant, D.N., Maqsood, T., 2014. GEM Technical Report 2014-X. Guidelines for Empirical Vulnerability Assessment, vol. 6. GEM Foundation, Pavia. www.globalquakemodel.org. GEM Technical Report 2010–X, 2014.

Rynn, J.M., Brennan, E., Hughes, P.R., Pedersen, I.S., Stuart, H.J., 1992. The 1989 Newcastle, Australia, Earthquake: the facts and the misconceptions. Bulletin of New Zealand National Society of Earthquake Engineering 25 (2), 77–144.

Schnabel, W., 1987. The accumulation of potential in Cyprus. Number 13: earthquake risk and insurance – e.g. in Cyprus. In: Conference Proceedings Nicosia 1987. Cologne Re., pp. 87–130

Sinadinovski, C., Greenhalgh, S., Love, D., 2006. Historical earthquakes: a case study for Adelaide 1954 earthquake. In: Australian Earthquake Engineering Society Conference, 2006, Canberra, ACT.

Standards Australia, 2007. Structural Design Actions: Part4: Earthquake Actions in Australia, second ed. Standards Australia, Sydney. AS 1170.4—2007.

Statistics NZ, 2015. Economic Metrics. http://www.stats.govt.nz/.

United Nations Economic Commission for Latin America and the Caribbean (ECLAC), 2003. Handbook for Estimating the Socio-economic and Environmental Effects of Disasters. Retrieved from: http://www.eclac.org/cgi-bin/getProd.asp?xml=/publicaciones/xml/4/12774/P12774.xml&xsl=/mexico/tpl-i/p9f.xsl&base=/mexico/tpl/top-bottom.xsl#.

World Bank, 2010. World Development Indicators Online Database. World Bank, Washington, DC.

CHAPTER

6

Indirect Loss Potential Index for Natural Disasters for National and Subnational Analysis

James E. Daniell, Bijan Khazai, Friedemann Wenzel

Geophysical Institute & Center for Disaster Management and Risk Reduction Technology (CEDIM), Karlsruhe Institute of Technology (KIT), Karlsruhe, Germany

INTRODUCTION

Direct economic consequences of earthquakes are often characterized by key sectors of loss; however, the long-term impacts of earthquakes and their indirect losses are less well known. The quantification of indirect loss potential requires in-depth analyses of past historical events and detailed review of public and private data. Traditionally, postevent indirect loss analysis is undertaken using postdisaster surveys, studies of business losses and changes in employment, input-output models, or supply chain loss analyses for different economic sectors.

In this study, an attempt is made to define indirect losses by reviewing existing data, literature, and methods. In addition, we present an index for indirect loss potential at both the national and subnational levels using components of the Global Earthquake Model (GEM) Socioeconomic Indicator Databases. The index provides measure for relative levels of indirect loss potential as an alternative to traditional business interruption models. The index assists where input-output models may not be feasible due to shortage of data. The index can also be useful for planning and comparative studies especially at the higher resolution town level where more complex models often fail due to lack of data or experience of town officials. We also foresee that the index can be used in the insurance and reinsurance industries in which large-area risk needs to be evaluated in a short amount of time.

BACKGROUND

There is no global standard for calculating the total economic loss from earthquakes. For example, HAZUS employs a method (Cochrane, 2004) that is different to the method

Risk Modeling for Hazards and Disasters
http://dx.doi.org/10.1016/B978-0-12-804071-3.00006-9

TABLE 6.1 A Few Existing Indirect Loss Definitions

Author	Description
ECLAC (2003)	Losses are "estimated from the economic flows resulting from the temporary absence of the damaged assets" (referring to the indirect losses).
Hallegatte (2014)	Indirect losses (also labeled "higher-order losses" in Rose, 2004) include all losses that are not provoked by the disaster itself, but by its consequences; they are spanning over a longer period of time than the event, and they affect a larger spatial scale or different economic sectors.

used by Middleman (2007), BTE (2001), or ECLAC (2003). Furthermore, terms such as damage, loss, or impact are used interchangeably (Okuyama and Sahin, 2009) and definitions of direct and indirect losses are often inconsistent (Table 6.1).

Different types of losses are accounted for in the economic analysis of natural disasters (Table 6.2). Double counting of losses may also occur if losses are split into direct and indirect losses. Furthermore, losses are often considered as a percentage of gross domestic products (GDP). GDP is however not easily split into its direct and indirect parts. For example, the cost of reconstruction paid by the public is unlikely to be equal to the actual impacts considered as part of the GDP (Cavallo and Noy, 2010). Standard macroeconomic methods, i.e., considering changes in stock values and flows, can however be used to describe the overall

TABLE 6.2 Types of Losses Accounted for in Economic Analysis of Natural Disasters (Loss = Replacement Costs) Using Components From Thompson and Handmer (1996, p. 58) and Daniell (2014)

Loss Type	Descriptor
Direct losses	• Property (private with residential and nonresidential, public infrastructure) • Infrastructure • Equipment, machinery, and contents • Production losses to industry and direct business losses
Indirect losses (negative)	• Business disruption (indirect damage), loss of wages • Loss of production/value added due to direct damage, slowdown • Loss of production due to transport/infrastructure/lifeline networks • Loss of production due to interindustry effects and supply chain effects (consumers, suppliers), input-output (forward or backward) • Restoration of households • Household alternative accommodation—additional costs (BTE, 2001) • Relief, clean up, and response costs (Labor etc.) with associated demand surge (Olsen and Porter, 2008) • Postponed impacts—cuts in household spending—time impacts of disaster
Intangible costs (negative)	• Fatalities, injuries, homelessness, health effects (debilitation) • Lost tourism and environmental, cultural, and historical assets • Damage to image of a company (or tangible-indirect) • Decreased competitiveness (or tangible-indirect) • Market disturbances (commodity price increases/decreases)

TABLE 6.3 Net Regional Losses and Gains Considered Separate From the Direct, Indirect, Intangible Components

Net Regional Losses/Gains (Negative/Positive)	• Rebuilding assistance, positive stimulus, survivor benefits, unemployment compensation, aid payments, node and network disruptions, bottleneck losses outside earthquake-affected area, systematic financial and institutional disruption losses, flows outside of the area in production, employment and income, income gains for commodities

process of loss of capital and GDP and herewith allows discriminating between direct and indirect loss components (Rose, 2004).

Negative losses such as the stimulation of the economy after a disaster are accounted for as indirect consequences when quantified. Regional losses may hence appear with negative trends. This is however unlikely at country scale with most disaster response and recovery initiatives funded through taxpayers or through special funds as taxpayers or funds might not feel stimulation of the economy in real time. The size of a country has a significant impact upon its recovery. Smaller and low-income economies see larger negative postevent loss impact than large high-income societies (Kousky, 2012). Disaster aid can reduce losses. The effect of disaster aid might however not be seen for weeks and months after an event. Quantifying the effects of aid is hence complex and requires considering long-term growth (Kim, 2010) (Table 6.3).

EXISTING METHODOLOGIES FOR QUANTIFYING INDIRECT LOSS IMPACT

Indirect losses are modeled using economical or statistical approaches. Indirect loss levels are often either considered a function of the direct damage level to the structures or are considered a function of the economic sectors (inputs). Various models allow calculating the potential indirect losses (Table 6.4). These models range from simple linear input-output models examining the interdependencies of sectors, to nonlinear general equilibrium-based systems that attempts to include all relevant changes to the economy after an event.

ECLAC (2003) developed a methodology examining direct and indirect losses resulting in their Post Disaster Assessment (PDNA). A methodology such as ECLAC (2003) requires an exhaustive dataset, which generally cannot be collected for the purpose of insurance modeling (which often has neither the time nor the exhaustive resources on the ground needed to do the exercise). Key sectors are analyzed in detail covering replacement- and reconstruction-costs using postdisaster economic loss calculation processes.

The Kobe earthquake was looked at in a "per capita GDP" impact over the 15 years following the earthquake since 1995 by the authors duPont and Noy (2015). It was found that a 13% decrease in GDP per capita had resulted in Hyogo Prefecture. Similarly Cavallo et al. (2010) used the method for analyzing the long-term impact of a disaster for a country's economy (but it is often difficult to split the effects of large-scale nondisaster economic

TABLE 6.4 Modeling Methods for Indirect Loss Estimation for Earthquakes in Addition to the Full GDP Methods Above

Modeling Method	Authors (Examples)	Description
Input-output/linear programming	Toyoda (2008); Cochrane (1974); Hallegatte and Ghil (2008); Hallegatte and Przyluski (2010)	A matrix of linear interdependencies between sectors is solved by looking at the ratios of inputs to outputs. See Appendix A for a full example. Generally overestimates the losses. Hallegatte and Ghil (2008) proposed the adaptive regional input-output (ARIO) model.
Input-output with supply/demand	FEMA (1999) – HAZUS	HAZUS combines a traditional input-output model with supply and demand disruptions. In this way, it also resembles a computable general equilibrium (CGE) or social accounting matrices (SAM) method but with a slightly different structure of inventories and flows.
Econometric	Kuribayashi et al. (1987)	These are statistical models that attempt to solve the problem by applying time-series data. This can be in particular sectors or the entire economy. They can be used stochastically and for forecasting but require much data.
Social accounting matrices	Cole (1995), Minnesota IMPLAN Group (1998)	A matrix of all economic transactions within an economy (outflows and inflows) from all actors. It can pick up interrupted flows in an economic network postdisaster.
General equilibrium (CGE)/hybrid CGE	Brookshire and McKee (1992), Rose (2004), Seung et al. (2000), Tatano and Tsuchiya (2008)	A system of nonlinear equations explaining each of the actors in the economic process (not just input-output) and also price changes to products and market fluctuations. Generally underestimates loss. They use SAMs and input-output.
Surveys	Tierney and Dahlhamer (1998)	Involves questionnaire techniques of the affected businesses to create a sample set for estimation.
Direct multiplication and indicator methods	Kundak and Dülger-Türkoğlu (2007), Hiete and Merz (2009), Wu et al. (2012)	Rule-of-thumb and observed system multiplication of direct losses via GDP or indicator systems looking at interdependencies.

impacts and the disaster impacts themselves). Insurance policies covering business interruption consider time limits as deductibles (loss paid after a certain amount of time, e.g., weeks) and/or whether losses arise directly from and along with a property loss or indirectly, i.e., due to supply chain network. Most policies (so far) exclude however indirect losses that are not a direct consequence of direct losses covered. Some long-run indirect loss calculations can show a reversed trend toward negative losses as opposed to results from short-run indirect loss calculation that in general come up with negative effect (Cavallo and Noy, 2010, see above). This makes it difficult to quantify the true impact of business interruption. In insurance, downtime is often defined as the time taken to replace or repair the business assets to become fully operational. Insurance is hence incapable of considering long-term increase in profit and hence negative losses as a result of a disaster. Significant deductibles are needed to make sure that the insured does share some risk.

The work of Kajitani et al. (2013) provides insight into sectoral-based losses for the Tohoku 2011 earthquake. The study examines capacity and the impact on industrial production losses for 21 manufacturing and 6 nonmanufacturing sectors. This includes the indirect supply-chain impact on car manufacturing as well as the impact of the earthquake on tourism (c.73% reduction). The study also considers the event's impact and the retail industry (around 2 months recovery). This data feed into generally used downtime functions that help to forecast business interruption for future events.

Below are the key drawbacks of most methodologies for insurance modeling:

1. Models follow a variety of different concepts describing reconstruction cost, demand surge, general sate of the economy, changes in consumer confidence (due to changing costs) or government subsidies, regulation, and mass production (lower costs). These data are not necessarily collected by the insurance industry and/or data are not "unique" and hence subject to large uncertainty.
2. Values of capital stock, infrastructure, utilities, and production change daily after a disaster. These items depend on the replacement and recovery time as well as cost inflation. Insurance companies might not have the time and resources to monitor these items. Lack of data can hence cause significant uncertainty in the modeled values.
3. The human element postdisaster in terms of employment, accommodation, politics, and the ability to work is poorly understood (and might depict large variability).
4. On a globalized world various levels of business networks interact resulting in unforeseen far-reaching consequences different from the assumptions generally considered in insurance models (which assume that the intensity of an event correlates directly with additional labor costs).

A good example of this last point is that three types of decomposition occur in input-output model matrices, where the (1) intraregional effects (domestic, M1), (2) interregional spill-over (international, M2), and (3) interregional feedback (international, M3) are examined. Okuyama (2009) studied these impacts for the 2004 Indian Ocean tsunami showing decreasing output and decreasing demand for each sector after the event.

CGE models incorporate input-output processes and can take into account supply constraints, changes in imports, and price changes. These models take into account changing equilibriums of systems after an event. Simple indicator models cannot take into account these changes (see below).

The ARIO model (Adaptive Regional Input-Output) of Hallegatte (2008) can handle complexities of indirect losses by looking at forward-backward linkages as well as postdisaster changes in production capacity. This model takes into account adaptive systems to the changes.

The CGE and ARIO models are deemed state of the art models for detailed indirect loss estimation. The simplified methodology described in this paper below, can either stand on its own or be complementary to more complex models. The methodology described below can serve as a fast, low-cost solution that can be used quickly after an event or prior to events as part of risk modeling, e.g., multiple countries or areas. The Chinese indirect loss standard method provides an example for these complementary tools. This method is detailed in the following section (Fig. 6.4 as well as Tables 6.4 and 6.5, see also Ye et al. (2011)). More elaborate models also capture detailed sectorial components not generally considered by a simple model.

Further key studies are summarized in the following table. This table give a deemed comprehensive distribution of the indirect loss indicator models around the globe. Most of these models come from China, Japan, and the USA.

ECONOMIC LOSSES FROM HISTORICAL EVENTS AND DIRECT STUDY RESULTS

The CATDAT Natural Disaster Socioeconomic loss database provides an overview of economic loss estimates from historical events and compares earthquake losses with losses from flood, cyclone/storm, and volcano disaster. A split between indirect and direct losses has been undertaken for each disaster where data were available. Although average direct and indirect losses might range in the same order of magnitude, there are no simple rules for how direct and indirect losses relate. The Ecuador 1987 earthquake suggests a total indirect to direct loss ratio of 5–7 whereas other earthquakes cause direct losses 10 times larger than indirect losses.

Additional work on the trends of indirect to direct loss ratios in developing economies however shows that higher indirect losses generally correlate with both higher direct losses and a decreasing development index. The latter seems counter-intuitive as business interruption exhibits a higher absolute value in developed countries. The lack of redundancies in the production sector in developing countries however causes the relative higher amount of indirect losses there. The larger the development index, the larger the economic loss; this suggests that the indirect to direct loss ratio has to follow a positive trend herewith correlating with increasing losses. This is however not visible in Fig. 6.2 (see below). Floods and cyclones appear to have higher indirect to direct loss ratios than earthquakes. This might be due to the larger areal extend (more likely to affect the economy and production) and often smaller average direct loss size per risk (building) of windstorms compared to earthquakes as well as the often larger concentration of infrastructure along the coast where tropical cyclones occur. The variability in the indirect to direct loss ratio leads (among others) from the fact that infrastructure losses and hence downtime of various sectors depend on highly critical and often complex and self-organized processes (Fig. 6.1).

Zhong and Lin (2004) considered losses from 20 earthquakes at 40 locations. The Tangshan earthquake in 1976 showed indirect to direct loss ratios of around 2.25 times for the city of Tangshan, whereas the Tianjin City ratio was around 0.86 in the same earthquake. Ratios for the 1996 Lijiang earthquake were around 0.83–1.09. Ratios range between 0.23 and 3.41 for the 20 earthquakes studied.

TABLE 6.5 Selected Indirect Losses Studies From the Literature Since 1980 (see also Daniell et al., 2015)

Author and Year	Location Studied	Key Indicators Examined
Ohtake et al. (2012a,b)	Kobe 1995, Tohoku 2011	Indirect loss data via GDP and employees
Porter and Ramer (2012)	Various	Downtime, reviews of repair cost
SISMA Ricostruzione (2012)	Emilia-Romagna 2012	No. of employers/companies, loss impact, GDP loss, sectors
Wu et al. (2012) and Ye et al. (2011)	Sichuan 2008	Different methods: First includes a subjective quantification of loss in terms of speed and size; general damage; infrastructure damage; urban function damage; recovery speed. Second uses input-output method using six parameters: architecture, industry and mining, chemical, transportation, agriculture, power generation, and supply
GB/T 27932, 2011 (2011)	Chinese Standard	Based on two methods as above; with regionalization factors for five settlement types
Giovinazzi and Nicholson (2010)	L'Aquila 2009	Connectivity, capacity, travel time, accessibility (network reliability)
Kuwata and Takechi (2010)	Kobe 1995	Production statistics (steel, manufacturing); industrial, production levels at damaged facilities, demand, no. of companies, water supply, recovery speed combining to utility of water in USD/yr/d/m^3—seismic resilience of the company, affected by delay
Heng et al. (2008)	China (Yunnan)	Rural area analysis (agricultural production)—direct and recovery time; Cobb–Douglas function—capital, labor, agricultural production, consumption via rural income, investment, speed of government relief
Nasserasadi et al. (2006)	Generalized through Izmit (1999), Tehran examples	Severity of the earthquake (important) Rehabilitation of factory (insignificant) Survived product storage capacity (insignificant) Mean time of household recovery (most important) Dependency of factory to household (most important)
BAPPENAS Joint Assessment Team (2006)	Indonesia	Key sectoral GDPs and employment; impact on revenues (but this is not a key value before)
Faizian et al. (2005)	Generalized	Loss of employment, loss of employees, loss of function, business interruption, consumption of materials and energy, macroeconomic impacts, and loss of heritage and the tourist industry; increase of investment production

(Continued)

TABLE 6.5 Selected Indirect Losses Studies From the Literature Since 1980 (see also Daniell et al., 2015)—cont'd

Author and Year	Location Studied	Key Indicators Examined
TCIP (2002)	Izmit 1999 and other	Nonperforming loans and bank deferrals; value-added loss; Emergency relief expenditures; job/fiscal changes
BTE (2001)	Edgecumbe 1987	Various parameters
Takashima and Hayashi (2000); Okuyama et al. (1999); Toyoda and Kochi (1997)	Kobe 1995	Electricity issues (consumption-based) index; gross regional product, impact area and supporting area (IoD), opportunity losses, I–O models; empirical production
Tierney and Dahlhamer (1998)	Loma Prieta 1989	Survey methods
Kawashima and Kanoh (1990)	Nihonkai-Chubu (1983), Japan	Agriculture (surveyed reduction), Sectoral Distribution of GDP and investment adjusted for value (production changes for production facilities, transportation and materials and goods)
Georgescu and Kuribayashi (1992)	Romania and Japan	Supplementary exports, imports, tourism receipts, lost production

These data are key for the Chinese standard of indirect loss assessment (GB/T 27932, 2011). Two methodologies are discussed in the Chinese standard including an empirical statistic and more complex economic method (using input-output processes). Industry losses are calculated in a matrix of products using consumption coefficients. This approach includes

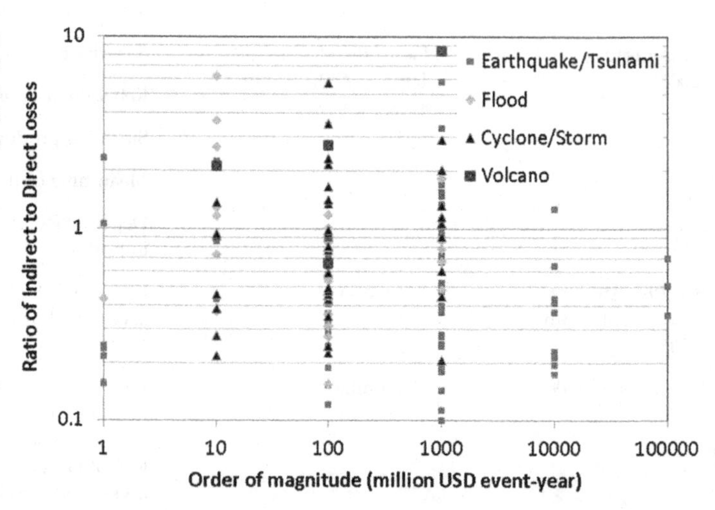

FIGURE 6.1 Ratio of indirect to direct losses versus overall earthquake loss for 150 events with similar Human Development Index value areas (Daniell et al., 2011). Data show large scatter with no obvious trend.

% of Direct Loss / Area GDP	<10%	10-50%	>50%
Western Rural	0.5 to 0.7	0.7 to 0.9	1.1 to 1.3
Western Urban	0.7 to 0.9	1.1 to 1.3	1.5 to 1.7
Eastern Rural	0.7 to 0.9	0.9 to 1.1	1.3 to 1.5
Eastern Urban	0.9 to 1.1	1.5 to 1.7	1.7 to 1.9
Megacity	1.2 to 1.6	1.6 to 2.0	2.0 to 2.5

FIGURE 6.2 Standard values for regional indirect economic losses (GB/T 27932, 2011).

traditional interindustry (in between industries) input-output procedures (initial investment, intermediate demand, and end products). This is undertaken for the primary, secondary, and tertiary sectors. The method includes the average daily values of an enterprise as well as the average daily values of reducing the amount of production along with the business downtime. The "Land loss equation" (GB/T 27932, 2011) includes the affected land area in m^2 and the reduction of the land per unit area in yuan per m^2 defined by the local land administration department. This is done along with a general downtime and reduced production calculation.

The empirical statistical model considered suggests that higher indirect losses correlate with increasing overall losses. Similarly, increased output per capita increases indirect losses. The scale of a disaster and the scale of urbanity are key parameters in the regional indirect loss calculation. Regional indirect losses are calculated by applying a factor to the direct losses in five cases combining the Western Rural and Urban settings, the Eastern Rural and Urban settings, and a Megacity standard.

The difficulty of the indirect loss potential however is that a direct relation of GDP is hard to see in the data outside of China. Different locations will have urban or rural settings; however, the relations of the 2011 standard do not seem to hold with respect to this study of the 150 events, where the larger events have scattered far exceeding that of the range.

Indirect losses may significantly increase for a company or state if international supply chains are considered. International supply chains include items shared across countries. Shared items can be cars being produced from car parts from two or more countries, or microchips being produced in locations other than those of the end products. The 1987 Ecuador earthquake provides an example of how downtime of destroyed pipelines can heighten losses for oil-dependent nations. A lack of redundancy in pipeline tracks meant that 33 km of destroyed pipelines with a pipeline repair cost of $122 million (1987 USD) caused a total of $766 million (1987 USD) losses including losses for missing supply and export. Supply chain losses absent of regional physical losses are considered contingent in the insurance industry. Insurance companies have shied away from covering contingent losses due to their potential systemic effects. This has changed more recently since oversupply in capital has created an increasingly competitive insurance market in which companies open up for more risky business. Other local or international business sectors such as tourism are often adversely affected by natural and manmade catastrophes. Examples include significant reduction in tourism in the Maldives after the 2004 tsunami.

Postdisaster employment changes drive changes in production (both increase and decrease) and govern further indirect gains and losses. Changes in unemployment rates may work as a key variable for this. Toyoda (2008) showed that larger firms are often more resilient to natural catastrophes, with smaller firms being more at risk to indirect losses resulting in larger indirect versus direct loss ratios. Higher levels of globalization may both provide more redundancies and hence decrease susceptibility to regional losses or can increase loss frequencies due to remote events affecting local businesses. Modeling global effects on indirect losses hence requires a methodology that both measures global events and supplies along with local impact.

Comparing sector losses as a percentage of preearthquake production helps approximating sector dependencies on earthquakes. Investigating historical events with ECLAC DaLa highlights differences in economic loss ratios for the different sectors. Productive sectors show higher susceptibility to indirect losses than administration sectors. Direct losses outweigh indirect losses by factors of between 2.5 and 3 for sectors including infrastructure and across sectors. Direct losses far outweigh indirect losses for the housing and building sectors (Fig. 6.3).

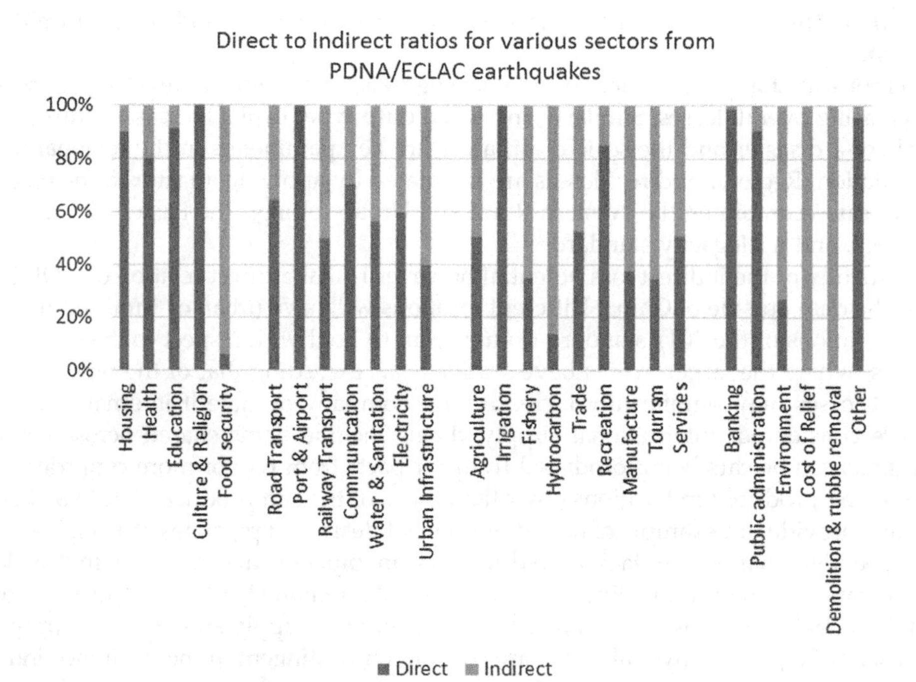

FIGURE 6.3 Direct versus indirect economic loss ratios from various sectors using over 25 earthquakes from various PDNAs in the ECLAC methodology.

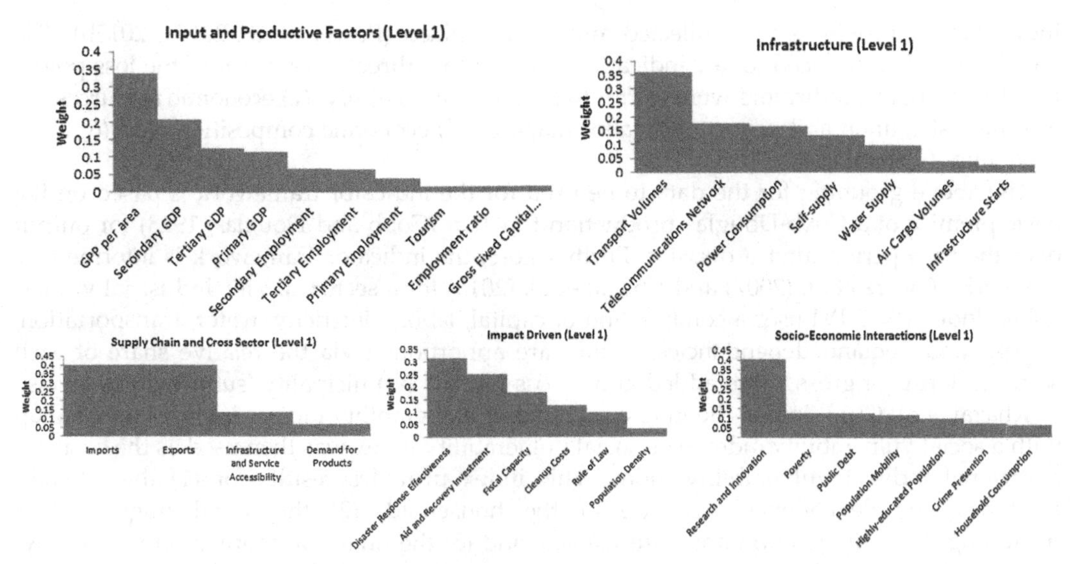

FIGURE 6.4 Level 1 weighting distributed over the various indicators (infrastructure).

Comparing sector losses for different historical events allows finding trends. Examples include the hydrocarbon (e.g., Ecuador 1987) sector as well as the agriculture and fishery industries. The above figure aggregates data from 25 events from the work of PDNAs. Kawashima and Kanoh (1990) did a direct and indirect loss study calculating both the indirect reduction as well as increase in products after large losses. In order to do this, results from modeled as well as actual losses were compared. Actual production losses for the construction sector do not include adjustments for lost stock and flows. After these adjustments, a small overall loss for the sector was changed into a negative loss due to an 11% increase in actual production after the event. Fisheries and forestry have some of the highest loss percentages in each sector.

Using analytical hierarchal processes (AHPs) and using indicators describing input-output linkages, so called vulnerability coefficient index models help studying loss impact, repair, and recovery phases. Japanese earthquakes such as the Nihonkai-Chubu Earthquake in 1983 and the Kobe (1995) earthquakes provide among the most comprehensive data and best-studied results of the world.

AN INDICATOR SYSTEM FOR VULNERABILITY TO INDIRECT LOSSES

An index for the vulnerability to potential indirect losses of countries and lower level impacts was undertaken as part of the GEM social vulnerability and integrated risk project (Daniell et al., 2015). It was decided that the indirect loss parameters would not be entirely split from the macroeconomic indicators, as all parameters are important when creating a vulnerability index. A global database of 1600 indicators for countries and up to 3000 indicators for provinces was collected and homogenized (Power et al., 2015a; 2015b). The dataset was then reduced to key indicators relevant to indirect assessment of the loss potential. The economic indicators were split into (1) economic activity; (2) economic resources; (3) income distribution and poverty; (4) labor markets; (5) economic composition; and (6) trade economics (Daniell et al., 2015).

The above grouping for the data to be used for the indicator framework is based on the basic premise of a Cobb-Douglas production function (Cobb and Douglas, 1928) for output over the time period after a disaster. Furthermore, the indicator framework is informed by the work of Merz et al. (2007) and Khazai et al. (2013) for a sector-specific industrial vulnerability index (IVI). IVI uses a combination of capital, labor, electricity, water, transportation, supply, and demand dependencies. Values are apportioned via the relative share of each sector in terms of gross value added and across NUTS3 (municipality/subprovince) levels.

Khazai et al. (2013) joined the industrial indirect vulnerability indices in order to come up with a social vulnerability index. The social vulnerability index was then used as the basis for a regional indirect vulnerability index. This index provides results for (1) the inherent fragilities—the dependency structure of the household, (2) the social fragmentation (including crime, integration and core values), and (3) the financial deprivation (unemployment and poverty). These three key sector results are then used to derive (a) the productive potential (training and innovation), (b) the service accessibility (transport and healthcare), and (c) the economic health (consumption level and disposable income).

Several adjustments are made to the approach. Lesser emphasis was given to social vulnerability as a basis for regional indirect vulnerability in favor of a more comprehensive set of indicators to describe the vulnerability of a country or a province to suffering indirect losses. The set of categories used commensurate with the taxonomy of the socioeconomic indicators of the GEM (Power et al., 2015a; Daniell et al., 2015). The set includes factors that describe the vulnerability of a country or province to indirect losses with respect to (1) production; (2) supply chain; (3) infrastructure; (4) socioeconomic factors; and (5) disaster impact factors (Table 6.6). The production factors including capital, labor, and material dependencies are

TABLE 6.6 Indirect Loss Indicator Groupings as Defined via the Literature

Primary Group	Secondary Groupings	Indicator Types
Productive factors	Capital dependency	Size and value of productive sectors (+)
	Labor dependency and intensity	Equipment (+)
	Material dependency	Size of economy (+)
		Specialization (+)
		Employment rates (+)
		Employment intensity (+)
		Amount of materials (+)

TABLE 6.6 Indirect Loss Indicator Groupings as Defined via the Literature—cont'd

Primary Group	Secondary Groupings	Indicator Types
Supply chain and cross-sector	Supply dependency	Production intensity (+)
	Supply and demand dependency	Simplification of sectors in terms of vertical integration (+)
	Supply and demand dependency	Clustering of the various sectors as well as population distribution (+)
	Demand dependency	Demand for certain products from customers (+)
Infrastructure	Water dependency	Infrastructure and service accessibility (+)
	Transport dependency	Network density (−)
	Power dependency	Transport volumes (+)
	Communication dependency	Importance of utilities (+)
		Consumption of utilities (+)
		Self-supply of utilities (−)
		Infrastructure starts (−)
Socio-economic interactions	Adaptive changes and measures	Household expenditure and income (+)
	Household factors	Ownership or willingness to Stay (+)
	Social fragmentation	Training and innovation (−)
	Productive potential	Poverty/GINI index (+)
	Financial deprivation	Crime rate (+)
		Education (−)
		Community expenditure (+)
Disaster impact-driven	Disaster magnitude/environment	Size of disaster (+)
	Aid payments and reserves	Prevalence of aftershocks (+)
	Budget systems	Budget effects (+)
	Population/exposure effects	Payment investment (−)
	Government decisions	Population distribution (degree of redundancy in systems) (+)
		Weather and temperature (+)

often components of methods described in Table 6.5. Economic loss data from historic events (see above) suggest higher trends of indirect loss to direct loss ratios for the infrastructure and production sectors as compared to the buildings and housing sector. Each of the resulting indicator types was given a "+" or "−" sign that describes their positive effect (gains) or negative effect (losses) (Table 6.6).

Each of the groupings takes into account components making up the indirect loss potential. Hiete and Merz (2009) note that industrial sectors with a high capital demand and material requirements have higher indirect losses than those with minor material requirements. Specialized machinery is difficult to replace and therefore has longer production downtimes. At a macro level, the China regional indices suggest that higher GDP per capita corresponds with (1) larger indirect losses, (2) increasing service levels, (3) larger industrial density, and (4) a smaller relative size of the agriculture sector.

The supply and demand method described above attempts to find indicators that represent (supply and demand) processes, their complexities, as well as interactions between their individual components. Yoshida and Deyle (2005) discuss the importance of these parameters for the indirect loss potential of the industrial sector dominating indirect losses. Zsidisin (2003) suggests that sector and/or population density as well as regional sourcing and geographic concentration of suppliers are also critical for indirect losses. The lesser a system depends on the supply of goods, the smaller its level of disruption caused by changes in the supply chain. The demand dependency components look at the customer proximity to the products as well as the interconnectivity of demand.

Our literature review suggests that the infrastructure components are also highly relevant for quantifying the indirect loss potential. Key factors include the following: (1) the degree of redundancy in the transport network along with the dependency of the economy on port and transport infrastructure and (2) the dependency of a system on the proximity of employers and employees. The road transportation network is important postdisaster, and the debris removal and reconstruction processes are included here. The power network and water supply systems are interlinked with critical infrastructure, productive sectors, services, and the industry. With greater power and water supply disruptions come higher indirect losses. The communication sector is becoming increasingly important with its mobile networks and the Internet that are strongly connected with the commercial (and increasingly private and industrial) sectors.

Many parameters are looked at in terms of social and human aspects of indirect losses. The key factors are outside of the industrial, commercial, and agriculture components and look at the system link back from the labor component (people) into the system. The ability of humans to withstand sudden shocks and the measures that are undertaken (resilience of communities) are key parameters, which govern this part of the index.

The above social factors however generally have a minor influence on the total indirect loss when quantified in a traditional sense (i.e., via the DaLa methodology [ECLAC, 2003]); however the implications of social decisions postdisaster in terms of purchasing power from households and other social factors are often partially interlinked with the indirect losses.

The productive potential of a community in terms of schooling and training is connected to the ability of people to go back to work after a large disruptive event. Consumption levels and accessibility are additional key parameters that overlap with the other sectors examined. Income (whether high or low) is not as such related to the indirect loss potential. For high-income locations, a greater loss in income of people will be reflected in higher absolute losses. Indirect losses however do not follow a similar trend as suggested earlier. Social stability and fragmentation are vulnerability components that play key roles for quantifying indirect losses. This is due to the fact that the ability of communities to recover their economic systems is faster if the society is better organized and hence more stable (Miller and Rivera, 2010). Crime rate, the level of integration, and inequality are all components that affect the social systems herewith influencing the labor component of production (WHO, 2005; Roy, 2010). Financial deprivation and household factors such as poverty and household size influence social vulnerability, which can affect indirect loss (Okuyama, 2008).

The final main component that impacts indirect losses postearthquake is the scale of the disaster itself along with how systems of governmental processes, financial processes, engineering processes, and both other public and private processes are arranged within the country. Aid payments and reserves are often a proxy for the speed of recovery along with investment done by the government in order to reduce indirect losses (Lindell, 2013). Investment in insurance, degree of building materials needed, engineers engaged, and other inflation measures and decisions made by the government involved in the disaster, contribute further to the indirect losses (Galloway and Hare, 2012; Chang et al., 2014). Aftershocks may increase indirect losses although aftershocks mostly affect governmental processes for occupational health and safety. Delays in recovery associated with aftershocks can lead to significant increase in indirect losses (Galloway and Hare, 2012). Climate and weather conditions during the recovery phase may play a major role in the recovery profile as well due to the effects on perishable goods from the agricultural sector.

RESULTS FOR COUNTRIES AND PROVINCES

The process for the framework behind the below described results is detailed in the GEM vulnerability to indirect losses report (Daniell et al., 2015). These results are used here as a tool for calibrating a business interruption model. Eight key steps for an indicator framework for an assessment were used from Nardo et al. (2005), however, with some slight changes as to the order for this analysis and methodology. The following eight-step process after Nardo et al. (2005) has been undertaken:

1. Developing a theoretical indicator framework
2. Data gathering for indicator datasets
3. Standardization of indicators
4. Selecting key indicator and subindicator groups
5. Final selection of indicators along with key subindicator groups
6. Weighting and aggregation
7. Sensitivity analysis
8. Visualization

The selection of indicators is contingent upon (1) availability of regional data and (2) availability of data within the above themes. Built upon literature review, 1600 national level indicators were reviewed and reduced with respect to their applicability to providing a useful basis for indirect losses. This resulted in a total of 200 indicators. The final selection of indicators was then undertaken considering the most relevant parameters for indirect losses. This was done using expert judgment via the parameters realized in the literature review as well as using quantitative parameters from ECLAC that are part of the makeup of existing indirect loss methodologies and historical reviews such as commerce or tourism aspects (see Table A.3 in appendix). The model applies the weighting based on the above methodology and the expert opinion using an AHP. The choice of weighting is related to the key results found through historical studies. This includes, e.g., increasing complexity leads to increased probability of indirect losses.

The indirect loss index globally correlates well with average historic event losses as expected. Singapore and Japan as centralized, highly industrialized, and network-oriented countries have very high potential for indirect losses. Similarly Netherlands, Belgium, and Taiwan as network-oriented economies with small areas also have high indirect loss potential as seen in events like Kaohsiung 2010 where chip manufacturing had huge downtime losses.

The results also concur with the Chinese indirect loss standard, which show increasing indirect losses for increasingly connected, high GDP regions, scaled with the size of the disaster (GB/T 27932, 2011).

The indirect loss potential index is built from 0 to 1. There is however no simple linear relationship between indirect losses and direct losses due to the various and complex interplay effects between different sectors. Further calibration is hence needed in case the index is used for insurance purposes.

The countries with the lowest potential for indirect economic (as opposed to potentially high direct) losses are those with low-level industrial systems and simple structures in their businesses with robust machinery. Examples include Mali and Niger (Fig. 6.5).

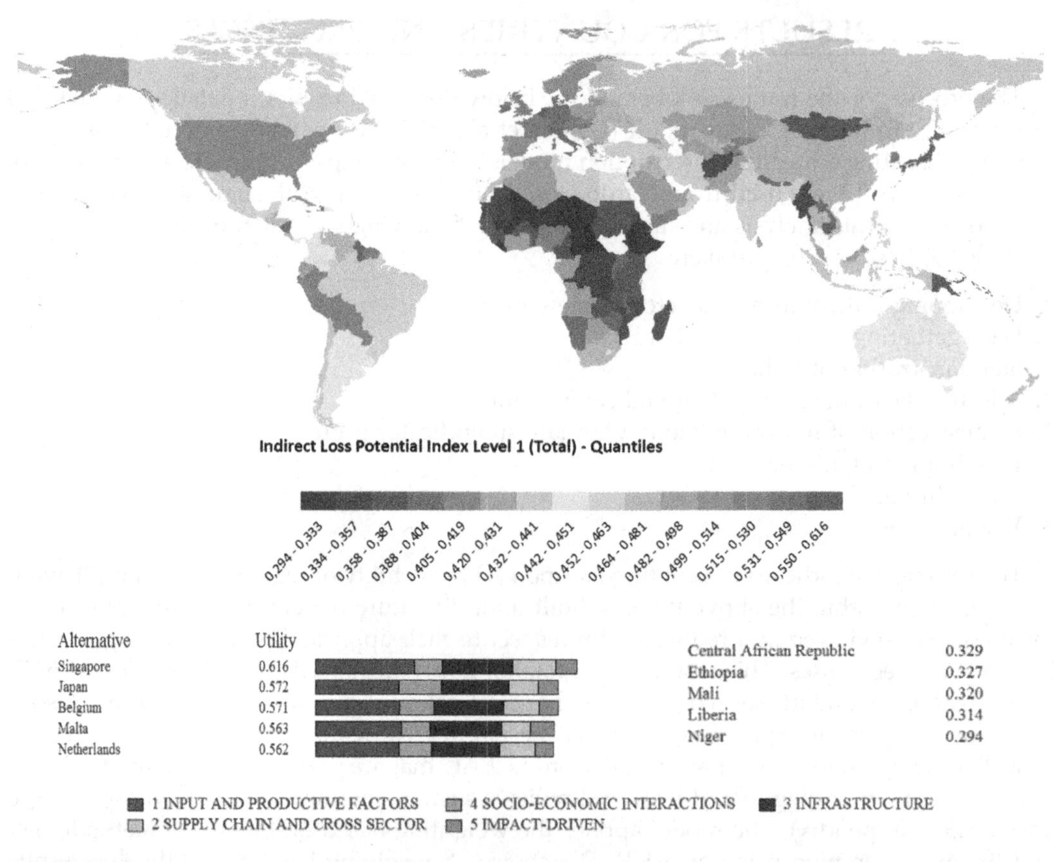

FIGURE 6.5 Level 1 indirect loss potential index: countries above 0.5 (i.e., high indirect economic loss vulnerability potential).

For a more detailed analysis on a regional level, the Asia-Pacific region was analyzed. Japan is analyzed in particular following their relatively large indirect losses in past events (as opposed to China for which direct economic losses dominated historical events). For Japan, NUTS Level 2 indicators were selected from a database of 3000 + indicators reduced to 200 indicators as described above. Country indicators may differ from NUTS Level indicators given the fact that provincial interactions may be different from interactions between countries.

Transformations for each parameter were made in line with the data and its potential correlation to indirect losses. Each transformation was done individually for each component on either a linear or lognormal (base 2 or 10) method, in line with the building of the Level 1 database. The Level 2 Indirect Economic Potential Loss Index denominates areas of potentially high indirect losses for Japan (Fig. 6.6). These locations show high industry and services concentrations, a large GDP per area production, dense networks of infrastructure, high supply chain, and cross-sector risk as well as high levels of government involvement and complex business processes.

The Input and Productive factors are the main component of the indirect loss potential index given their direct relevance for the calculation of losses. Higher GDP often (depending on the makeup of the GDP and exposure to earthquakes) correlates to higher economic losses (e.g., China Standard.). The impact ranking portrayed depends on how much dominant industries in locations are contingent on infrastructure and also Osaka outranks Tokyo in terms of infrastructure ranking (Fig. 6.7). In the Kobe earthquake, Osaka's business was highly affected by the road closures between Osaka and Kobe along with other infrastructure outages.

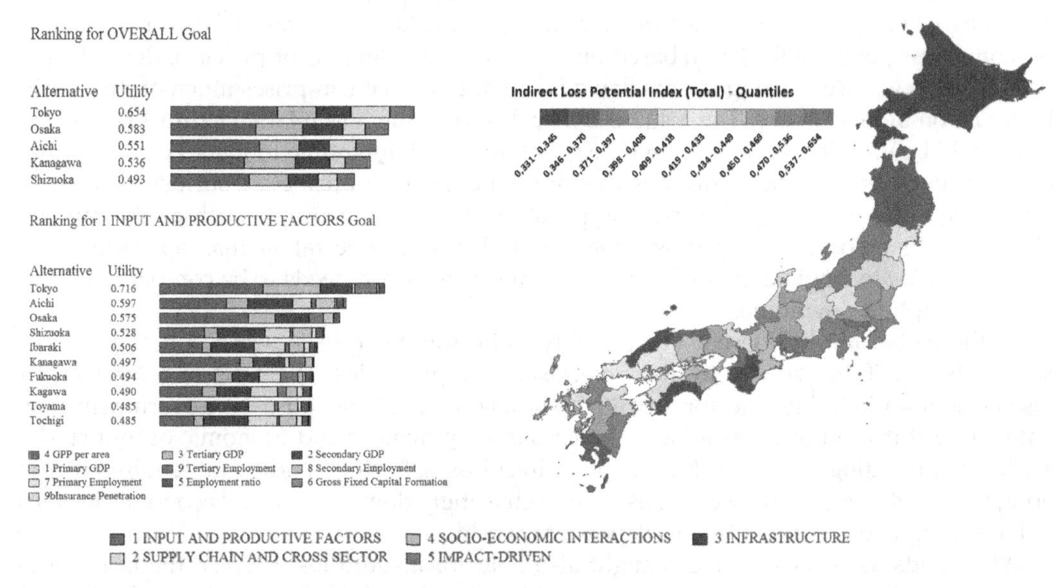

FIGURE 6.6 Level 2 indirect loss potential index–by quantile.

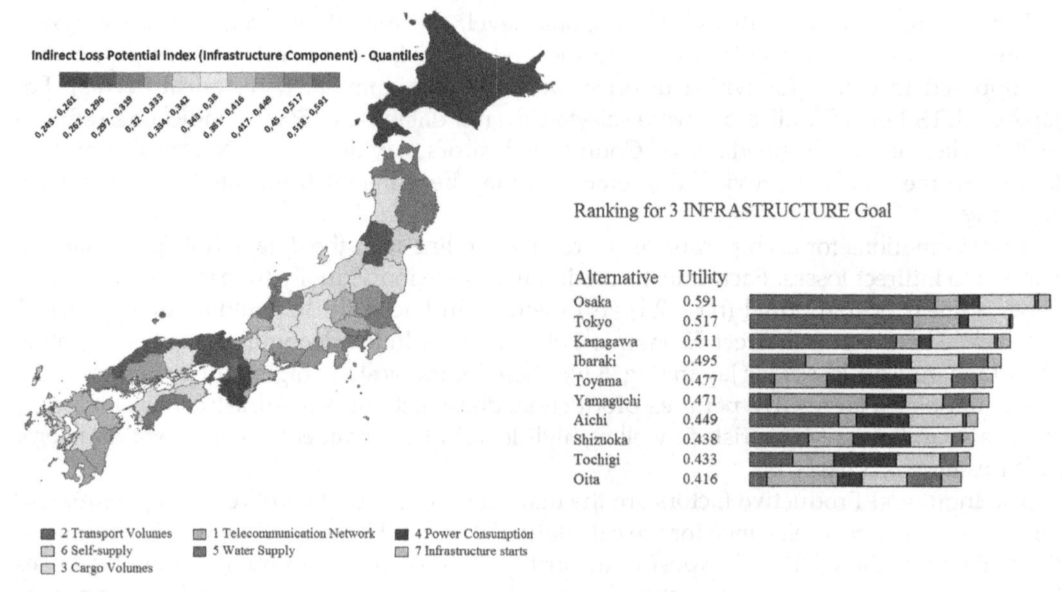

FIGURE 6.7 Level 2 indirect loss potential index–infrastructure quantiles.

SUMMARY

There is no simple method to calculate indirect losses. However, key components of indirect losses can be examined rapidly using an indicator framework. The framework introduced in this study allows relative ranking of specific portfolios (e.g., selection of firms), countries, or provinces. A selection of suitable indicators has been undertaken for the indirect economic loss potential for Japan based on both a country (level 1) or province (level 2) comparison. This was done using the available Japan dataset that comprises much of the specific literature on indirect losses. The result is a map that denominates relative indirect losses. This map could be used in combination with e.g., a traditional downtime model to provide differentiation between various businesses that are dependent on business from others. In addition, it provides a framework for the application of various concepts such as the building characteristics associated with downtime. The index is relative rather than absolute and in order to calculate tangible losses business income information needs to be compared to business interruption downtime.

At the global level, different vulnerability indicators have been selected based on local level indictors. These adhere to the same taxonomy principles. The scale of a disaster, the resilience of an individual sector, employment changes, and postdisaster government mandates control the indirect loss totals. Postdisaster employment and economic output studies are key to providing accurate estimates of indirect losses. Further studies are required considering the fact that indirect losses are likely to increasingly dominate overall economic losses in a developing and increasingly interdependent world.

What holds for economic losses might also hold for insured losses given the fact that insurance policies have become more comprehensive increasingly covering infrastructure and supply chain losses (see recent earthquakes in Chile and NZ). Infrastructure losses were deemed largely uninsurable in the past due to their potentially significant system risk involved. The use of more detailed supply chain and CGE models have hence their place in detailed insurance analyses and are becoming increasingly prevalent (Hallegatte and Ghil, 2008) for analyzing the effects in particular settings and locations.

The complex interplay of factor and decisions after an event makes it difficult to predict indirect losses using only one model. This means that modeling the evolution of indirect losses in detail demands continuous model adjustment and learning. Having an index for the indirect loss potential before a disaster allows for planning, systems optimization, and mitigation of potential indirect losses as well as creating awareness of key factors, which drive indirect losses.

We suggest that the above methodology requires further expert discussions in the selection of potential variables and their weights along with further calibration for use within portfolio settings. Nevertheless, the current index provides a first-order model for obtaining a ranking of the vulnerability of countries and provinces to indirect loss potential informed by the detailed studies of indirect losses of past historic events and typical inputs used in other more complex methodologies.

References

Albala-Bertrand, J.M., 2008. Localization and Social Networking, Background Paper for the Global Facility for Disaster Reduction and Recovery, the World Bank.

BAPPENAS, the Provincial and Local Governments of D.I.Yogyakarta, the Provincial and Local Governments of Central Java, and International Partners (2006) "Preliminary Damage and Loss Assessment: Yogyakarta and Central Java Natural Disaster", prepared for the 15th Meeting of the Consultative Group on Indonesia, Jakarta, Indonesia.

Benson, C., 2012. Indirect Economic Impacts of Disasters. Commissioned review, Foresight, Government Office for Science, UK.

Brookshire, D., McKee, M., July 1992. Other indirect costs and losses from earthquakes: issues and estimation. In: Indirect Economic Consequences of a Catastrophic Earthquake. National Earthquake Hazards Reduction Program, Federal Emergency Management Agency, pp. 267–325.

Bureau of Transport Economics (BTE), 2001. Economic Costs of Natural Disasters in Australia. Bureau of Transport Economics Report 103. Canberra, Australia.

Cavallo, E., Noy, I., 2010. The Economics of Natural Disasters. IDB Working Paper Series No. IDB-WP-124.

Cavallo, E., Galiani, S., Noy, I., Pantano, J., 2010. Catastrophic Natural Disasters and Economic Growth. Research Department Publications No. 4671. Retrieved from: http://ideas.repec.org/p/idb/wpaper/4671.html.

Chang, S.E., Taylor, J.E., Elwood, K.J., Seville, E., Brunsdon, D., Gartner, M., 2014. Urban disaster recovery in Christchurch: the central business district cordon and other critical decisions. Earthquake Spectra 30 (1), 513–532.

Cobb, C.W., Douglas, P.H., 1928. A theory of production. American Economic Review 18 (Supplement), 139–165.

Cochrane, H.C., 1974. Predicting the economic impact of earthquakes. In: Social Science Perspectives on the Coming San Francisco Earthquake. Natural Hazards Research Paper, (25).

Cochrane, H.C., 2004. Economic loss: myth and measurement. Disaster Prevention and Management 13 (4), 290–296.

Comerio, M.C., 2006. Estimating downtime in loss modelling. Earthquake Spectra 22 (2), 349–365.

Cole, S., 1995. Lifelines and livelihood: a social accounting matrix approach to calamity preparedness. Journal of Contingencies and Crisis Management 3 (4), 228–246.

Daniell, J.E., 2014. Development of Socio-Economic Fragility Functions for Use in Worldwide Rapid Earthquake Loss Estimation Procedures (Doctoral thesis). Karlsruhe Institute of Technology, Karlsruhe, Germany.

Daniell, J.E., Khazai, B., Wenzel, F., Vervaeck, A., 2011. The CATDAT damaging earthquakes database. Natural Hazards and Earth System Sciences 11 (8), 2235–2251. http://dx.doi.org/10.5194/nhess-11-2235-2011.

Daniell, J.E., Khazai, B., Power, C., 2014a. Earthquake Vulnerability Indicators for the Determination of Indirect Loss Potential. CEDIM, 73p.

Daniell, J.E., Simpson, A., Gunasekara, R., Baca, A., Schaefer, A., Ishizawa, O., Murnane, R., Tijssen, A., Deparday, V., Forni, M., Himmelfarb, A., 2014b. Review of Open Source and Open Access Software Packages Available to Quantify Risk from Natural Hazards. Understanding Risk Report. GFDRR.

Daniell, J.E., Khazai, B., Power, C., 2015. Earthquake Vulnerability Indicators for the Determination of Indirect Loss Potential, GEM Socioeconomic Module Technical Report.

duPont IV, W., Noy, I., 2015. What happened to Kobe? A reassessment of the impact of the 1995 earthquake in Japan. Economic Development and Cultural Change 63 (4), 777–812. University of Chicago Press (published as a 2012 working paper also).

Faizian, M., Schalcher, H.R., Faber, M.H., 2005. Consequence assessment in earthquake risk management using damage indicators. In: Proceedings of the 9th International Conference on Structural Safety and Reliability (ICOSSAR 05). Rome, Italy, pp. 19–23.

Federal Emergency Management Agency (FEMA), 1999. Multi-Hazard Loss Estimation Methodology. Earthquake Model. Hazus®–Latest Version: MH MR5. Technical Manual. Retrieved from: www.fema.gov/plan/prevent/hazus.

Federal Emergency Management Agency (FEMA), 2013. Multi-hazard Loss Estimation Methodology. Earthquake Model. Hazus®–MH MR5. Technical Manual. Retrieved from. www.fema.gov/plan/prevent/hazus.

Galloway, B.D., Hare, H.J., 2012. A review of post-earthquake building control policies with respect to the recovery of the Christchurch CBD. Bulletin of the New Zealand Society for Earthquake Engineering 45 (3), 105–116.

GB/T27932-2011, 2011. Assessment of Indirect Losses (地震灾害间接经济损失评估方法) (in Chinese).

Georgescu, E.S., Kuribayashi, E., 1992. Study on seismic losses distribution in Romania and Japan. In: Proceedings of the 10th World Conference on Earthquake Engineering. Taylor & Francis, Madrid, Spain.

Giovinazzi, S., Nicholson, A., 2010. Transport network reliability in seismic risk analysis and management. In: 14th ECEE European Conference on Earthquake Engineering. Ohrid, Macedonia, 30 August–03 September 2010.

Green, T., Feser, E., 2007. Business Interruption Loss Modeling by a Modified HAZUS Approach. Department of Urban and Regional Planning, University of Illinois at Urbana Champaign, Urbana, Illinois through Steelman and Hajjar, 2008.

Gurenko, E., 2006. Earthquake Insurance in Turkey: History of the Turkish Catastrophe Insurance Pool. World Bank Publications.

Hallegatte, S., 2008. An adaptive regional input-output model and its application to the assessment of the economic cost of Katrina. Risk Analysis 28 (3), 779–799.

Hallegatte, S., 2014. The Indirect Cost of Natural Disasters and an Economic Definition of Macroeconomic Resilience. SDRFI Impact Appraisal Project.

Hallegatte, S., Ghil, M., 2008. Natural disasters impacting a macroeconomic model with endogenous dynamics. Ecological Economics 68, 582–592.

Hallegatte, S., Przyluski, V., 2010. The Economics of Natural Disasters: Concepts and Methods. World Bank. © World Bank. https://openknowledge.worldbank.org/handle/10986/3991. License: Creative Commons Attribution CC BY 3.0.

Ham, H., Kim, T.J., Boyce, D., 2005. Assessment of economic impacts from unexpected events with an interregional commodity flow and multimodal transportation network model. Transportation Research Part A: Policy and Practice 39 (10), 849–860.

Hayashi, T., 2005. Issues on Recovery Funds, Report on Comprehensive Evaluation of Recovery of 10 Years after. Hyogo Prefecture vol. II, 372–445 (in Japanese).

Heng, L., Chao, L., Shujun, C., 2008. Study on earthquake indirect economic loss in rural area. In: 15th WCEE, Beijing.

Hiete, M., Merz, M., 2009. An indicator framework to assess the vulnerability of industrial sectors against indirect disaster losses. Nr. 131. In: Landgren, J., Nulden, U., Van der Walle, B. (Eds.), Proceedings of the 6th International ISCRAM Conference. Gothenburg, Sweden.

Kajitani, Y., Chang, S.E., Tatano, H., 2013. Economic impacts of the 2011 Tohoku-oki earthquake and tsunami. Earthquake Spectra 29 (s1), S457–S478.

Kawashima, K., Kanoh, T., 1990. Evaluation of indirect economic effects caused by the 1983 Nihonkai-chubu, Japan, earthquake. Earthquake Spectra 6 (4), 739–756.

Khazai, B., Merz, M., Schulz, C., Borst, D., 2013. An integrated indicator framework for spatial assessment of industrial and social vulnerability to indirect disaster losses. Natural Hazards 67 (2), 145–167.

Kim, C.-K., 2010. The effects of natural disasters on long-run economic growth. The Michigan Journal of Business 41, 15–49.

Kousky, C., 2012. Informing Climate Adaptation: A Review of the Economic Costs of Natural Disasters, Their Determinants, and Risk Reduction Options. Resources for the Future, pp. 12–28. Discussion Paper.

Kundak, S., 2004. Economic loss estimation for earthquake hazard in Istanbul, 2004. 44th European Congress of the European Regional Science Association. R, Porto Portugal, 8/25/2004–8/29/2004.

Kundak, S., Dülger-Türkoğlu, H., 2007. Evaluation of earthquake risk parameters in the historical site of Istanbul. ARI (The Bulletin of the Istanbul Technical University) 55 (1), 53–66.

Kuribayashi, E., Ueda, O., Tazaki, T., 1987. An econometric model of long-term effects of earthquake losses. In: 13th Joint Meeting of US-Japan Panel on Wind and Seismic Effects, Tsukuba, Japan.

Kuwata, Y., Takechi, J., 2010. The effect of the industrial water outage on manufacturing business after the Kobe earthquake. In: Proceedings of the 14th European Conference on Earthquake Engineering. Paper No. 1262.

Leontief, W.W., 1986. Input Output Economics. Oxford University Press.

Li, Z., 2010. Historic Chinese Earthquake Descriptions. Retrieved from. http://data.earthquake.cn/data/zhenli/wjf/html/zhenli087.htm.

Lindell, M.K., 2013. Recovery and reconstruction after disaster. In: Encyclopedia of Natural Hazards. Springer, Netherlands, pp. 812–824.

MAEviz, 2008. "MAEviz Software," Mid-America Earthquake Center. University of Illinois at Urbana-Champaign, Urbana, Illinois. http://maeviz.ce.uiuc.edu/software and tools/maeviz.html.

Merz, M., Hiete, M., Bertsch, V., Rentz, O., 2007. Decision support for managing interruptions in industrial supply chains. In: Proceedings des 8. Forums Katastrophenvorsorge–Katastrophenvorsorge im Klimawandel.

Middleman, 2007. Natural Hazards in Australia: Identifying Risk Analysis Requirements. Geoscience Australia, Canberra, Australia.

Miller, D.S., Rivera, J.D., 2010. Community Disaster Recovery and Resiliency: Exploring Global Opportunities and Challenges. CRC Press.

Minnesota IMPLAN Group, 1998. Elements of the Social Accounting Matrix. MIG, Inc, Stillwater, MN.

Nardo, M., Saisana, M., Saltelli, A., Tarantola, S., Hoffman, A., Giovannini, E., 2005. Handbook on Constructing Composite Indicators.

Nasserasadi, K., Mechler, R., Ashtiany, M.G., 2006. Seismic indirect loss model for industrial facilities. In: Proceeding of 1th European Conference on Earthquake Engineering and Seismology, Geneva, Switzerland.

Ohtake, F., Okuyama, N., Sasaki, M., Yasui, K., 2012a. Measuring the Longlasting Effect of Natural Disasters: The Case of the 1995 Hanshin–Awaji Earthquake. Paper presented at the East West Center, December.

Ohtake, F., Okuyama, N., Sasaki, M., Yasui, K., 2012b. Impacts of the Great Hanshin–Awaji Earthquake on the labor market in the disaster areas. Japan Labor Review 9 (4), 42–63.

Okuyama, Y., 2008. Critical Review of Methodologies on Disaster Impacts Estimation. Background paper for EDRR report.

Okuyama, Y., 2009. Impact Estimation of Higher Order Effects. Background Paper for EDRR Report. GFDRR, Washington, DC.

Okuyama, Y., Sahin, S., 2009. Impact estimation of disasters: a global aggregate for 1960 to 2007. In: Policy Research Working Paper, Vol. 4963. The World Bank, Washington, DC.

Okuyama, Y., Hewings, G.J.D., Sonis, M., 1999. Economic impacts of an unscheduled, disruptive event: a Miyazawa multiplier analysis. In: Hewings, G.J.D. (Ed.), Understanding and Interpreting Economic Structure. Springer, Berlin, pp. 113–144.

Olsen, A., Porter, K.A., 2008. A Review of Demand Surge Knowledge and Modelling Practice. Willis Research Network. White Paper.

Pachakis, D., Kiremidjian, A., 2003. Ship traffic modeling methodology for ports. Journal of Waterway, Port. Coastal, and Ocean Engineering 129 (5), 193–202.

Porter, K.A., Ramer, K., 2012. Estimating earthquake-induced failure probability and downtime of critical facilities. Journal of Business Continuity & Emergency Planning 5 (4), 352–364.

Power, C., Daniell, J.E., Khazai, B., Burton, C., Oberacker, C., 2015a. Social and Economic Vulner- Ability Global Indicator Database Handbook. CEDIM, 112 p.

Power, C., Daniell, J.E., Khazai, B., Burton, C., Oberacker, C., 2015b. Social and Economic Vulner- Ability Eastern Asia and Paci C Sub-National Indicator Database Handbook. CEDIM, 112 p. 130 p.

Rose, A., 2004. Economic principles, issues, and research priorities in hazard loss estimation. Modeling Spatial and Economic Impacts of Disasters 13–36.

Rose, A., Benavides, J., Chang, S.E., Szczesniak, P., Lim, D., 1997. The regional economic impact of an earthquake: direct and indirect effects of electricity lifeline disruptions. Journal of Regional Sciences 37, 437–458.

Roy, S., 2010. The Impact of Natural Disasters on Violent Crime (Unpublished Doctoral Dissertation). University of Canterbury. Retrieved: June 1, 2014.

Seung, C.K., Harris, T.R., Englin, J.E., Netusil, N.R., 2000. Impacts of water reallocation: a combined computable general equilibrium and recreation demand model approach. The Annals of Regional Science 34 (4), 473–487.

Steelman, J.S., Hajjar, J.F., September 2008. Capstone Scenario Applications of Consequence-Based Risk Management for the Memphis Testbed. Mid-American Earthquake Center, Department of Civil and Environmental Engineering, University of Illinois at Urbana-Champaign, Urbana, Illinois.

Takashima, M., Hayashi, H., 2000. Indirect loss estimation using electricity consumption index. In: WCEE 2000.

Tatano, H., Tsuchiya, S., 2008. A framework for economic loss estimation due to seismic transportation network disruption: a spatial computable general equilibrium approach. Natural Hazards 44 (2), 253–265.

Thompson, P., Handmer, J., 1996. Economic Assessment of Disaster Mitigation. Center for Resource and Environmental Studies, ANU.

Tierney, K.J., Dahlhamer, J.M., 1998. Earthquake Vulnerability and Emergency Preparedness Among Businesses. University of Delaware Disaster Research Center.

Tirasirichai, C., Enke, D., 2007. Case study: Applying a regional CGE model for estimation of indirect economic losses due to damaged highway bridges. The Engineering Economist 52 (4), 367–401.

Toyoda, T., Kochi/Kawauchi, A., 1997. Estimation of industrial losses caused by the great Hanshin-Awaji earthquake. Kokumin Keizai Zasshi 176 (2), 1–15 (in Japanese).

Toyoda, T., 2008. Economic impacts of Kobe earthquake: a quantitative evaluation after 13 years. In: Fiedrich, F., Van de Walle, B. (Eds.), Proceedings of the 5th International ISCRAM Conference. Washington, DC, USA, pp. 606–617.

United Nations Economic Commission for Latin America and the Caribbean (ECLAC), 2003. Handbook for Estimating the Socio-Economic and Environmental Effects of Disasters. Retrieved from: http://www.eclac.org/cgi-bin/getProd.asp?xml=/publicaciones/xml/4/12774/P12774.xml&xsl=/mexico/tpl-i/p9f.xsl&base=/mexico/tpl/top-bottom.xsl#.

WHO, 2005. Violence and Disasters. http://www.who.int/violence_injury_prevention/publications/violence/violence_disasters.pdf.

Wu, J., Li, N., Xie, W., 2012. Interregional economic impact analysis of the Wenchuan earthquake, China. In: Proceedings of the 20th International Input-Output Association (IIOA) Conference, Bratislavia, Slovakia.

Xie, W., Li, N., Wu, J.D., Hao, X.L., 2014. Modeling the economic costs of disasters and recovery: analysis using a dynamic computable general equilibrium model. Natural Hazards and Earth System Science 14 (4), 757–772.

Yamano, N., Kajitani, Y., Shumuta, Y., 2007. Modeling the regional economic loss of natural disasters: the search for economic hotspots. Economic Systems Research 19 (2), 163–181.

Ye, S., Zhai, G., Hu, J., 2011. Damages and lessons from the Wenchuan earthquake in China. Human and Ecological Risk Assessment: An International Journal 17 (3), 598–612. http://dx.doi.org/10.1080/10807039.2011.571086.

Yeh, C.-H., Loh, C.-H., Tsai, K.-C., 2006. Overview of Taiwan earthquake loss estimation system. Natural Hazards 37 (1–2), 23–37.

Yoshida, K., Deyle, R.E., 2005. Determinants of small business hazard mitigation. Natural Hazards Review 6, 1–12.

Zhong, J., Lin, Q., 2004. Indirect economic losses from the earthquake research. Natural Disasters 12 (4), 88–92 (in Chinese).

Zsidisin, G.A., 2003. A grounded definition of supply risk. Journal of Purchasing and Supply Management 9 (5/6), 217–224.

Further Reading

Daniell, J.E., 2003–2014. The CATDAT Damaging Earthquakes Database. Digital Database, Karlsruhe, Germany.

Daniell, J.E., 2011. The CATDAT Damaging Earthquakes Database - 2010-The Year in Review. CEDIM Research Report 2011-01 (CEDIM Loss Estimation Series). Karlsruhe, Germany. Retrieved from: http://quakesos.sosearthquakesvz.netdna-cdn.com/wp-content/uploads/2011/03/CATDAT-EQ-Data-1st-Annual-Review-2010-James-Daniell-03-03-2011.pdf.

Daniell, J.E., Vervaeck, A., 2013. The CATDAT Damaging Earthquakes Database - 2012-The Year in Review. CEDIM Research Report 2013-01 (CEDIM Loss Estimation Series). Karlsruhe, Germany. Retrieved from: http://earthquake-report.com/2013/01/07/damaging-earthquakes-2012-database-report-the-year-in-review/.

Daniell, J.E., Khazai, B., Wenzel, F., Vervaeck, A., 2012. The worldwide economic impact of earthquakes. Paper No. 2038. In: Proceedings of the 15th World Conference of Earthquake Engineering, Lisbon, Portugal.

GADM v1 and v2, 2011–2013. GADM v1 and v2. Retrieved from: www.gadm.org.
Kouno, T., 2011. The Lessons of the Great Tohoku Earthquake and its Effects on Japan's Economy (Part 4). The Effects of Power Shortages on Japan's Economy. Analysis Based on Inter-Industry Table: Effects of Blackouts Alone Could Surpass ¥1 Trillion. Retrieved from: http://jp.fujitsu.com/group/fri/en/column/message/2011/2011-04-14.html.
Mechler, R., 2003. Macroeconomic Impacts of Natural Disasters. World Bank Report.
Merz, M., Hiete, M., Bertsch, V., 2009. Multicriteria decision support for business continuity planning in the event of critical infrastructure disruptions. International Journal of Critical Infrastructures 5 (1/2), 156–174.
Okuyama, Y., Chang, S.E. (Eds.), 2004. Modeling Spatial and Economic Impacts of Disasters. SpringerVerlag.
Przyluski, V., Hallegatte, S., 2011. Indirect costs of natural hazards. In: CONHAZ Report WP02 Supported by the European Community's Seventh Framework Program Through the Grant to the Budget of the Coordination Action CONHAZ. Contract 244159, SMASH-CIRED, France, 41 pp.
Thieken, A.H., Ackermann, V., Elmer, F., Kreibich, H., Kuhlmann, B., Kunert, U., Maiwald, H., et al., 2008. Methods for the evaluation of direct and indirect flood losses (2009 report, 2008). In: 4th International Symposium on Flood Defense: Managing Flood Risk, Reliability and Vulnerability. Toronto, Ontario, Canada.

APPENDIX A: ELECTRONIC SUPPLEMENT

Table A.1 shows the chosen parameters used for Japan as part of the indirect loss potential index.

TABLE A.1 Level 2 Indicators for Japan Chosen for the Indirect Loss Potential Index

Primary Group	Indicator Types Along Level 1 Categories	Level 2 Indicators Chosen	Code	+/−
Input and productive factor	Primary GDP	Primary GDP per capita	ECOEAC791-794	+
	Secondary GDP	Secondary GDP per capita	ECOEAC795	+
	Tertiary GDP	Tertiary GDP per capita	ECOEAC796-806	+
	GPP per area	Gross prefectural production value (millions of yen)	ECOEAC788	+
	Employment ratio	Employment ratio prefecture (%)	ECOLAMH66	+
	GFCF	Gross fixed capital formation (real) (millions of yen)	ECOEAC828	+
	Primary employment	Primary industry employment ratio (%)	ECOEAC833	+
	Secondary employment	Secondary sector employment ratio (%)	ECOEAC845	+
	Tertiary employment	Tertiary sector employment ratio (%)	ECOEAC864	+
	Insurance penetration	Auto insurance penetration (objective) (%)	INFTCO216	−

(*Continued*)

TABLE A.1 Level 2 Indicators for Japan Chosen for the Indirect Loss Potential Index—cont'd

Primary Group	Indicator Types Along Level 1 Categories	Level 2 Indicators Chosen	Code	+/−
Supply chain and cross-sector	Industrial area	Industrial and semiindustrial area ratio (%)	ECOEREA65	+
	Clustering of sectors	Number of municipalities	GICGEF062	+
	Demand	Financial capability index (financial state)	INXXXX005	+
	Accessibility	Income from outside the prefecture (nominal) (millions of yen)	ECOEAC823	−
Infrastructure	Telephones	Number of telephone subscribers (per 1000 population)	INFTCO284	+
	Transport volumes	The average traffic volume roads (/12 h units)	INFTCO292	+
	Air cargo	The amount of air cargo transportation (kg) per capita	ECOERE869	+
	Power consumption	Electric power consumption (per household) (kwh)	INFEWS392	+
	Water supply	Water supply water supply population ratio (%)	INFEWS386	−
	Self-supply of utilities	Per capita health expenditure (1000 yen)	ECOIDP363	−
	Infrastructure starts	Construction building construction costs estimated amount (US $ million)	ECOEAC845	−
Social/human-economy interaction	Household consumption	Consumption expenditure (all households) (¥)	ECOIDP158	+
	Staying potential	Home ownership rate (%)	ECOEREA34	+
	Training and innovation	Per capita expenditure on education (1000 yen)	ECOIDP367	−
	Poverty/inequality	Gini coefficient of annual income (households of two or more persons) (10,000 yen)	ECOIDP305	+
	Crime rate	Number of road traffic law violation arrests (per 1000 population)	INFTCO228	+
	Education	Percentage of university graduates to total employment (%)	ECOLAMH81	−
	Community expenditure	Public debt expense (finance municipalities) (US $ thousands)	ECOERE664	+

TABLE A.1 Level 2 Indicators for Japan Chosen for the Indirect Loss Potential Index—cont'd

Primary Group	Indicator Types Along Level 1 Categories	Level 2 Indicators Chosen	Code	+/−
Impact-driven	Government effectiveness	Number of fire-fighting water supply (plants) (per 100,000 population)	GICGEF168	−
	Rule of law	Number of legislators per state per capita	GICPSC003	+
	Payment investment	Civil costs per capita [total municipal finance department] (1000 yen)	ECOIDP364	−
	Population distribution	Population density per $1k^2$ inhabitable land (people)	ECOERE875	+
	Budget effects	Prefectural income per capita (US $ thousand)	POPPPSJ38	+
	Cleanup costs	Number of residential premises is in contact with the road width of less than 6 m (residential) (% of total)	GICGEF062	+

Table A.2 shows various studies of indirect losses with their key indicators as seen in the literature.

TABLE A.2 Indirect Losses Study of Literature From Various Studies Since 1980 (Daniell et al., 2015)

Author and Year	Location Studied	Key Indicators Examined
Comerio (2006) and Comerio et al. (2010)	Loma Prieta (1989)	Quantifying downtime via start and complete repairs
HAZUS (1999)	USA	Full indirect economic I-O matrix created
Ohtake et al. (2012a,b)	Kobe (1995), Tohoku (2011)	Indirect loss data via GDP and employees
Porter and Ramer (2012)	Various	Downtime, reviews of repair cost
SISMA Ricostruzione (2012)	Emilia-Romagna (2012)	No. of employers/companies, loss impact, GDP loss, sectors
Wu et al. (2012)	Sichuan (2008)	Direct damage extent in terms of sectors and capital—thus bottlenecks in production; environmental instability; product price increase, labor supply shortages, raw material costs; reconstruction funding

(Continued)

TABLE A.2 Indirect Losses Study of Literature From Various Studies Since 1980 (Daniell et al., 2015)—cont'd

Author and Year	Location Studied	Key Indicators Examined
Ye et al. (2011)	Sichuan (2008)	Different methods: First includes a subjective quantification of loss in terms of speed and size; general damage; infrastructure damage; urban function damage; recovery speed. Second uses input-output method using six parameters: architecture, industry and mining, chemical, transportation, agriculture, power generation and supply
Li (2010) and Xie et al. (2014)	General	Amplification (socioeconomic, sociopolitical), safety and planning laws, housing quality, and resident lifestyle for industrial production systems
Giovinazzi and Nicholson (2010)	L'Aquila (2009)	Connectivity, capacity, travel time, accessibility (network reliability)
Kuwata and Takechi (2010)	Kobe (1995)	Production statistics (steel, manufacturing); industrial, production levels at damaged facilities, demand, no. of companies, water supply, recovery speed combining to utility of water in USD/yr/d/m^3—seismic resilience of the company, affected by delay
Okuyama and Sahin (2009)	Indonesia, Thailand, Sri Lanka	Using the 2000 Asia input-output table
Heng et al. (2008)	China (Yunnan)	Rural area analysis (agricultural production)—direct and recovery time, (Cobb–Douglas function—capital, labor, agricultural production), consumption via rural income, investment, speed of government relief
Albala-Bertrand (2008)		Negative and positive impact
Tirasirichai and Enke (2007)	USA	Bridge model for indirect losses, full transportation module
BAPPENAS Joint Assessment Team (2006)		Key sectoral GDPs and employment; impact on revenues (but this is not a key value before)
Gurenko (2006)	Turkey	Nonperforming loans and bank deferrals; value-added loss; Emergency relief expenditures; job/fiscal changes
Ham et al. (2005)	USA	Commodity flows for transportation network
Kundak (2004)	Istanbul scenario	No. of employees and GDP analysis with parameters
Yamano et al. (2007)	Japan	Power substations, interregional input-output tables
Pachakis and Kiremidjian (2003)		Business interruption and lost income
Yoshimura (2003)	Japan	Manufacturing: number of employees and value of fixed assets; Construction: difficulty of demand and supply in terms of the lost assets adding the positive effects on capital recovery; Essential utilities: days out in addition to the number of added value of output; Wholesale trade: capital stock, assets, employees (and loss); Traffic and

TABLE A.2 Indirect Losses Study of Literature From Various Studies Since 1980 (Daniell et al., 2015)—cont'd

Author and Year	Location Studied	Key Indicators Examined
		communication: also positive vs. negative effects; Services: similar
BTE (2001)	Edgecumbe (1987)	Various as shown in Background section; Table 6.2
Takashima and Hayashi (2000)	Kobe (1995)	Electricity issues (consumption based) index; gross regional product, impact area and supporting area (IoD), opportunity losses
Taipei (1999)		% downtime and % GDP in sectors
Tierney and Dahlhamer (1998)	Loma Prieta (1989)	Survey methods
Hanshin-Asawi EQ Committee – Hayashi (2005, 1997)	Kobe (1995)	Estimate via empirical means including GDP percentage per location per sector
Okuyama et al. (1997, 1999)	Kobe (1995)	Input-output model
Rose et al. (1997)/Rose and Benavides (1998)	Tennessee, Mid-America	Electricity disruption costs
Toyoda and Kawauchi (1997)	Kobe (1995)	Survey of firms
Brookshire and McKee (1992)	USA—Northridge (1994)	General equilibrium (CGE)/hybrid CGE
Kawashima and Kanoh (1990)	Nihonkai-Chubu (1983), Japan	Agriculture (surveyed reduction), sectoral distribution of GDP and investment adjusted for value (production changes for production facilities, transportation, and materials and goods)
Kuribayashi et al. (1987)	Japan	Cobb-Douglas: Labor, capital, adjustment factors
Georgescu and Kuribayashi (1992)	Romania and Japan	Supplementary exports, imports, tourism receipts, lost production
FEMA (2013) via Wilson (1982)	USA	Input-output mode sectors

Author	Calculations
National Research Council (1999)	Indirect costs increase as a proportion of total disaster costs with the size of a disaster
Kates (1965)	Indirect damage as percentage of direct damage for commercial buildings (37%), industrial buildings (45%) but clean-up costs as direct costs.
Smith et al. (1979)	1974 Lismore flood—18.5% of direct for commercial, 36% for industrial—business disruption contributed 67%–71% of the indirect losses
SMEC (1975)	35% for commercial and 65% for industrial buildings
Smith et al. (1990)	55% of direct for commercial and industrial buildings
Sturgess and Associates (2000)	25%–35% of direct
BTE (2001)	No simple relationships but 25%–40% of direct costs for floods are a reasonable estimate

The culmination of the various parameters from the methodology as seen from around 20 South and Central American earthquakes over a period of 30 years, as well as expert opinion are shown in the following summary of the four volume document giving the key provisions needed for analysis (Table A.3).

Insurance-Based Modeling

Insurance modeling of indirect loss in addition to the methodologies mentioned above generally centers around supply chain disruption, downtime (business interruption), and employment reduction. The key component is the damage calculation via vulnerability functions of the built infrastructure postevent whether it be power outages, utilities, building loss, or road closures. This is commonly the driving factor for indirect losses and thus is a required part of any analysis. Structural damage often influences indirect losses via the complexity of the rebuild causing delays to business. This can often be calculated from empirical relations or by expert-influenced analytical modeling. Buildings, contents, and income of an affected business are often modeled in conjunction with the downtime calculation and business income, policy conditions, civil authority influence, limits, and deductibles to create a final business interruption.

At the simplest level, mean damage ratio versus business interruption downtime data is often available from insurance companies for historic events, which can then be used as a first estimate in the analysis of distributed portfolios. The stochastic modeling of power distribution outages due to an earthquake via random substation outages has inherent complexities. This provides the next level of complexity as utility disruption for a nondependent business is also a key component. The level of redundancy and resilience of their business processes determine the speed of recovery.

However, in the case of a multilocation or single business, dependence diagrams and scorecards are often required to see whether local or regional resources are used, contract terms with partners, resilience of each branch, and the overall continuity probability postdisaster. This dependency of a business to losses via other businesses for production is one of the most difficult calculations. Thus, business interruption models have differing complexity with businesses with only physical damage and/or utility (local/country) or location (authority) disruptions requiring much less calculation time than the addition of dependent businesses as input distributors or producers for other businesses. The delay times are often calculated in supply chain systems or input-output processes as described above with the reduction in productivity produced temporally as well as spatially. Distances between businesses often play a major role in these supply chain models. Damage matrices of dependent businesses can be set up for modeling the potential losses from stochastic event sets with varying magnitudes of events. Concurrently a CGE model and/or ARIO model may be set up to capture complexities. In conjunction with such methodologies as well as the specific policy clauses in terms of definition of downtime and calculation of background rates of economic outlook, the premium can then be set by using the AAL + inherent variability and costs.

Market recovery rates, macroeconomic modeling, government mitigation action modeling, and repair speeds are often added by expert opinion or separate modeling outside of the system analysis as influencing factors for adjusting a premium.

TABLE A.3 Indirect Loss Provisions Needed as Examined by ECLAC as Part of the DaLa Methodology (Daniell et al., 2015)

Topic Theme	Provision Necessary for Quantifying Indirect Losses
All encompassing	Local availability of materials
	Capacity of the construction sector
	Availability of equipment
	Timeframe of reconstruction
	Land acquisition
	Demolition and removal of all existing material
	Construction of temporary buildings
	Replacement of contents for use in the time before full replacement
	Losses in terms of rent required for premises
Housing	Temporary water and sanitation and electricity services setup
	Cost of food
	Loss of rental income
Education	Provision of education for students and the cost of temporary education services
	Overtime costs of health workers (overlap with demand surge)
	Repairing of overused facilities that were used as shelters
	Revenue losses to private schools
	Accelerated training to replace deceased teachers
Health	Cost increases for medical and health care
	Costs of monitoring outbreaks of diseases
	Overtime costs of health workers (overlap with demand surge)
	Provision of medical supplies and medicines above that of normal use
	Psychological and physical rehabilitation of affected people
	Transport costs to unaffected health centers
	Environmental disposal of radioactive/hazardous materials
	Outreach and advertising programs for prevention and mitigation of diseases, vaccination and outbreaks
	Decreases in number of patients due to interruption of access
Transportation	Alternate route provision and time costs associated
	Alternate transport mode provision
	Volume flows of traffic (reduction or increase) to be compared with the predisaster traffic quantities

(Continued)

TABLE A.3 Indirect Loss Provisions Needed as Examined by ECLAC as Part of the DaLa Methodology (Daniell et al., 2015)—cont'd

Topic Theme	Provision Necessary for Quantifying Indirect Losses
	Higher operating costs for the time period of reconstruction/rehabilitation (3 months–5 years generally depending on damage)—type of vehicle, terrain, and road quality
	Salary changes and overtime
	Depreciation and monetary value changes of the transport corridors
	Transportation mode-wear of vehicle ratios (i.e., fuel, repairs, tires) as determined by conservation of surface
Electricity	Replacement machinery and equipment required (outside of the materials)
	Labor cost changes
	Electricity demand changes (in line with other sector reconstruction components—residential, agricultural, industrial, commercial)—higher operational costs
	Lower operational revenues (associated with the above demand changes) or other costs due to replacement power plants or power imports
Water and Sanitation	Water demand changes (from destroyed/damaged or affected locations) and higher operational costs or overuse of unaffected provisions—*This will not occur until the housing and full reconstruction has occurred.*
	Lower operational revenues (associated with the above demand changes) or other costs due to replacement water or sanitation services; partial supply of services
	Nonprovision of water affected sectors
Commerce	Destruction of stocks of goods to sell or flow effects of these raw goods
	Stoppage of sales due to lack of electricity, water
	Unavailability or shortages of labor
	Disruption of goods availability and inflow from needed components
	Future unavailability of goods to sell in various sectors (i.e., livestock)
	Working capital issues at commercial enterprises
	Demand changes due to income declines or increases or construction material demands
Tourism	Tourist trust in the location (fear or lack of information)
	Environmental assets and cultural site visits
	Foreign advertising and promotion campaigns (and domestic)
	Permits and licenses as well as bureaucratic issues
	Demand changes
	Other supply changes (contingent on bed capacity and the all-encompassing construction effects)
	Overtime payments to staff
	Lack of luxury provisions

Existing Software Incorporating Methods for Quantifying Indirect Loss Impact

Table A.4 shows the parameters for indirect loss analysis and the dependencies within the indirect loss index.

Aspects of indirect losses are quantified within HAZUS and MAEviz (now ERGO) following an input-output methodology and through fiscal impact analysis; however, there

TABLE A.4 Defining Parameters for Indirect Loss Analysis (Combining in Components of Macroeconomic Analysis) and the Dependency as Part of the Indirect Loss Potential Index

Parameter	Definition	Dependency
Perishable goods	Perishable foods are any that will spoil or rot within days or hours. Most perishables will not even last a few hours without refrigeration or freezing. In the case of the Limon 1991 earthquake, bananas waiting for shipment spoiled due to an inability to ship	Social
Business interruption	The lack of ability of a business to function postdisaster. Business interruption is the revenue losses, perturbations, and adjustments to a business postdisaster	Input, Supply Chain
Diminished production due to paralysis of activities	This refers to the disruption of production of goods or services due to direct paralysis of ongoing activities	All
Budgetary reorientation	Adjustment of the budget (albeit national, council or regional) due to a disaster	Impact
Income reduction	Postdisaster income reduction is often an indirect loss associated with the decision process of governments or businesses after the earthquake, i.e., the loss of future harvests may reduce incomes in that field	Input, Social
Relief costs	The costs associated with the relief process such as food packages and other aid processes postearthquake	Impact
Clean-up costs	These are the nondebris removal costs of clean-up such as removing silt after liquefaction or removing fallen trees or such	Impact, Infrastructure
Flow-on effects of disasters (traffic congestion etc.)	Additional effects that follow the earthquake including traffic congestion due to blocked roads, lack of public transport services etc.	Infrastructure
Linkage effect production losses (backward or forward)	This is the forward and backward modeling of production processes, i.e., a backward linkage (demand reductions) is referring to the growth of an industry that leads to the growth of the industries that supply inputs to it; for a phone manufacturer, this would mean that the parts manufacturers and chip manufacturers that supply them. Forward linkages (earthquake induced supply shortages) are then the industries that the phone manufacturer supports, i.e., stores, multinational onsellers etc.	Supply Chain

(Continued)

TABLE A.4 Defining Parameters for Indirect Loss Analysis (Combining in Components of Macroeconomic Analysis) and the Dependency as Part of the Indirect Loss Potential Index—cont'd

Parameter	Definition	Dependency
Gross investment and consumption	"The impact of a disaster on levels of consumption and investment depends on a range of demand and supply factors, most importantly access to sufficient finance by the private sector and individual households to replace lost assets, continue to implement prior investment plans, and maintain levels of consumption. The consequences, in turn, for levels of consumption and investment will help shape the pace of economic recovery and thus the scale and nature of other indirect and secondary effects of the disaster." (Benson, 2012)	Input, Social
Balance of payments	The balance of payments is also affected through adverse effects of earthquakes. These are often due to the greater costs via budgetary reassignment	Impact
Inflationary effects	Inflation can occur as a result of major earthquakes due to financial changes which in turn require the raising of interest rates or prices	Impact, Input
Employment losses and sector estimates	The employment losses as part of an earthquake due to industrial, service, or agricultural sector disruption	Input
Intangible and psychological effects	The effects to the human psyche and the psychological injuries from earthquakes as well as those that cannot be quantified such as life costing	Partially in social
Debris removal and demolition	This is the clearing of debris and destroyed components postearthquake that often are rubble blocking streets or causing indirect damage through disruption. The demolition costs of badly damaged buildings are often included in indirect loss estimates	Impact
Retrofitting and stabilizing work	The stabilizing work postearthquake is thought of as an indirect loss given that it is as a result of an earthquake that stabilization is required and thus associated costs indirect. Retrofitting and shoring of buildings and the associated costs are indirect	Input, Social
Compulsory land acquisition	These are the government costs, where land is declared unusable postdisaster or land is required for reconstruction or resettlement projects. One of the largest land acquisitions was post-Christchurch 2011 earthquake	Impact
Temporary housing units	Costs associated with the building and supplying of shelter postearthquakes albeit built out of canvas or semipermanent concrete structures. In some cases, temporary housing units have been used for over 20 years, i.e., in Armenia after the 1988 earthquakes	Impact, infrastructure
Other temporary structures and detours	Roads and other infrastructure may need to be built post-earthquake as well as building detour options that are only for a few years such as longer roads, or lower standard roads	Infrastructure

TABLE A.4 Defining Parameters for Indirect Loss Analysis (Combining in Components of Macroeconomic Analysis) and the Dependency as Part of the Indirect Loss Potential Index—cont'd

Parameter	Definition	Dependency
Temporary leases	The leases over the period postdisaster where they are required due to earthquake losses can be counted under indirect	Infrastructure
Cultural heritage	The cultural value of an asset is something that cannot generally be quantified in a tangible way. The indirect losses associated with the loss of the object, spiritual or cultural patterns, tourism losses, and historical impacts contribute to the total loss	Not taken into account
Disaster mitigation	Mitigation postdisaster includes the work and costs made to improve or change the earthquake risk of an object	Impact
Increased costs for health care and services	The strain on health care and services often lead to increased pricing due to either a lack of doctors and care workers in the health industry, a lack of resources, or outages or perturbations in the services sector	Social
Temporary supply (of electricity) and demand	The rewiring and reworking of electricity supply postdisaster as well as the demand surge or change due to lack of electricity and the cost change associated with it	Infrastructure
Alternative accommodation	The cost of hotel stays as well as additional costs to families supporting those in a disaster contribute to indirect losses	Impact
Population losses and indirect impacts	The mortality rates and the effect on families and loss of income over long term as well as the health effects postdisaster in terms of diseases and injuries contribute to costing. Life costing can be looked at as a difference over many years	Impact
Lost records	There are often high operational costs due to destruction of inventories such as health files in a hospital	Social
Contract compliance issues	There are often variation clauses in contracts which an earthquake can disrupt, causing additional fees and penalties to be paid	Impact
Postearthquake information services	Advertising postearthquake required to inform the public of changes as well as additional situation systems or radio/TV network provisions are included are indirect losses	Infrastructure
Disaster repercussions via third parties	Losses or gains from external factors and external people who are not direct victims and beneficiaries of the disaster	Social
Deteriorating competitiveness	A major issue is market gap analysis after a disaster. In some cases it can be seen that certain markets are not needed or are irrelevant to the economy meaning that they become less competitive	Supply chain
Productivity reduction due to aftershocks	Fear and ability of government protocols to deal with production in conjunction with aftershocks. Often machinery shutdowns or additional checks lead to huge delays	Impact, Input
Seasonal variations in demand	In productive sectors where there are seasonal variations, the time of the disaster can influence strongly the indirect loss potential	Impact

are limited options outside of this methodology for indirect loss analyses in existing software packages. FEMA (1999) defines the indirect losses in an I-O framework as a forward-linked model (output-purchasing) or a backward-linked model (input-supplied). Many parameters of the GDP sectoral makeup as well as other linkages are required to create the input-output ratios in these models.

Input-output models (Leontief, 1986; FEMA, 1999) are based on the premise that the usual trend of growth will be maintained without a disaster and that the difference between the direct losses to production and the trended growth is the loss due to indirect causes. The reduced availability of goods, given the removal of the affected production, governs the process of business interruption. Two key links propagate business interruption losses: if the regional customers purchase their output, or if regional suppliers provide their inputs, then they have a greater chance of losses. HAZUS (FEMA, 1999) allows for calculation of downtime functions based on expert and empirical data, in various sectors based on US conditions.

MAEviz (2008) has a few modules that are associated with indirect loss calculations, including the business interruption module, fiscal impact module, and other such modules. The fiscal impact module looks at the impact on tax revenue (Steelman and Hajjar, 2008), looking at property taxes levied on the building stock and calculating the effects on the building stock. There is an overlap of macroeconomic and indirect loss aspects in this component. The business interruption module for the building stock is similar to the HAZUS-MH modules (Green and Feser, 2007). Warehouses and heavy industrial facilities are most greatly affected in their case and this is provided based on interruption loss per area. The work of EQvis and ERGO have integrated these as part of their software.

TELES (Taiwan) (Yeh et al., 2006) is another software that integrates indirect economic losses in their software. As present, insurance pricing, tax calculations, and other macroeconomic effects are calculated; however, a HAZUS style indirect loss calculation is envisaged. With the proliferation of various open source risk software packages, a number of indirect loss methodologies are likely to be available in the future (Daniell et al., 2014a,b).

Internal models within insurance companies or modeling firms have differing complexity when it comes to implementing clause differences in terms of the time over which business disruption is paid (i.e., 72 h after an event until 6 months, or recovery of the assets in the question).

The key parameters often looked at are supply shortages, sales decline, opportunity costs, and the general economic loss. To break the methodology down further, the key components are represented in the following interactive effects between

- The goods and services shipped from industry to industry
- The shipments to consumers (goods and services), businesses (investment in plant and equipment and retained inventories), government (goods, services and equipment), to other regions (exported goods and services)
- Gross value Added in various sectors: wages and salaries to employees/labour; capital creation (rents, interest); business taxes from the government; royalties from natural resources (i.e., agriculture, mining, construction, transportation, trade, finance, services, government)
- Imports to each producing sector from other regions.
- Clean up, repair, and downtime where alternative measures and temporary accommodation of previous services are undertaken.

The complexity of models not directly implemented in open software frameworks (Tatano and Tsuchiya, 2008) similarly proposes a full CGE model to analyze long-term and short-term equilibrium in the system post- and preearthquake. The components of their analysis include all industries, transport conditions, business, labor, capital, household goods, and prices varied depending on the sectors examined.

CHAPTER

7

Probability Gain From Seismicity-Based Earthquake Models

Kristy F. Tiampo[1], *Robert Shcherbakov*[2], *Paul Kovacs*[3]

[1]**Department of Geological Sciences and Cooperative Institute for Research in Environmental Sciences, University of Colorado, Boulder, CO, USA;** [2]**Department of Earth Sciences, Department of Physics and Astronomy, Western University, London, ON, Canada;** [3]**Institute for Catastrophic Loss Reduction, Toronto, ON, Canada**

INTRODUCTION

Natural hazards represent a significant risk to the people and economy of every country in the world and are most often a future liability that is difficult to quantify or predict. Although societies make significant efforts to protect themselves from disasters, recent data suggest that their impacts have been rising. Since the 1970s, the average damage from natural disasters has increased at an alarming rate of 11.4% per year. The insurance industry estimates that insured catastrophic losses average $60 billion (USD) per year, and that these, along with total losses, have increased steadily since 1980 (Munich Re, 2013). This increase is likely due to a number of complicated factors but generally is attributed to climate change and associated human factors such as population growth (Cutter and Emrich, 2005; Munich Re, 2013).

Urbanization and the projected 35%–55% growth in Earth's population over the next 50 years will result in sustained, rapid growth in our urban centers and associated risk to their populations (United Nations Office for Disaster Risk Reduction, 2015). The potential for future disasters arising from inevitable natural hazards is exacerbated by population growth and concentration of exposure (Aster, 2012). It is estimated that the occurrence of a large earthquake in one of the world's megacities could result in 12 million deaths and monetary losses exceeding $100 billion (USD) for various earthquake scenarios around the globe (Bilham, 2009). In 2011, losses from the Tohoku earthquake in Japan totaled more than $309 billion (USD), with insured losses reaching more than $40 billion (USD) (AonBenfield, 2011).

The impact to life and property from large earthquakes is potentially catastrophic. The M7.0 earthquake in Haiti, 2010, was the fifth deadliest earthquake in history, with fatality estimates that range from tens to hundreds of thousands (Cavallo et al., 2010). Direct economic

Risk Modeling for Hazards and Disasters
http://dx.doi.org/10.1016/B978-0-12-804071-3.00007-0

damage from the M8.8 earthquake that struck Chile in 2010 reached $30 billion (USD), ~18% of Chile's annual economic output (Kovacs, 2010).

Insurance companies estimate premiums and allocate capital to cover extreme events such as large earthquakes by integrating over risk curves and calculating the likelihood of losses in the tails of those distributions. Most of the tools and models used today are based on long-term hazard, most often derived from risk assessment products offered by the same or a small selection of model providers. Only recently, in an increasingly competitive market, have insurance companies asked for improved insights regarding short-term changes in hazard and risk. If those improved, time-dependent estimates provide even a small change in risk, when implemented over several deals in a portfolio, the resulting change in tail risk and price can be the definitive factor when assessing whether a company will assume or reject the associated risk. Today, time-dependent estimates of earthquake hazard should be considered when assessing the tail risk of a large, extreme seismic event.

As a result of their potential regional and national impact, research into earthquake prediction has been ongoing in various forms for almost 100 years (Kanamori, 1981; Geller et al., 1997; Wyss, 1997; Jordan, 2006). The controversy surrounding earthquake forecasting is exemplified by the *Nature* debates in the late 1990s (Main, 1999). These debates focused on the nature of earthquakes themselves and whether they might be intrinsically unpredictable. Although many of the underlying questions remain unanswered, it marked a turning point in the field such that today earthquake forecasting, or the assessment of time-dependent earthquake hazard, is the standard in earthquake predictability research (Tiampo and Shcherbakov, 2012b).

A wealth of seismicity data at progressively smaller magnitude levels has been collected over the past 40 years, in part the result of the digitization of seismic records and in part out of recognition that there is still much to learn about the underlying process, particularly after the Parkfield prediction window passed without an earthquake (Bakun et al., 2005). Although it has long been known that temporal and spatial clustering exists in natural seismicity, much of the research on those patterns in the early years focused on a relatively small fraction of the events, primarily at the larger magnitudes (Kanamori, 1981). Examples include, but are not limited to, Mogi donuts and precursory quiescence (Mogi, 1969; Yamashita and Knopoff, 1989; Wyss et al., 1996), aftershock sequences (Gross and Kisslinger, 1994; Nanjo et al., 1998), characteristic earthquakes and seismic gaps (Swan et al., 1980; Haberman, 1981; Bakun et al., 1986; Ellsworth and Cole 1997), temporal clustering (Frohlich, 1987; Dodge et al., 1996; Eneva and Ben-Zion, 1997; Jones and Hauksson, 1997), pattern recognition (Keilis-Borok and Kossobokov, 1990; Kossobokov et al., 1999), stress transfer and triggering over large distances (King et al., 1994; Deng and Sykes, 1996; Stein, 1999; Brodsky, 2006), and time-to-failure analyses (Bufe and Varnes, 1993; Bowman et al., 1998; Jaumé and Sykes, 1999). Although these efforts provided invaluable insights, they were hampered by the poor statistics associated with the small numbers of moderate-to-large events available for analysis at that time.

Today, the availability of higher resolution data sets over a broader range of magnitudes and major computational advancements have provided new material for earthquake forecasting (Jordan, 2006; Tiampo and Shcherbakov, 2012a,b). In 2002, the first prospective forecast using small magnitude earthquake data was published (Rundle et al., 2002; Tiampo et al., 2002). The resulting renewed interest in seismicity-based forecasting methodologies engendered new efforts aimed at better defining and testing these techniques. In particular, landmark initiatives in the earthquake forecasting validation and testing arena include the working group on Regional Earthquake Likelihood Models (RELMs) as well as the Collaboratory on the Study of Earthquake Predictability (CSEP), both founded after 2000 (Field, 2007; Gerstenberger and Rhoades, 2010; Zechar et al., 2010).

It should be noted that a number of other precursory phenomena have been studied over the years, including hydrologic phenomena, tilt and strain precursors, chemical emissions, precursory seismic velocity changes, stress transfer studies, and electromagnetic signals (Turcotte, 1991; King et al., 1994; Stein, 1999; Scholz, 2002; Crampin and Gao, 2010). Important discussion exists elsewhere on the appropriate standard for a testable earthquake forecast (Jackson and Kagan, 2006; Jordan, 2006) and the various forecast pros and cons of testing methodologies and their evaluation (Vere-Jones, 1995; Field, 2007; Zechar et al., 2010; Gerstenberger and Rhoades, 2010). In addition, Tiampo and Shcherbakov (2012b) provide a complete review of different intermediate-term seismicity-based earthquake forecasting techniques. In general, these methods assume a particular physical mechanism is associated with the generation of large earthquakes and their precursors and uses the instrumental and/or historic catalogue to isolate those precursors.

Here we review the current status of these seismicity-based forecasting methodologies and their value over various time scales, from days to years. Again, we limit the discussion to methodologies, which rely on the instrumental catalogue for their data source and which attempt to produce forecasts which are limited in both space and time in some quantifiable manner. In the next section we provide an overview of time-dependent earthquake forecasts that use seismicity data to improve our knowledge of seismic hazard. In Sections Short-Term Seismicity-Based Earthquake Forecasting and Intermediate-Term Seismicity-Based Earthquake Forecasting we describe short- and intermediate-term earthquake forecasting. We provide examples of each and discuss the associated probability gains. We conclude with a brief discussion of the potential future of these methods.

TIME-DEPENDENT SEISMICITY-BASED EARTHQUAKE FORECASTS

Today, earthquake hazard forecasting can be divided into three major categories: long-, intermediate-, and short-term estimation (Tiampo and Shcherbakov, 2012b; Jordan, 2014). Long-term hazard forecasting encompasses those earthquake reoccurrence processes that are the foundation of the probabilistic seismic hazard mapping (PSHA) (Cornell, 1968; McGuire, 2004, 2008). These processes generally are presumed to be time independent, and the long-term average seismic hazard rate is assumed to remain constant over time scales ranging from decades to centuries (Cornell, 1968; Woo and Marzocchi, 2014). PSHA remains the standard for long-term modeling and has been extraordinarily successful in providing reliable building code guidelines, infrastructure and land-use planning, disaster and emergency response planning for hazard response, and risk mitigation (Field et al., 2014). However, as discussed earlier, spatial and temporal clustering exists in local and regional seismicity records (Kanamori, 1981; Jordan and Jones, 2010; Field et al., 2015). As a result, significant research into earthquake forecasting has resulted in incorporation of a time-dependent component into the most recent version of the Uniform California Earthquake Rupture Forecast, Version 3 (UCERF3) by the 2014 Working Group on California Earthquake Probabilities (WGCEP14) (Field et al., 2015). Again, both of these approaches result in hazard estimates for earthquake and ground-shaking probabilities on the order of decades to centuries (Field et al., 2014, 2015).

The earthquake process has significant temporal variability in the seismic rate, and these variations are much larger than the random variability associated with a completely time-independent process. Although there is a wide range of spatial and temporal scales associated with this variability, the focus in recent years has evolved along two separate but related lines of research based on the first-order physics. Most methods designed for short-term forecasting are based on the premise that any earthquake alters the physical conditions surrounding it in a way, potentially triggering nearby earthquakes (Woo and Marzocchi, 2014). Intermediate-term forecasting recognizes this paradigm but is concerned with the complex series of interactions that occurs on the larger scales associated with fault systems (Tiampo and Shcherbakov, 2012b).

One of the primary drivers for new initiatives aimed at quantifying time-dependent seismic hazard was the postmortem analysis following the 2009 L'Aquila, Italy earthquake. On April 6, 2009, the L'Aquila earthquake, M6.3, struck killed more than 300 people, injured more than 1500, and destroyed approximately 20,000 buildings, displacing more than 65,000 local residents. The long-term seismic hazard of the area was well understood. PSHA model of Italy identified the region as one of the country's most seismically dangerous regions. However, beginning in January of the same year, a number of small, felt earthquakes occurred, which did result in school evacuations, among other preparedness measures. Retrospectively, it is clear now that these were foreshocks to the main event. However, at the time there was insufficient evidence for such a determination, and the government agency with responsibility for hazards and risk evaluation did not designate these small magnitude events as precursors to a larger earthquake (Hall, 2011; Jordan et al., 2011). However, subsequent analysis and evaluation has resulted in new efforts to quantify short-term time-dependent seismic hazard in an attempt to effectively and responsibly implement operational earthquake forecasting (OEF).

OEF, whether short term or intermediate term, is primarily aimed at providing earthquake forecasts that can be used to implement effective preparedness, mitigation, and response actions (Jordan, 2014). Given what has been learned over the past 40 years about the statistical nature of the earthquake sequence, it is possible to formulate local and regional statistical models of observed earthquake behavior. These models can, in turn, estimate the changing probability of large events with time (Jordan and Jones, 2010). For example, the nature of aftershock sequences, which occur over days and weeks after a large event, is well understood, and the associated probability of a large aftershock after a main event can be estimated for a given region (Utsu, 1961; Utsu et al., 1995; Gerstenberger et al., 2005, 2007; Shcherbakov, 2014). Although short-term forecasts such as this have operational utility, the interpretation and implementation of such forecasts is still being developed because, although the increase in local probability may be as much as orders of magnitude greater than the long-term probability, the absolute probability of an event remains very low, typically less than 1% per day. As a result, time-dependent earthquake forecasting from seismicity-based models is restricted to a low-probability environment. (Jordan et al., 2014). Fig. 7.1 illustrates the issue schematically. Because the likelihood of occurrence of a large event on the time scales of days or weeks is so small to start with, even an increase of 1000-fold in probability results in forecast probabilities ranging from 0.01 to 1%. However, despite these limitations, ongoing research is

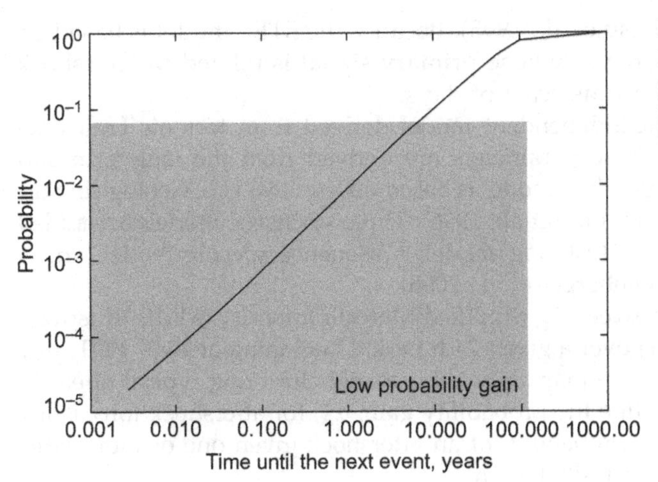

FIGURE 7.1 Probability of occurrence for large earthquakes for a typical tectonic environment. The *purple region* (*light gray in print versions*) denotes areas of low probability for an event.

focused on both the potential usefulness and means of improving short- and intermediate-term earthquake forecasts. In the next two sections we provide examples of both. We conclude with a discussion of potential implications.

SHORT-TERM SEISMICITY-BASED EARTHQUAKE FORECASTING

Short-term earthquake forecasting is generally defined as time-dependent estimates of earthquake occurrence on time scales of days to weeks (Jordan, 2014). In practice today, short-term earthquake forecasting falls into one of two categories: aftershock forecasting after mainshocks and mainshock forecasting using foreshocks. The first are based on well-known statistics of aftershock sequences (Utsu, 1961; Utsu et al., 1995). The second are less well understood, but are based on long-standing observations that some large events are preceded by smaller sequences (Jones and Molnar, 1979; Yamashita and Knopoff, 1989; Agnew and Jones, 1991; Dodge et al., 1996; Christophersen and Smith, 2008; Michael, 2012).

Aftershock Forecasting

Time-dependent models that use seismic data alone to track the spatiotemporal development of aftershock sequences have been tested and, in some cases, implemented in a number of regions (Gerstenberger et al., 2005; Marzocchi and Lombardi, 2009; Marzocchi et al., 2014; Woo and Marzocchi, 2014). The short-term forecasting models implemented to date fall into two categories: the short term earthquake probability (STEP) model (Gerstenberger et al., 2005, 2007) and those that employ stochastic epidemic-type aftershock sequence (ETAS) models (see Ogata, 1987, 1988, 1989; Zhuang et al., 2004).

The STEP model was inaugurated in 2005 at http://pasadena.wr.usgs.gov/step (Gerstenberger et al., 2005). STEP is a method that employs a universal seismicity law (in this case the modified Omori aftershock law) and historic and instrumental data to create a time-dependent forecast (Utsu et al., 1995). Because the STEP model is based on the Omori law it is a short-term forecast whose primary signal is related to aftershock sequences, producing forecasts on a time scale of days.

The STEP model combines a time-independent model derived from tectonic fault data with stochastic clustering models whose parameters are derived from the long-term and recent catalogue data. The time-independent model is based on the 1996 U.S. Geological Survey (USGS) long-term hazard maps (Frankel et al., 1996). Three stochastic models are added to that background model: a generic clustering model, a sequence specific model, and a spatially heterogeneous model (Gerstenberger et al., 2005).

Fig. 7.2 shows a STEP forecast for exceeding Modified Mercalli Intensity (MMI) VI (strong shaking) (Wood and Neumann, 1931) over a given 24-h period beginning at 14:07 PDT, July 28, 2004 (Gerstenberger et al., 2005). The map shows the spatial clustering typical of earthquake aftershock sequences. If we define the probability gain, α_A, for aftershock forecasting to be the ratio of the time-dependent probability of an aftershock given one or more large event, $P(A)_{STEP}$, to the long term hazard, P(C), then

$$\alpha_A = P(A)_{STEP}/P(C) \tag{7.1}$$

The probability gain, α_A, for the short-term probability of an event (1-day) in those local areas is approximately five times that of background and the absolute probability of exceeding MMI VI is ~0.02 (Gerstenberger et al., 2005).

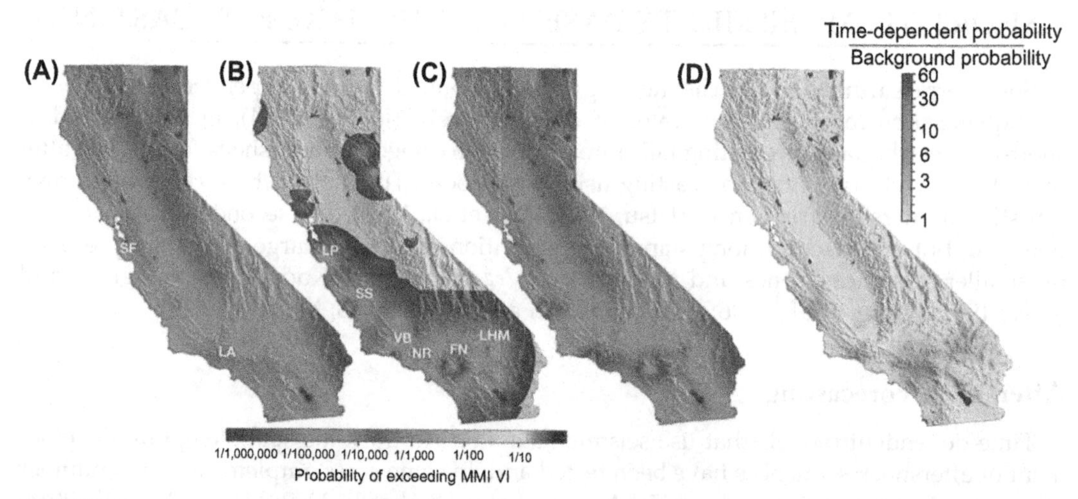

FIGURE 7.2 Short term earthquake probability for exceeding Modified Mercalli Intensity (MMI) VI over a given 24 h period starting at 14:07 PDT, July 28, 2004. (A) Time-independent hazard based on 1996 USGS hazard maps for California. SF and LA are San Francisco and Los Angeles, respectively; (B) time-dependent hazard exceeding background including contributions from several events: December 22, 2003, San Simeon (SS, Mw = 6.5), an M4.3 earthquake 4 days earlier near Ventura (VB), an M3.8 event that near San Bernardino (FN), the 1999 Hector Mine, M7.1 earthquake (LHM), and the 1989 M6.9 Loma Prieta (LP) earthquake and (C) the combination of (A) and (B), representing the total forecast of the likelihood of ground shaking in the next 24-h period; (D) ratio of the time-dependent contribution to the background. *Modified from Figure 1, Gerstenberger, M.C., Wiemer, S., Jones, L.M., Reasenberg, P.A., 2005. Real-time forecasts of tomorrow's earthquakes in California. Nature 435, 328–331.*

Originally formulated by Ogata (1987, 1988, 1989), the ETAS model primarily has been used to forecast the evolution of aftershock sequences. However, ETAS is not simply a model for aftershock sequences; it is fundamentally a model of triggered interacting seismicity in which all events have identical roles in the triggering process. In ETAS, every earthquake is regarded as both triggered by earlier events and a potential trigger for subsequent earthquakes. Every event is a potential aftershock, mainshock, or foreshock, with its own aftershock sequence. To reproduce natural seismicity, a background term with a random component is added to the branching model. In the intervening years, the model has been used in many studies to describe the spatiotemporal distribution and features of actual seismicity (Ogata, 1988, 1999, 2005; Ma and Zhuang, 2001; Helmstetter and Sornette, 2002, 2003a,b; Console et al., 2003; Zhuang et al., 2004, 2005; Ogata and Zhuang, 2006; Vere-Jones, 2006, among others). For a thorough review of the development of ETAS, see Ogata (1999) and Helmstetter and Sornette, (2002). Fig. 7.3 shows a 1-day earthquake forecast for Italy based on an ETAS model (Falcone et al., 2010).

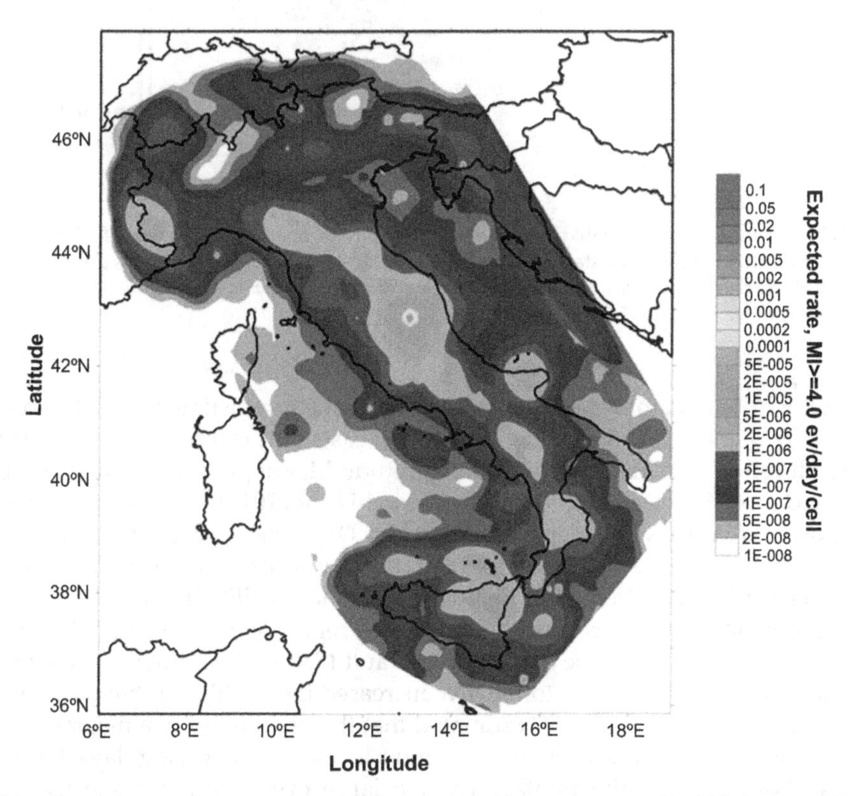

FIGURE 7.3 Short-term occurrence-rate density (events $M \geq 4.0$, per day per cell of 0.1 degree $\times$ 0.1 degree) for the Italian territory, for a 24-h period starting at 00:00 UTC on March 21, 2009, from an ETAS-type model which propagates seismicity using a) the Omori law. *Modified from Falcone, G., Console, R., Murru, M., 2010. Short-term and long-term earthquake occurrence models for Italy: ETES, ERS and LTST. Annals of Geophysics 53. http://dx.doi.org/10.4401/ag-4760.*

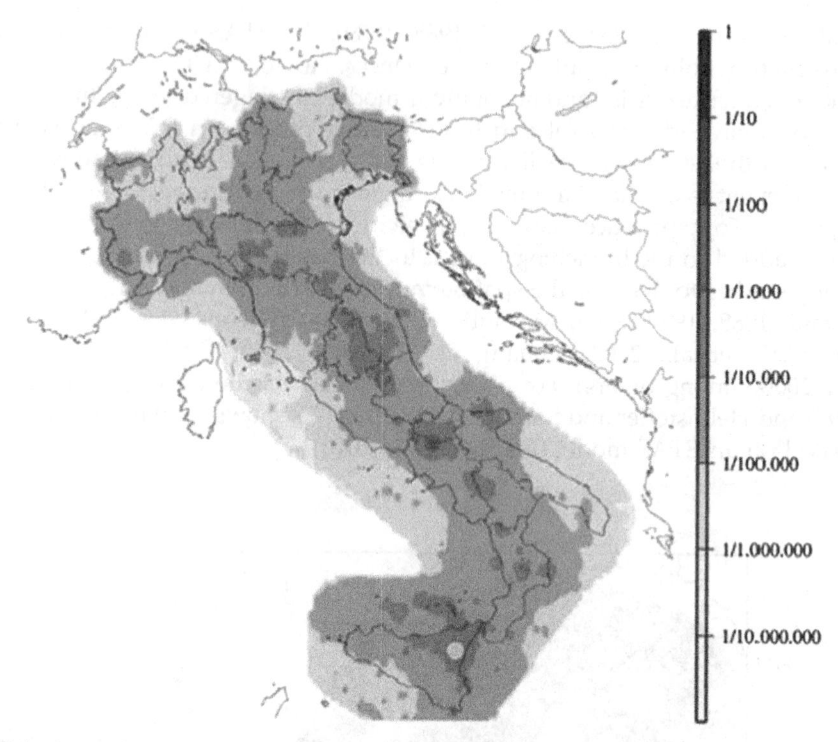

FIGURE 7.4 Spatial forecast for Italy showing the probability of one or more events of $M \geq 4$ in each cell of 0.1 degrees by 0.1 degrees in the week starting from December 31, 2013. *Modified from Marzocchi, W., Lombardi, A.M., Casarotti, E., 2014. The establishment of an operational earthquake forecasting system in Italy. Seismological Research Letters 85 (4), 961–969. http://dx.doi.org/10.1785/0220130219.*

Today, OEF in Italy employs a short-term earthquake forecasting model based on a combination of ETAS and STEP forecasts. Weekly forecasts that estimate the number of events that in the *i*th cell of a spatial grid with a magnitude M_i are produced. These are combined and converted to MMI exceedance values (Marzocchi et al., 2014). Fig. 7.4 shows one example of these short-term forecasting maps for Italy from December of 2013.

Short-term forecasts have been used for operational purposes in California. In March 2009, the California Earthquake Prediction Evaluation Council (CEPEC) evaluated a sequence of more than 50 small earthquakes clustered in a small region south the Salton Sea and located near the southern extension of the San Andreas Fault (SAF). The sequence includes an M4.8. CEPEC determined that proximity to the SAF increased the likelihood that they could trigger a large earthquake ($M \geq 7.0$). CEPEC estimated that the probability of a major earthquake on the SAF had increased to values between 1% and 5% over the following days, based on existing methods for assessing the probability of a large event from foreshocks on the SAF (CEPEC Preliminary Report to Cal-EMA, 2009).

In Agnew and Jones (1991), it was shown that, for earthquake forecasting based on foreshock occurrence, if P(F) is the probability that an event is a foreshock and P(C) is, again, the a priori probability of a mainshock, then the probability gain from a foreshock prediction α_F is equal to

$$\alpha_F = P(F)/P(C) \tag{7.2}$$

Agnew and Jones (1991) estimated the background 1-day probability of a large event in California is 5.5×10^{-6} (Agnew and Jones, 1991). Given that, the single day probability gain, α_F, is on the order of 1000 s, although recent studies suggest that the probability of an event is several orders of magnitude higher for southern California (Field et al., 2014).

More recently, retrospective studies of the M $\geq$ 8.5 Tohoku, Japan earthquake of 2011 provide more quantitative estimates of the probability gain α_F from foreshock modeling. Fig. 7.5 shows a spatial forecast map for Japan, based on the methodology of that shown in Fig. 7.4, for the 1-year period of 2011 (Lombardi and Marzocchi, 2011; Woo and Marzocchi, 2014). The scale shows the annual probability for an M $\geq$ 8.5 earthquake in each grid cell of 0.1 degrees by 0.1 degrees. Also shown is the location of the Tohoku earthquake. A strong seismic sequence occurred a few days before the mainshock, where the largest foreshock was M7.2. Fig. 7.6 shows a forecast map for the same location for the period between March 9 through 16, 2011 (Woo and Marzocchi, 2014). The weekly probability for a region of

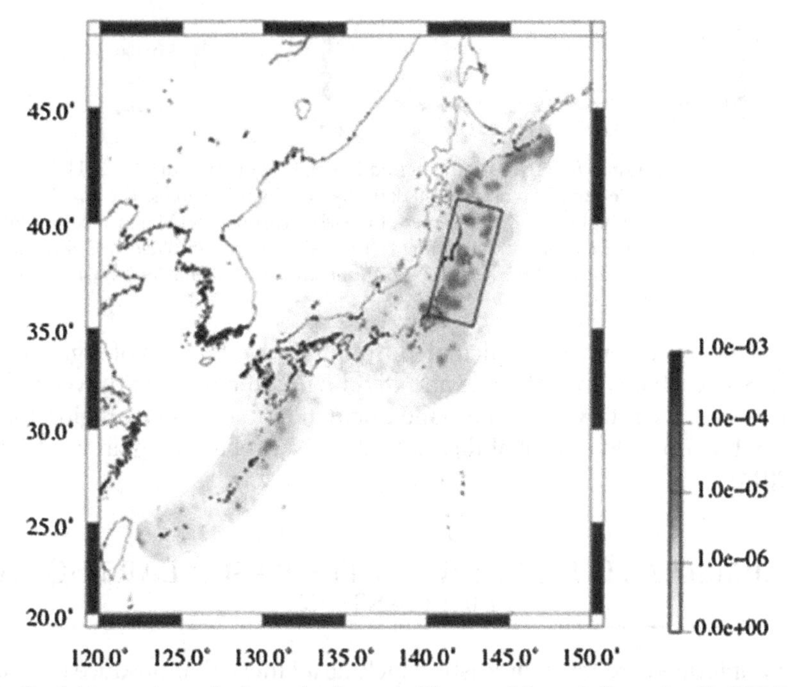

FIGURE 7.5 Spatial forecast map for Japan for the period January 1 through December 31, 2011. The scale shows the annual probability for an M $\geq$ 8.5 earthquake in each cell of 0.1 degrees by 0.1 degrees. The box shows the location of the Tohoku earthquake, and the star is the epicenter. *Modified from Woo, G., Marzocchi, W., 2014. Operational Earthquake Forecasting and Decision-making. In: Wenzel, F., Zschau, J. (Eds.), Early Warning for Geological Disasters. Springer, Berlin, pp. 353–367. http://dx.doi.org/10.1007/978-3-642-12233-0_18.*

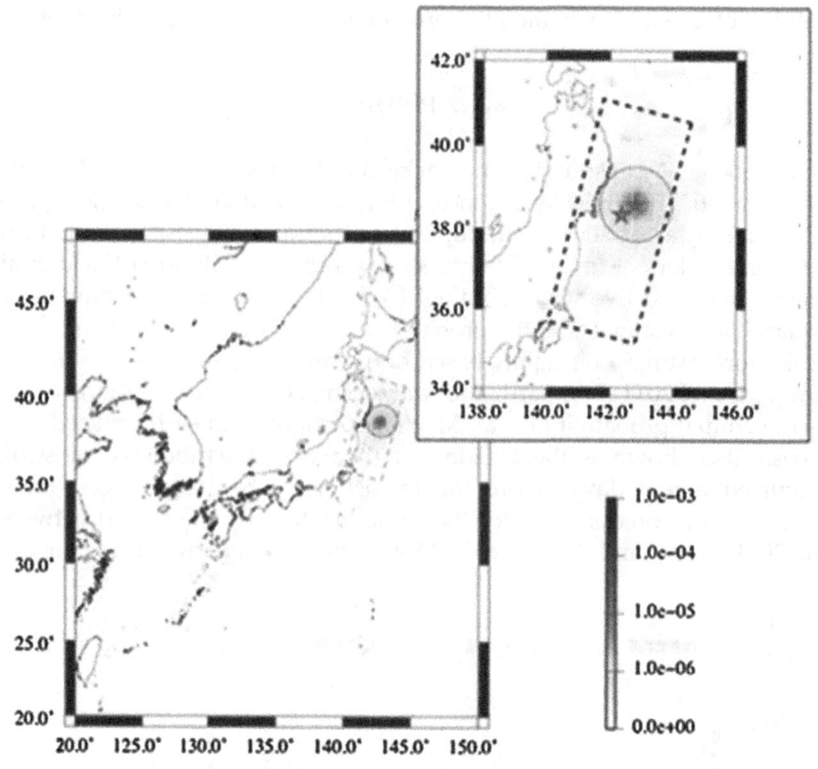

FIGURE 7.6 Spatial forecast map for Japan for the period March 9 through March 16, 2011. The scale shows the annual probability for an $M \geq 8.5$ earthquake in each cell of 0.1 degrees by 0.1 degrees. The box shows the location of the Tohoku earthquake; the circle highlights the epicenter of the M7.2 earthquake, March 9, 2011. *Modified from Woo, G., Marzocchi, W., 2014. Operational Earthquake Forecasting and Decision-making. In: Wenzel, F., Zschau, J. (Eds.), Early Warning for Geological Disasters. Springer, Berlin, pp. 353–367. http://dx.doi.org/10.1007/978-3-642-12233-0_18.*

100 km around the epicenter can be calculated from the original map of Fig. 7.5 by summing the probabilities in each box in that circle and dividing by the number of weeks in each year. That a priori probability is 0.0012%. The same calculation for the same region for the 1-week forecast map of Fig. 7.6 yields a probability of 0.12% or a probability gain α_F of 100 (Woo and Marzocchi, 2014).

INTERMEDIATE-TERM SEISMICITY-BASED EARTHQUAKE FORECASTING

Methods for intermediate-term forecasting include all those that produce estimates on time scales of months to years. These range from models based on physical processes, such as accelerating seismic moment release (Bufe and Varnes, 1993; Bowman et al., 1998; Brehm and Braile, 1998; Jaumé and Sykes, 1999; Ben-Zion and Lyakhovsky, 2002; Mignan and Di Giovambattista, 2008; Mignan, 2011) to those that are primarily based on smoothed models of statistical seismology, such as those based on ETAS (Ogata, 1987, 1988, 1989; Console et al., 2003; Helmstetter et al., 2005, 2006, 2007; Falcone et al., 2010; Lombardi and Marzocchi, 2011, among others). Another interesting development is the introduction of the hybrid Coulomb/statistical model for earthquake forecasting based on stress transfer models (Steacy et al., 2014). A complete review of these methods can be found in Tiampo and Shcherbakov (2012b). Here we present one method that bridges these two classes as an example of the potential probability gain from intermediate-term forecasting.

The Pattern Informatics (PI) index is an analytical method for quantifying the spatiotemporal seismicity rate changes in historic seismicity (Rundle et al., 2002; Tiampo et al., 2002; Holliday et al., 2006). The PI method measures the local change in seismicity relative to the long-term background seismicity that has been used to forecast large earthquakes, or P(C). The method identifies spatiotemporal patterns of anomalous activation or quiescence that serve as proxies for changes in the underlying stress that may precede large earthquakes and can be related to the location of large earthquakes that occur in the years following their formation (Tiampo et al., 2002, 2006). Again, theory suggests that these seismicity structures are related to changes in the underlying stress level (Dieterich, 1994; Dieterich et al., 2002; Tiampo et al., 2006).

The PI index is calculated using historic catalog data from seismically active areas. The seismicity data are gridded by location into boxes. In California, a grid box size of 0.1 degrees in latitude and longitude was successful, but that may vary with tectonic area. Time series are created for each of these gridded locations. The mean is removed from each time series and divided by the standard deviation (Tiampo et al., 2002). Dividing by the constant standard deviation normalizes the regional seismicity by its background and illuminates small, local fluctuations in seismicity. The result is averaged over all possible years. For any given catalog and time period, the PI index is the power associated with that value, $\Delta S'(x_i,t_1,t_2)$, such that $\Delta P(x_i,t_1,t_2) = \{\Delta S'(x_i,t_1,t_2)\}^2 - \mu_p$ where μ_p is the spatial mean of $\{\Delta S'(x_i,t_1,t_2)\}^2$ or the time-dependent background (Tiampo et al., 2002).

In 2002, Rundle et al. published a prospective forecast for California for the period of 2000–10, inclusive. Fig. 7.7 reproduces that forecast. Fig. 7.7 shows the normalized seismicity rate change for California, 2000–10 inclusive, in each cell of 0.1 degrees by 0.1 degrees. Color scale is logarithmic, where the number represents the exponent. Circles identify those events of magnitude $M \geq 5$ that occurred during the forecast period. Thirty-nine events occurred between 2000 and 2010, and 37 fell on or within one box size (~11 km) of an anomaly (successful forecasts), the margin of error for this forecast (Holliday et al., 2006, 2007), and two did not (missed events).

Note that the long-term probability of events $M \geq 5$ in the region is approximately 1.25 events per year (Field et al., 2014). Given that, the average number of events of $M \geq 5$ during that time should have been 13 or 14, not 39. This is an example of the time-dependent clustering in earthquake seismicity, even on large regional scales. If we estimate the total uniform probability in each 0.1-degree by 0.1-degree grid cell for every box in the forecast in Fig. 7.7 (10,100), the individual 10-year probability for each location is 1.2×10^{-4}. Retrospective studies of Fig. 7.7 suggest that the map is best interpreted as a threshold probability—anything above a very small value should be considered as having equal probability of an event as any other nonzero site. Those with zero probability are extremely unlikely to have an event. Here, we assume that all sites with PI > 0 are equally probable, whereas those

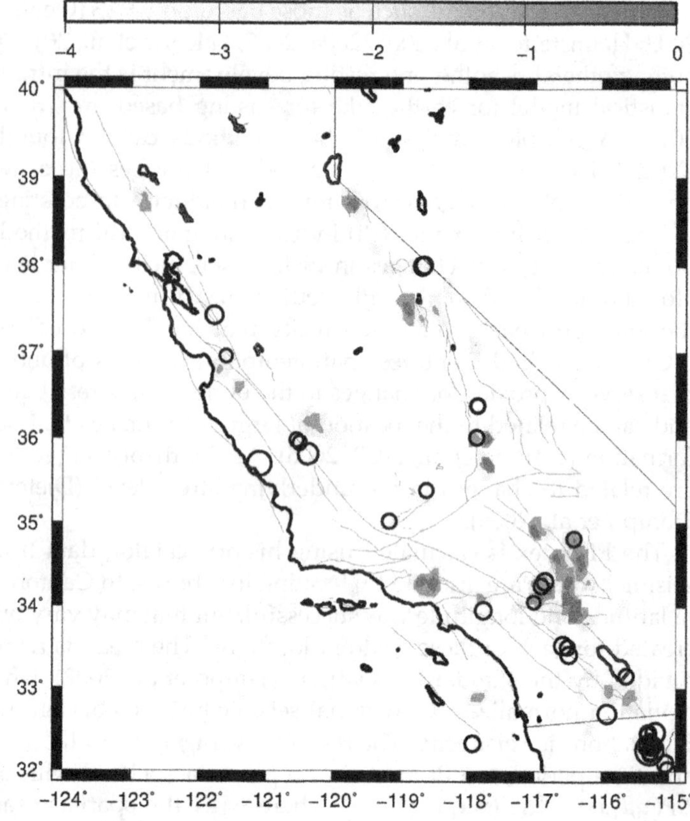

FIGURE 7.7 Normalized seismicity rate change for California, 2000–2010 inclusive, in each cell of 0.1 degrees by 0.1 degrees. *Color scale (gray in print versions)* is logarithmic, where the number represents the exponent. Black circles are events that occurred from 2000 to 2010, $M \geq 5$ (circle size also scales with magnitude).

with PI = 0 have zero probability. Note that those events without where PI = 0 provide important information on where locations are not likely to occur. There are 197 sites with PI > 0. Therefore, the probability of an event in each box is 0.0063 over 10 years. The resulting probability gain for intermediate-term forecasting using the PI method, α_{PI}, is 52.

Fig. 7.8 shows a PI forecast for $M \geq 5$ for eastern Canada. Again, we use historic seismicity data for the forecast, $M \geq 4$, 1902 through 2011, to forecast the next 11 years, 2012 through 2022 inclusive. Here we map the binary probabilities for grid boxes of 0.05 degrees by 0.05 degrees, where any location in red is equally likely to experience a large event.

There are 16,281 grid sites in the region shown in Fig. 7.8. The number of earthquakes of $M \geq 5$ for an 11-year period in eastern Canada is approximately 1.5 (Tiampo et al., 2010; Fereidoni, 2014). The probability of an event in each of the 16,281 boxes over those 11 years is 9.2×10^{-5}. However, the total number of boxes with PI ≥ 0 is 183, and the 11-year probability for each of those boxes is 0.0082, a probability gain for each box alone, α_{PI}, of ~89.

Seismicity is clustered in space, and we can estimate the probability gain for a particular smaller region, which can be useful as well. Fig. 7.9 shows the same PI forecast for Montreal alone. If we sum the total probability of an event over the regions shown, 45 degrees to 46 degrees latitude and −73 degrees to −74.5 degrees longitude, the probability of an $M \geq 5$ in the

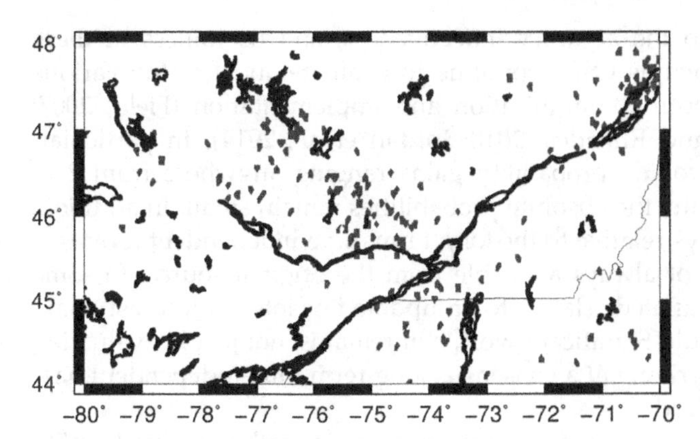

FIGURE 7.8 PI binary forecast for eastern Canada for the time period 2012 through 2022, inclusive, calculated on a grid of cell of 0.05 degrees by 0.05 degrees. The forecast threshold is set at zero, so that all boxes with a PI values > 0 are contoured as *red (gray in print versions)* alarm cells.

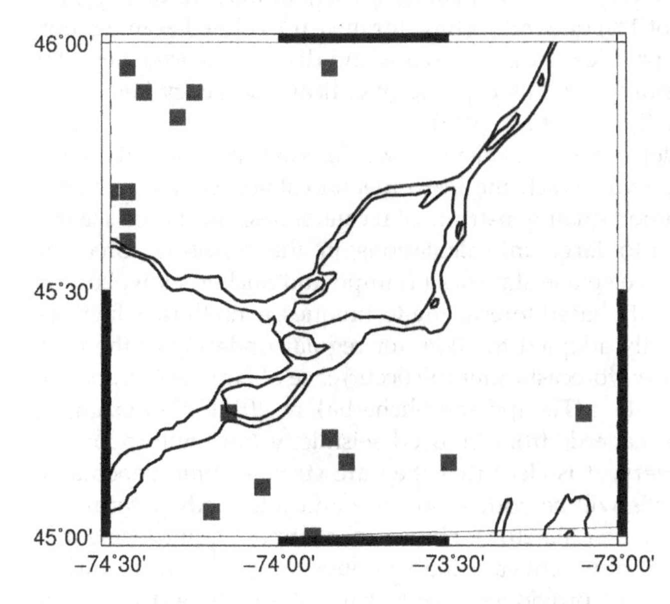

FIGURE 7.9 PI binary forecast for Montreal for the time period 2012 through 2022, inclusive, calculated on a grid of cell of 0.05 degrees by 0.05 degrees. The forecast threshold is set at zero, so that all boxes with a PI values > 0 are shown as *red (gray in print versions)* alarm cells.

next 11 years in the Montreal area is 0.14. The probability of a forecast using the average probability for the entire region is 0.055, or an α_{PI} of ~2.5.

CONCLUSIONS

Recent progress in the field of statistical seismology, in conjunction with the availability of large quantities of seismic data at smaller scales, has led to significant advances in the study of time-dependent earthquake processes. As a result, new developments in the field of intermediate- and short-term seismicity-based earthquake forecasting have resulted in implementation of OEF methods in real time.

Significant questions remain as to the accuracy, effectiveness, and usefulness of these methods. Although organizations such as CSEP continue to evaluate and test the various methods, issues remain regarding both communication and implementation (Field, 2007; Zechar et al., 2010; Gerstenberger and Rhoades, 2010; Jordan et al., 2014). In particular, although many of these methods provide probability gains ranging anywhere from 2 to 5000 over time periods of days to years, the absolute probabilities remain small. In addition, the probability gains are almost always relative to the long-term, time-independent forecasts, which can be complicated and are not always available from the original source. In some cases, those probability gains are calculated relative to an updated or interim forecast, making comparison and evaluation difficult. Significant work still remains, not just in evaluating the methods themselves but also in arriving at a consensus long-term time-independent hazard map for use as a universal standard.

However, forecasts that provide small but measurable increases in earthquake probability can still be useful. An increase from a very small probability (orders of magnitude less than 1%) to simply unlikely (on the order of 1%) does allow time for measures aimed at mitigation and readiness, including review and practice of disaster plans and drills, enhanced alerts for emergency personnel and facilities, storage and stockpiling of critical emergency items, and reevaluation of reinsurance coverage (Jordan et al., 2014).

It should be noted that all time-dependent seismicity-based forecasts are limited by the quality of the data and time periods over which the instrumental catalog contains reliable data, highlighting the importance of good quality instrumental catalogues, and dense seismic networks. These errors propagate into large miscalculations in the resulting forecasts, particularly for large events, which have sparse statistics (Tiampo and Shcherbakov, 2012b).

One important advantage of seismicity-based forecasting techniques, even those which are time independent, is that they are easily adapted to allow for regular updating of the forecasts as well as the provisions for revised forecasts after the occurrence of large events, potentially capturing the dynamics of the system (Tiampo and Shcherbakov, 2012b). For example, as we come to better understand the hazards from induced seismicity from anthropogenic processes such as hydrocarbon recovery, it is clear that they are strongly time dependent. Forecasts associated with these hazards will be both short term and intermediate term and will require frequent updating (Atkinson et al., 2015). However, continued improvement of time-dependent seismic hazard estimates presents a valuable opportunity for a wide variety of users, ranging from emergency hazard providers to financial institutions and insurance companies, to better assess and mitigate the associated risk. Future work must include studies into both more efficient means of updating these time-dependent forecasts and more effective methods of communicating their meaning and the appropriate response to public, private, and government entities.

Acknowledgments

The research of KFT was made possible by an NSERC Discovery Grant and the NSERC CSRN Strategic Network Grant and the ongoing support of ICLR. The research of RS was supported by the NSERC Discovery grant and by an NSERC CRD grant. We would like to thank Professor Dr. Friedemann Wenzel for his valuable review. Several images were plotted with the help of GMT software developed and supported by Paul Wessel and Walter H.F. Smith.

References

Agnew, D.A., Jones, L.M., 1991. Prediction probabilities from foreshocks. Journal of Geophysical Research 96, 11959–11971.

AonBenfield, 2011. Tohoku Earthquake and Tsunami Event Recap Report. Impact Forecasting.

Aster, R., 2012. Expecting the unexpected: black swans and seismology. Seismological Research Letters 83. http://dx.doi.org/10.1785/gssrl.83.1.5.

Atkinson, G.M., Ghofrani, H., Assatourians, K., 2015. Impact of induced seismicity on the evaluation of seismic hazard: some preliminary considerations. Seismological Research Letters 86, 1009–1021. http://dx.doi.org/10.1785/0220140204.

Bakun, W.H., King, G.C.P., Cockerham, R.S., 1986. Seismic slip, aseismic slip, and the mechanics of repeating earthquakes on the Calaveras fault, California. In: Das, S., Boatwright, J., Schlotz, C.H. (Eds.), Earthquake Source Mechanics, Geophysical Monograph Series, 37. AGU, Washington, D.C, pp. 195–208.

Bakun, W.H., Aagaard, B., Dost, B., Ellsworth, W.L., Hardebeck, J.L., Harris, R.A., Ji, C., Johnston, M.J.S., Langbein, J., Lienkaemper, J.J., Michael, A.J., Murray, J.R., Nadeau, R.M., Reasenberg, P.A., Reichle, M.S., Roeloffs, E.A., Shakal, A., Simpson, R.W., Waldhauser, F., 2005. Implications for prediction and hazard assessment from the 2004 Parkfield earthquake. Nature 437, 969–974.

Ben-Zion, Y., Lyakhovsky, V., 2002. Accelerated seismic release and related aspects of seismicity patterns on earthquake faults. Pure and Applied Geophysics 159, 2385–2412.

Bilham, R., 2009. The seismic future of cities. Bulletin of Earthquake Engineering. http://dx.doi.org/10.1007/s10518-009-9147-0.

Bowman, D.D., Ouillon, G., Sammis, C.G., Sornette, A., Sornette, D., 1998. An observational test of the critical earthquake concept. Journal of Geophysical Research 103, 24,359–24,372.

Brehm, D.J., Braile, L.W., 1998. Intermediate-term earthquake prediction using precursory events in the New Madrid Seismic Zone. Bulletin of Seismological Society of America 88, 564–580.

Brodsky, E.E., 2006. Long-range triggered earthquakes that continue after the wave train passes. Geophysical Research Letters 33, L15313. http://dx.doi.org/10.1029/2006GL026605.

Bufe, C.G., Varnes, D.J., 1993. Predictive modeling of the seismic cycle of the greater San Francisco Bay region. Journal of Geophysical Research 98, 9871–9883.

Cavallo, E., Powell, A., Becerra, O., 2010. Estimating the Direct Economic Damage of the Earthquake in Haiti. Inter-American Development Bank working paper series, No. IDB-WP-163.

CEPEC Preliminary Report to Cal-EMA, 2009. The M4.8 Bombay Beach Earthquake Swarm of March 2009. http://response.scec.org/node/75.

Christophersen, A., Smith, E.G., 2008. Foreshock rates from aftershock abundance. Bulletin of Seismological Society of America 98, 2133–2148. http://dx.doi.org/10.1785/0120060143.

Console, R., Murru, M., Lombardi, A.M., 2003. Refining earthquake clustering models. Journal of Geophysical Research 108, 2468. http://dx.doi.org/10.1029/2002JB002130.

Cornell, C.A., 1968. Engineering seismic risk analysis. Bulletin of Seismological Society of America 58 (5), 1583–1606.

Crampin, S., Gao, Y., 2010. Earthquakes can be stress-forecast. Geophysical Journal International 180, 1124–1127. http://dx.doi.org/10.1111/j.1365-246X.2009.04475.x.

Cutter, S.L., Emrich, C., 2005. Are natural hazards and disaster losses in the U.S. Increasing? EOS Transactions 86 (381), 388–389.

Deng, J.S., Sykes, L.R., 1996. Triggering of 1812 Santa Barbara earthquake by a great San Andreas shock: implications for future hazards in southern California. Geophysical Research Letters 23, 1155–1158.

Dieterich, J., 1994. A constitutive law for rate of earthquake production and its application to earthquake clustering. Journal of Geophysical Research 99, 2601–2618.

Dieterich, J.H., Cayol, V., Okubo, P., 2002. The use of earthquake rate changes as a stress meter at Kilauea volcano. Nature 408, 457–460.

Dodge, D.A., Beroza, G.C., Ellsworth, W.L., 1996. Detailed observations of California foreshock sequences: implications for the earthquake initiation process. Journal of Geophysical Research 101, 22,371–22,392.

Ellsworth, W.I., Cole, A.T., 1997. A test of the characteristic earthquake hypothesis for the San Andreas Fault in central California. Seismological Research Letters 68, 298.

Eneva, M., Ben-Zion, Y., 1997. Techniques and parameters to analyze seismicity patterns associated with large earthquakes. Journal of Geophysical Research 102, 17,785–17,795.

Falcone, G., Console, R., Murru, M., 2010. Short-term and long-term earthquake occurrence models for Italy: ETES, ERS and LTST. Annals of Geophysics 53. http://dx.doi.org/10.4401/ag-4760.

Fereidoni, A., 2014. Seismicity Processes in the Charlevoix Seismic Zone, Canada (Ph.D. thesis). University of Western Ontario.

Field, E.H., 2007. Overview of the working group for the development of regional earthquake likelihood models (RELM). Seismological Research Letters 78, 7–16.

Field, E.H., Arrowsmith, R.J., Biasi, G.P., Bird, P., Dawson, T.E., Felzer, K.R., Jackson, D.D., Johnson, K.M., Jordan, T.H., Madden, C., Michael, A.J., Milner, K.R., Page, M.T., Parsons, T., Powers, P.M., Shaw, B.E., Thatcher, W.R., Weldon II, R.J., Zeng, Y., 2014. Uniform California earthquake rupture forecast, version 3 (UCERF3) - the time-independent model. Bulletin of Seismological Society of America 104, 1122–1180. http://dx.doi.org/10.1785/0120130164.

Field, E.H., Arrowsmith, R.J., Biasi, G.P., Bird, P., Dawson, T.E., Felzer, K.R., Jackson, D.D., Johnson, K.M., Jordan, T.H., Madden, C., Michael, A.J., Milner, K.R., Page, M.T., Parsons, T., Powers, P.M., Shaw, B.E., Thatcher, W.R., Weldon II, R.J., Zeng, Y., 2015. Long-term time-dependent probabilities for the third uniform California earthquake rupture forecast (UCERF3). Bulletin of Seismological Society of America 105, 511–543. http://dx.doi.org/10.1785/0120140093.

Frankel, A., Mueller, C., Barnhard, T., Perkins, D., Leyendecker, E., Dickman, N., Hanson, S., Hopper, M., 1996. National Seismic Hazard Maps, June 1996. U.S. Geological Survey, Openfile Report 96–532.

Frohlich, C., 1987. Aftershocks and temporal clustering of deep earthquakes. Journal of Geophysical Research 92, 13,944–13,956.

Geller, R.J., Jackson, D.D., Kagan, Y.Y., Mulargia, F., 1997. Enhanced: earthquakes cannot be predicted. Science 275, 49–70.

Gerstenberger, M.C., Rhoades, D.A., 2010. New Zealand earthquake forecast testing centre. Pure and Applied Geophysics 167, 877–892.

Gerstenberger, M.C., Wiemer, S., Jones, L.M., Reasenberg, P.A., 2005. Real-time forecasts of tomorrow's earthquakes in California. Nature 435, 328–331.

Gerstenberger, M.C., Jones, L.M., Wiemer, S., 2007. Short-term aftershock probabilities: case studies in California. Seismological Research Letters 78, 66–77.

Gross, S.J., Kisslinger, C., 1994. Tests of models of aftershock rate decay. Bulletin of Seismological Society of America 84, 1571–1579.

Haberman, R.E., 1981. Precursory seismicity patterns: stalking the mature seismic gap. In: Simpson, D.W., Richards, P.G. (Eds.), Earthquake Prediction: An International Review, Maurice Ewing Series, vol. 4. AGU, Washington, DC, pp. 29–42.

Hall, S.S., 2011. Scientist on trial: at fault? Nature 477, 264–269.

Helmstetter, A., Sornette, D., 2002. Subcritical and supercritical regimes in epidemic models of earthquake aftershocks. Journal of Geophysical Research 107. http://dx.doi.org/10.1029/2001JB001580.

Helmstetter, A., Sornette, D., 2003a. Importance of direct and indirect triggered seismicity in the ETAS model of seismicity. Geophysical Research Letters 30, 1576. http://dx.doi.org/10.1029/2003GL017670.

Helmstetter, A., Sornette, D., 2003b. Predictability in the Epidemic-Type Aftershock Sequence model of interacting triggered seismicity. Journal of Geophysical Research 108, 2482. http://dx.doi.org/10.1029/2003JB002485.

Helmstetter, A., Kagan, Y.Y., Jackson, D.D., 2005. Importance of small earthquakes for stress transfers and earthquake triggering. Journal of Geophysical Research 110. http://dx.doi.org/10.1029/2004JB003286.

Helmstetter, A., Kagan, Y.Y., Jackson, D.D., 2006. Comparison of short-term and time-independent earthquake forecast models for southern California. Bulletin of Seismological Society of America 96, 90–106.

Helmstetter, A., Kagan, Y.Y., Jackson, D.D., 2007. High-resolution time-independent grid-based forecast for $m \geq 5$ earthquakes in California. Seismological Research Letters 78, 78–86.

Holliday, J.R., Rundle, J.B., Tiampo, K.F., Klein, W., Donnellan, A., 2006. Systematic procedural and sensitivity analysis of the Pattern Informatics method for forecasting large M>5 earthquake events in southern California. Pure and Applied Geophysics. http://dx.doi.org/10.1007/s00024-006-0131-1.

Holliday, J.R., Chen, C.-C., Tiampo, K.F., Rundle, J.B., Turcotte, D.L., Donnellan, A., 2007. A RELM earthquake forecast based on Pattern Informatics. Seismological Research Letters 78, 87–93.

Jackson, D.D., Kagan, Y.Y., 2006. The 2004 Parkfield earthquake, the 1985 prediction, and characteristic earthquakes: lessons for the future. Bulletin of Seismological Society of America 96. http://dx.doi.org/10.1785/012005082.

Jaumé, S.C., Sykes, L.R., 1999. Evolving towards a critical point: a review of accelerating seismic moment/energy release prior to large and great earthquakes. Pure and Applied Geophysics 155, 279–306.

Jones, L.M., Hauksson, E., 1997. The seismic cycle in southern California: precursor or response? Geophysical Research Letters 24, 469–472.

Jones, L.M., Molnar, P., 1979. Some characteristics of foreshocks and their possible relationship to earthquake prediction and premonitory slip on faults. Journal of Geophysical Research 84.

Jordan, T.H., 2006. Earthquake predictability, brick by brick. Seismological Research Letters 77, 3–6.

Jordan, T.H., 2014. The Prediction Problems of Earthquake System Science (GSA Presidential Medal Lecture).

Jordan, T.H., Jones, L.M., 2010. Operational earthquake forecasting: some thoughts on why and how. Seismological Research Letters 81.

Jordan, T.H., Chen, Y.-T., Gasparini, P., Madariaga, R., Main, I., Marzocchi, W., Papadopoulos, G., Sobolev, G., Yamaoka, K., Zschau, J., 2011. Operational earthquake forecasting, state of knowledge and guidelines for utilization, international commission on earthquake forecasting for civil protection final Report. Annals of Geophysics 54, 4. http://dx.doi.org/10.4401/ag-5350.

Jordan, T.H., Marzocchi, W., Michael, A.J., Gerstenberger, M.C., 2014. Operational earthquake forecasting can enhance earthquake preparedness. Seismological Research Letters 85, 955–959.

Kanamori, H., 1981. The nature of seismicity patterns before large earthquakes. In: Earthquake Prediction: An International Review. AGU Monograph. AGU, Washington, DC, pp. 1–19.

Keilis-Borok, V.I., Kossobokov, V.G., 1990. Times of increased probability of strong earthquakes M $\geq$ 7.5 diagnoes by algorithm M8 in Japan and adjacent territories. Journal of Geophysical Research 95, 12,413–12,422.

King, G.C.P., Stein, R.S., Lin, J., 1994. Static stress changes and the triggering of earthquakes. Bulletin of Seismological Society of America 84, 935–953.

Kossobokov, V.G., Romashkova, L.L., Keilis-Borok, V.I., 1999. Testing earthquake prediction algorithms: statistically significant advance prediction of the largest earthquakes in the Circum-Pacific, 1992–1997. Physics of the Earth and Planetary Interiors 111, 187–196.

Kovacs, P., 2010. Reducing the Risk of Earthquake Damage in Canada: Lessons from Haiti and Chile. ICLR Research Paper Series, 49.

Lombardi, A.M., Marzocchi, W., 2011. The double branching model for earthquake forecast applied to the Japanese seismicity. Earth Planets Space 63, 187–195.

Ma, L., Zhuang, J., 2001. Relative quiescence within the Jiashi swarm in Xinjiang, China: an application of the ETAS point process model. Journal of Applied Probability 38, 213–221.

Main, I., (moderator), 1999. Is the reliable prediction of individual earthquakes a realistic scientific goal? Debate in Nature, (moderator)www.nature.com/nature/debates/earthquake/equake_frameset.html.

Marzocchi, W., Lombardi, A.M., 2009. Real-time forecasting following a damaging earthquake. Geophysical Research Letters 36, L21302. http://dx.doi.org/10.1029/2009GL040233.

Marzocchi, W., Lombardi, A.M., Casarotti, E., 2014. The establishment of an operational earthquake forecasting system in Italy. Seismological Research Letters 85 (4), 961–969. http://dx.doi.org/10.1785/0220130219.

McGuire, R., 2004. Seismic Hazard and Risk Analysis. Earthquake Engineering Research Institute, Oakland, California.

McGuire, R., 2008. Probabilistic seismic hazard analysis: early history. Earthquake Engineering and Structural Dynamics 37, 329–338.

Michael, A.J., 2012. Fundamental questions of earthquake statistics, source behavior, and the estimation of earthquake probabilities. Bulletin of Seismological Society of America 102, 2547–2562.

Mignan, A., 2011. Retrospective on the accelerating seismic release (ASR) hypothesis: controversy and new horizons. Tectonophysics 505, 1–16.

Mignan, A., Di Giovambattista, R., 2008. Relationship between accelerating seismicity and quiescence, two precursors to large earthquakes. Geophysical Research Letters 35, L15306. http://dx.doi.org/10.1029/2008GL035024.

Munich Re, 2013. NatCatSERVICE, Geo Risks Research.

Mogi, K., 1969. Some features of recent seismic activity in and near Japan 2, Activity before and after large earthquakes. Bulletin of the Earthquake Research Institute, the University of Tokyo 47, 395–417.

Nanjo, K., Nagahama, H., Satomura, M., 1998. Rates of aftershock decay and the fractal structure of active fault systems. Tectonophysics 287, 173–186.

Ogata, Y., 1987. Long-term dependence of earthquake occurrences and statistical models for standard seismic activity in Japanese. In: Saito, M. (Ed.), Mathematical Seismology, vol. II. Inst. of Stat. Math., Tokyo, pp. 115–125.

Ogata, Y., 1988. Statistical models for earthquake occurrence and residual analysis for point process. Journal of the American Statistical Association 83, 9–27.

Ogata, Y., 1989. Statistical model for standard seismicity and detection of anomalies by residual analysis. Tectonophysics 169, 159–174.

Ogata, Y., 1999. Seismicity analysis through point-process modeling: a review. Pure and Applied Geophysics 155, 471–507.

Ogata, Y., 2005. Synchronous seismicity changes in and around the northern Japan preceding the 2003 Tokachi-oki earthquake of *M*8.0, 110, B08305. http://dx.doi.org/10.1029/2004JB003323.

Ogata, Y., Zhuang, J., 2006. Space-time ETAS models and an improved extension, 413, 13–23.

Rundle, J.B., Tiampo, K.F., Klein, W., Sá Martins, J., 2002. Self-organization in leaky threshold systems: the influence of near mean field dynamics & its implications for earthquakes, neurobiology and forecasting. Proceedings of the National Academy of Sciences of the United States of America 99 (Suppl. 1), 2463.

Scholz, C.H., 2002. The Mechanics of Earthquakes and Faulting, second ed. Cambridge Univ. Press, Cambridge.

Shcherbakov, R., 2014. Bayesian confidence intervals for the magnitude of the largest aftershock. Geophysical Research Letters 41, 6380–6388. http://dx.doi.org/10.1002/2014GL061272.

Steacy, S., Gerstenberger, M., Williams, C., Rhoades, D., Christopheresen, A., 2014. A new hybrid Coulomb/statistical model for forecasting aftershock rates. Geophysical Journal International 196, 918–923.

Stein, R.S., 1999. The role of stress transfer in earthquake occurrence. Nature 402, 605–609.

Swan, F.H., Schwartz, D.P., Cluff, L.S., 1980. Recurrence of moderate to large magnitude earthquakes produced by surface faulting on the Wasatch fault zone, Utah. Bulletin of Seismological Society of America 70, 1431–1462.

Tiampo, K.F., Shcherbakov, R., 2012a. Optimization of seismicity-based forecasts. Pure and Applied Geophysics. http://dx.doi.org/10.1007/s00024-012-0457-9.

Tiampo, K.F., Shcherbakov, R., 2012b. Seismicity-based earthquake forecasting techniques: ten years of progress. Tectonophysics. http://dx.doi.org/10.1016/j.tecto.2011.08.019.

Tiampo, K.F., Rundle, J.B., McGinnis, S., Gross, S., Klein, W., 2002. Mean-field threshold systems and phase dynamics: an application to earthquake fault systems. Europhysics Letters 60, 481–487.

Tiampo, K.F., Rundle, J.B., Klein, W., 2006. Premonitory seismicity changes prior to the Parkfield and Coalinga earthquakes in southern California. Tectonophysics 413, 77–78.

Tiampo, K.F., Klein, W., Li, H.-C., Mignan, A., Toya, Y., Kohen-Kadosh, S.L.Z., Rundle, J.B., Chen, C.-C., 2010. Ergodicity and earthquake catalogs: forecast testing and resulting implications. Pure and Applied Geophysics 167, 763. http://dx.doi.org/10.1007/s00024-010-0076-2.

Turcotte, D.L., 1991. Earthquake prediction. Annual Review of Earth and Planetary Sciences 19, 263–281.

United Nations Office for Disaster Risk Reduction, 2015. Sendai Framework for Disaster Risk Reduction 2015–2030. UNISDR, Geneva.

Utsu, T., 1961. A statistical study on the occurrence of aftershocks. Geophysical Magazine 30, 521–605.

Utsu, T., Ogata, Y., Matsu'ura, R.S., 1995. The centenary of the Omori formula for a decay law of aftershock activity. Journal of Physics of the Earth 43, 1–33.

Vere-Jones, D., 1995. Forecasting earthquakes and earthquake risk. International Journal of Forecasting 11, 503–538.

Vere-Jones, D., 2006. The development of statistical seismology: a personal experience. Tectonophysics 413, 5–12.

Woo, G., Marzocchi, W., 2014. Operational earthquake forecasting and decision-making. In: Wenzel, F., Zschau, J. (Eds.), Early Warning for Geological Disasters. Springer, Berlin, pp. 353–367. http://dx.doi.org/10.1007/978-3-642-12233-0_18.

Wood, H.O., Neumann, F., 1931. Modified Mercalli intensity scale of 1931. Bulletin of Seismological Society of America 21, 277–283.

Wyss, M., 1997. Cannot earthquakes be predicted? Science 278, 487–490.

Wyss, M., Shimaziki, K., Urabe, T., 1996. Quantitative mapping of a precursory seismic quiescence to the Izu-Oshima 1990 M6.5 earthquake, Japan. Geophysical Journal International 127, 735–743.

Yamashita, T., Knopoff, L., 1989. A model of foreshock occurrence. Geophysical Journal International 96, 389–399.

Zechar, J.D., Schorlemmer, D., Liukis, M., Yu, J., Euchner, F., Maechling, P., Jordan, J., 2010. The collaboratory for the Study of Earthquake Predictability perspective on computational earthquake science. Concurrency and Computation Practice and Experience 22, 1836–1847.

Zhuang, J., Ogata, Y., Vere-Jones, D., 2004. Analyzing earthquake clustering features by using stochastic reconstruction. Journal of Geophysical Research 109, B05301. http://dx.doi.org/10.1029/2003JB002879.

Zhuang, J., Chang, C., Ogata, Y., Chen, Y., 2005. A study on the background and clustering seismicity in the Taiwan region by using point process models. Journal of Geophysical Research 110, B05S18. http://dx.doi.org/10.1029/2004JB003157.

Further Reading

Swiss Re, 2015. Natural Catastrophes and Man-Made Disasters in 2014. Sigma, No. 2.

CHAPTER

8

Big Data Challenges and Hazards Modeling

Kristy F. Tiampo[1], Seth McGinnis[2], Yelena Kropivnitskaya[3], Jinhui Qin[3], Michael A. Bauer[3]

[1]Department of Geological Sciences and Cooperative Institute for Research in Environmental Sciences, University of Colorado, Boulder, CO, USA; [2]Institute for Mathematics Applied to Geosciences, National Center for Atmospheric Research, Boulder, CO, USA; [3]Department of Earth Sciences, Department of Computer Sciences, Western University, London, ON, Canada

INTRODUCTION

Over the past few decades, technological advancements have led to an avalanche of data from a variety of domains, including scientific sensors, health-care information, and financial, Internet, and supply chain system data. While the growing number of data sets is challenging our ability to analyze and understand those data, most qualitative definitions of these emerging data issues emphasize the difficulties associated with large volumes (Kouzes et al., 2009; Ma et al., 2015; Miller et al., 2011; Moore et al., 1998). For example, today we can generate vast databases of DNA and RNA sequencing data each week at a cost of less than US$5,000, a rate that also is matched by other technologies, such as real-time imaging and mass spectrometry (Schadt et al., 2010). Individual, user-generated data are adding to the volume daily because, as noted in the Harvard Business Review, "each of us is now a walking data generator" (McAfee and Brynjolfsson, 2012). The term used today to capture the essence of this trend is "big data." However, big data is not simply about large volume. The structure, quality, and end-user requirements of big data, including real-time analysis, require new and innovative techniques for acquisition, transmission, storage, and processing (Hu et al., 2014).

Big data is revolutionizing a wide variety of industries today. For example, the McKinsey Global Institute estimates incorporation of big data into decision-making in the US health-care system could result in 100 billion dollars in revenue and savings annually. In particular, effective use of big data will result in identification of new potential drug candidates

Risk Modeling for Hazards and Disasters
http://dx.doi.org/10.1016/B978-0-12-804071-3.00008-2

and more rapid development and approval of new medicines for pharmaceutical companies. For example, incorporation of the large volume and a variety of existing molecular and clinical data into complex models for advanced analysis could help identify new candidate drugs with an increased probability of successful development of safe and effective treatments (Cattell et al., 2013).

Two reports from the U.S. Department of Energy (DOE) have identified a new era of data-intensive computing (U.S. Department of Energy, 2004, 2013). Hilbert and López (2011) estimated that, as of 2007, the world's capacity to store information was 2.9×10^{20} optimally compressed bytes, while our ability to communicate those data was almost 2×10^{21} bytes. Finally, humankind was able to carry out 6.4×10^{18} instructions per second on general-purpose computers, which corresponds to the number of nerve impulses executed per second in the human brain. Between 1986 and 2007, the world's technological information processing capacities grew at exponential rates, while general-purpose computing capacity grew at a rate of 58% per year and globally stored information grew at 23% per year (Hilbert and López, 2011).

The more recent report from the U.S. Department of Energy (2013) identified four points critical to the success of data-intensive science. First, because data-intensive science relies on the collection, analysis, and management of big data, investments in exascale systems will be necessary to analyze those massive volumes of data. Second, integration of data analytics with exascale simulations is essential for scientific advancement. In the past, large-scale simulation or data assimilation efforts were followed by online data analyses and visualizations. Today's exascale simulations require that in-situ analysis and visualization occur while data are still resident in memory. Third, there is an urgent need to simplify analysis and visualization workflows for data-intensive science. Fourth, we need to increase the pool of computational scientists trained in big data and exascale computing to effectively advance the goals of today's data-intensive science.

In order for publicly funded science to provide credible, salient, and legitimate information for policy making, we need to better understand the opportunities for and constraints to scientific use of data and to identify factors that constrain or foster data usability, particularly as they relate to data acquisition, storage, and processing. Data usability is a function of the context of potential use and almost always involves iterative interaction between data producers and data users (Dilling and Lemos, 2011). Big data has an important role to play in that process and can thus be viewed as a boundary object. Boundary objects, a concept originally defined by Star and Griesemer (1989), are abstract or physical constructs that reside at the interface between two different social worlds, such as science and policy. Their primary function is to bridge differences between these communities and facilitate cooperation through mutual understanding (Karsten et al., 2001). One of the most common types of boundary object is information, often in the form of data repositories that are used in different ways by different users. Such repositories have a structure that is understandable in all worlds, adaptable to the local needs and constraints of different users, and robust enough to maintain a common identity (Guston, 2001; Star and Griesemer, 1989).

The next generation of remote earth observation sensors, coupled with those previously archived and currently operational, will collect a large number of data sets covering large geographical areas over many spatial scales and with time steps that range from minutes to hours to decades. As a result, satellite imagery is increasingly applied to the study of natural and anthropogenic hazards. This can include forecasting, assessment, and response for phenomena such as floods, wind, landslides, tornadoes, and earthquakes, as well as longer-term perils such as climate change. Applications include urban management, disaster response, adaptation, and mitigation. For example, disaster managers are beginning to integrate remote sensing information into characterizations of physical and social vulnerability for both planning and postdisaster response. In the period immediately following a disaster, humanitarian relief teams urgently require information to help them to assess the situation and plan relief operations. Providing rapid and appropriate access to timely and usable big data available from earth observation satellites is critical to effective communication between scientists and users (Gamba et al., 2011).

In this paper, we detail the challenges presented by big data for the hazard community today (Section Big Data). In Sections Lessons From Climate Science and Lessons from Solid Earth Science we present new approaches to solving two issues that will not be solved by improved hardware and storage capabilities. Finally, we conclude (Section Conclusions) with strategies and opportunities for the future.

BIG DATA

Manyika et al. (2011) defines big data as those data sets that are too large for the ability of standard database software to capture, store, manage, and analyze. The Large Hadron Collider at CERN, which generates around 15 petabytes of data every year, is an example of a big data endeavor that challenges the state of the art in computation, networking, and data storage.

However, "big data" is a popular term whose definition varies both within and between fields, and it is frequently used to refer to more than just very large volume. The National Institute of Standards and Technology defines big data as data sets having volume, acquisition velocity, or data representation that interferes with the use of traditional methods or architectures to analyze it (Cooper and Mell, 2012; Hu et al., 2014). For example, high-resolution global Earth observation era satellites produce several terabytes of remote sensing data per day, an order of magnitude less than CERN, but those data streams must be transmitted through downlink channels at a rate of only gigabits per second, presenting new challenges in transmission, storage, and analysis (Gamba et al., 2011; Ma et al., 2015). Data sets from clinical trials are even smaller, but present big data challenges because their significance creates a need to preserve, reuse, and recombine them with other data sets in a variety of different contexts and representations (Lynch, 2008; Marx, 2013).

Because data sets change in size, software capabilities advance over time, and "big" is defined relative to standard methods and architecture, the question of whether a given data set is big data is both contextual and time-dependent. Fundamentally, this question can only be answered by assessing the differentiating features of the data set in a given use context.

In 2001, Laney introduced the "3 V's" of big data, in which he defined the three main components of big data as volume, velocity, and variety (Laney, 2001). Since then, other characteristics have been added to the initial list of three. Depending on the particular emphasis, different authors have added value, veracity, variability, validity, viscosity, and volatility to the original list, expanding it to the "4 V's," "5 V's," or "7 V's" of big data, depending on the combination of interest (Biehn, 2013; Gantz and Reinsel, 2011; Khan et al., 2014; McNulty, 2014; Meijer, 2011; Vorhies, 2013; Zikopoulos and Eaton, 2011). The following is a brief discussion of each.

Volume refers to the massive amounts of data generated today, in a variety of different ways. In the 1970s and 1980s the world's data storage requirements went from megabytes to gigabytes; in the late 1980s we graduated to terabytes of data storage needed; and in the late 1990s we went from terabytes to petabytes. Today we are faced with storage needs of exabytes (Hu et al., 2014). For example, the European Bioinformatics Institute (EBI), one of the world's largest data repositories for biology data, currently stores 20 petabytes alone. The genomic data portion of that, ~2 petabytes, doubles every year (Marx, 2013). In the context of hazards research, the Earth Observing System Data and Information System (EOSDIS) provides end-to-end capabilities for managing NASA's Earth science data from various sources: satellites, aircraft, field measurements, and various other programs. The NASA network transports the data to the science operations facilities and generates and archives higher-level science data products for EOS missions. EOSDIS currently stores 8292 unique data collections distributed to more than 2 million users each year with a total archive volume greater than 9 petabytes, increasing at a rate of 6.4 terabytes per day (NASA, 2015). Note that while storage capability increases exponentially every year, data storage needs to grow in response; these data sets have come into existence because the capacity to store them made previously infeasible research possible.

Data velocity refers to the rate of data acquisition or usage, such as a stream of readings taken from a sensor (e.g., a seismometer, tweets, unmanned vehicle, or satellite) operating in real time. This stream of incoming data needs to be captured, stored, and analyzed. Timeliness (rate of capture or usage) and latency (lag time) can also be important, both for data acquisition and data usage. Another aspect of data velocity is the lifespan of the data, i.e., whether it remains valuable over time or loses its meaning and importance. The last important aspect of velocity is the speed with which data must be stored and retrieved. For example, when someone visits sophisticated content websites such as Amazon or Yahoo, specific ads have been selected based on the capture, storage, and analysis of current and past web visits and associated analytics. The associated architecture needs to support real-time analysis for thousands of visitors each minute. The same operational requirements exist for real-time hazard analysis (Vorhies, 2013).

Variety describes structural heterogeneities in big data collections. Data can come in different forms (textual, numeric, relational, graphical, geospatial, sensory), have different degrees of structure (unstructured text, semistructured email, semantically marked-up documents), have different formats (plain-text file, .csv, fixed-length fields, Excel spreadsheet, HTML table), come from different sources, come from different kinds of sources (human-generated, automated sensor logging, scientific instruments, simulations), and mean different things. Data representing the same information in different ways constitute another important form of variety and are often due to different sources, different intended uses, and incomplete or missing standards and specifications. The variety in big data presents two major challenges. First, there is a need to store and retrieve these different data types quickly, efficiently, and at a low cost. Second, it is necessary to align and integrate different representations to analyze the target correctly. Note that variety presents a unique challenge and opportunity. For many years there has been a significant amount of attention and effort focused on improving storage capabilities to handle large data volumes, but incorporation of both structured and unstructured data variety into management and analytics presents a new challenge for data integration and computation architecture (Vorhies, 2013).

Value is an important characteristic of big data in the business and tech sectors, where big data efforts frequently involve mining for insight in data sets collected for other purposes. In such cases, it is important to evaluate whether the results have a meaningful relationship to the motivating enterprise (this issue is also sometimes referred to as the "viability"). In contrast, in the scientific realm, data sets are created and stored for analysis because they are presumed to have value. What makes scientific data sets big is the fact that their "big V" captures value that would otherwise be inaccessible. Data with big volume capture value that is either rare (found in a much smaller and difficult-to-predict subset), or diffuse (found with statistical analysis that requires a very large input set to achieve useful results). Data with big velocity captures value found on very short timescales. Data with big variety captures value that comes from relationships between very different sources of data. Extracting the unique value in a big data set therefore requires addressing the other "V"s that it possesses, which can be a challenge. For example, a researcher working with a big climate model data set might be interested in studying some localized and comparatively rare phenomena such as hurricanes, atmospheric rivers, or ice storms. Simply finding these events within the data set can itself be a large research project. This aspect of big data is particularly relevant to hazards research because rarity is one of the things that make powerful events hazardous; powerful but common events are not hazardous because we have plenty of opportunities to adapt to them. Note also that big data frequently has latent value that will not become apparent until later. Given an ever-widening array of ways to use and analyze big data and the decreasing costs of storage, many data providers conclude that the potential benefits of having a big data set in the future outweigh the costs of collecting and storing it in the present (Biehn, 2013; Vorhies, 2013).

Veracity addresses the inherent trustworthiness of big data. It is important to understand the provenance of the data, the reliability of the original source, and the accuracy and completeness of the data. Concerns regarding ambiguity, inconsistency, and incompleteness can become major obstacles to forming reliable conclusions from big data. As a result, adequate and transparent quality control (QC), provenance tracking, and data management and governance practices are critical when handling big data (Vorhies, 2013). The importance of veracity depends critically on value and needs to be assessed accordingly.

Validity is related to veracity but concerns the accuracy and correctness of the data in the context of a particular usage. For example, the intensity of a storm could be gauged using satellite imagery or social media posts. Both data sets may have no veracity problems, but the former would be valid for evaluating the quality of a numeric weather prediction model where the latter would not, and vice versa for assessing the impact of the storm on the local populace.

Variability refers to data whose meaning is constantly changing. For example, this is the case for data related to language processing. Because words do not have static definitions, their meaning can vary wildly with context. Variability also can be introduced as a result of data availability, sampling processes, and changes in the characteristics of an instrument or other data source. It can be a difficult analytic problem to distinguish new and valuable signals in the data from artifacts because of variability (Vorhies, 2013; McNulty, 2014). While understanding variability is critical to relevant and worthwhile data analytics, it is also related to data value at some level.

Viscosity and volatility are both related to velocity. Data viscosity describes data velocity relative to the timescale of the event being studied or described (Vorhies, 2013). Volatility refers to the rate of data loss and the stable lifetime of the data. Scientific data often have a practically unlimited lifetime, but data coming from social media and business sources may vanish after some finite period of time if actions are not taken to preserve it. Reliable and credible analytics, including real-time hazards assessments, depend on data that are stable and consistent through time (Khan et al., 2014).

Of the characteristics described above, volume, velocity, and variety are the most important. While there is substantial overlap between them, each of the three primarily affects storage, transmission, and access of that data, respectively. The other V's are entrained with value and emerge in the context of a particular use or analysis of the data. Understanding and evaluating these characteristics is essential to the design of scalable big data systems and to advancing the methods and technologies used for storage, retrieval, and analysis of big data (Hu et al., 2014).

Big data analytics applications generally can be classified as either batch processing or streaming processing. In the traditional batch-processing paradigm, big data is first stored and analysis is done separately at a later time. This is most common when data generation is constrained (for example, because a computationally expensive model cannot be run on demand, or an instrument has limited availability) and the analysis is not known a priori. Batch data processing usually benefits significantly from parallelism. Batch processing is most effective when structured so that the most reductive steps happen early, to minimize the total amount of processing. Because computation costs less than storage and storage costs less than data transmission, it is also beneficial to push processing as far "upstream" as possible; many data providers are shifting toward a paradigm of "moving the computation to the data," and allowing data consumers to perform computation on the cyberinfrastructure that houses the data to minimize the amount of data that must be transmitted. This model is applied most often today in the fields of bioinformatics, web mining, and machine learning (Hu et al., 2014).

Streaming processing is most commonly applied when the greatest potential value of data depends on its timeliness. Here, data are analyzed as soon as possible to produce results as quickly as possible. It is most suitable for applications in which data are available as a live online stream and immediate processing is necessary. Necessarily, the analysis must be known beforehand or constructed on the fly. Because it arrives continuously with high velocity and volume, only a small portion of the data can be stored in local memory, and the analysis must be computationally inexpensive. Streaming processing can produce rapid results for a variety of applications, such as near-real-time hazard assessment (Hu et al., 2014).

Although volume is clearly one of the most common aspects of big data, solutions to handling large volume are primarily technological. In the next two sections, we provide two examples of how to deal with the variety and the velocity aspects of big data.

LESSONS FROM CLIMATE SCIENCE

Big data with a large variety component is especially difficult to work with because variety puts limits on automation. Automation depends on structural homogeneity, which is to say, on minimizing variety. When data variety cannot be reduced and heterogeneity must be retained, the analysis must be performed by a human, which in turn requires reducing any other "V"s (usually volume) in the data.

Variety is usually found in data sets that are cross-disciplinary or that in some way cross a boundary between data sources. Unified data sources have strong incentives and low costs for creating homogenous data, while independent data sources frequently have different data structures because they aim to support different uses of the data. Big data with high variety often results from joining data sets from multiple sources in support of "long tail" cross-domain uses and consequently is often encountered by smaller user communities.

In this section, we present strategies based on the experiences of the climate science community for both reducing big data variety and for retaining it; both approaches are important because perfection in either is unattainable.

Reducing Variety: Lessons From NARCCAP, CORDEX, and CMIP

Catastrophe losses and inadequate risk management can reduce profitability of insurance companies. It is important to assess potential losses from large events to ensure proper risk management and appropriate pricing. As a result, climate and weather models have become a critical element for organizations that deal with catastrophe risk (Dlugolecki et al., 2009). Even small changes, implemented for a large risk portfolio, can result in a change in price or tail risk that might make all the difference as to whether a company assumes or rejects that risk.

Because both climate models and weather forecasts have become routine inputs for hazard assessment and catastrophe modeling, it is important to both understand and appropriately integrate big data sets into those models. In particular, those risk models routinely used by insurance companies must be designed to handle the large volumes associated with big data.

Numeric weather prediction and climate models simulate atmospheric dynamics by dividing the atmosphere up into many small elements and calculating the fluxes between them based on physical laws. Storing the results at a usefully high spatial and temporal resolution generates big data with high volume. The data are even bigger when the results come from multiple models, as in the various Climate Model Intercomparison Projects (CMIPs, e.g., Taylor et al., 2013) used in the IPCC assessment reports, or the North American Regional Climate Change Assessment Program (NARCCAP, Mearns et al., 2009) and COordinated Regional climate model Downscaling EXperiment (CORDEX, Giorgi et al., 2009) projects, which provide high-resolution climate change scenario data for impacts.

The output from any individual model in such a project is typically fairly structurally homogeneous, but there can be significant differences between models, and this variety makes direct comparison of the results difficult. The climate science community reduces variety through the use of community-developed standards and formats.

One of the most important ways in which climate modelers standardize output is through the use of the NetCDF file format. NetCDF is a self-describing, machine-independent data format for array-oriented scientific data (Rew and Davis, 1990). The format's self-describing structure embeds metadata within the file, which enables the development of smart software that can make decisions about how to process a data file based on its contents. Embedded metadata is much harder to lose track of and also enables automated provenance tracking. The NetCDF file format is significantly enhanced by widespread adoption of the CF (Climate and Forecasting) metadata standard (Gregory, 2003), which standardizes the "use metadata" in NetCDF files. The CF standard provides rules and controlled vocabulary for structuring and recording the scientific meaning of content in NetCDF files, further increasing the feasibility of automated processing.

In addition to making use of CF-conformant NetCDF files for storage, model intercomparison projects have specification documents that standardize the experimental design of the simulations, the set of archived model outputs, and the organization of the archives where the data sets are stored. For the CMIP projects, most modelers were able to incorporate a standard software interface (CMOR, the Climate Model Output Rewriter) into their models to ensure these standards were met. This approach has proved less feasible in regional modeling projects such as NARCCAP and CORDEX, which must instead rely on stringent QC.

The NARCCAP Archive Specification (Gutowski and McGinnis, 2007) is based on the *IPCC Standard Output Contributed to the PCMDI Archive* (Taylor, 2005). Its first requirement is that all outputs be stored in NetCDF format and conform to version 1.0 or later of the CF Conventions metadata standard. CF conformance dictates to a large degree the structure of the file contents, as well as a number of accompanying metadata attributes describing the contents. The archive specification, or spec, also requires that each output variable be stored separately, on the model's native simulation grid, in files spanning at most 5 years and ending at 00 UTC on January 1 of a year ending in 5 or 0. This last requirement was necessary to stay below the 2-GB file size limit inherent to older, 32-bit versions of the NetCDF library that were still prevalent at the time. We found that splitting the data in this way created the opportunity for errors to occur because there could be gaps or overlaps between segments or metadata could be inconsistent from one segment to the next. For this reason, we recommend a principle of maximizing atomicity by aiming for a one-to-one correspondence between files and model output fields from a single simulation when constructing archives.

The NARCCAP spec requires that dimensions of array data be ordered time, level, latitude, longitude, and that the data be stored from north to south and west to east. It specifies the numeric precision of different quantities and the values used to represent missing data. It also requires or recommends a number of attributes containing metadata about the model configuration, the driving boundary conditions, the individuals or group that performed the simulation, and information about the modification history of the data.

The Output Requirements portion of the NARCCAP spec contains a list of output fields to be included in the archive and specifies for each the name of the variable, the units it should be recorded in, the "long_name" attribute (a human-readable string frequently used by plotting software), and the "standard_name" attribute, which comes from a controlled vocabulary associated with the CF standard and identifies the geophysical quantity represented by the data. The spec also lists the frequency of reporting for each variable, how the associated temporal coordinates should be stored (e.g., when the "day" begins and ends for daily average values), whether a variable should be recorded as an instantaneous or time-averaged value, the sign convention for directional quantities such as wind speeds and radiation fluxes, and, where applicable, the height above the surface. The spec prescribes an algorithm for constructing filenames from the variable stored in the file, its temporal coverage, and standardized names for the models generating the data.

In NARCCAP, modeling groups from different institutions performed simulations, postprocessed the raw model output from those simulations according to the spec documents, and submitted the results to a central archive at NCAR, the National Center for Atmospheric Research. The data management group at NCAR performed QC on the submitted data sets, added them to the archive, and published them through the Earth System Grid data portal website. We designed the QC process to ensure that files published through the data portal adhered to the specifications and contained model output for the variables in question that was free of gross errors. To ensure that the files met the CF standard, we used the stand-alone CF Checker software tool developed by Rosalyn Hatcher. We checked the metadata prescribed by the NARCCAP spec by a semiautomated process of checking one exemplar from each group of files by hand, and then checking that any differences from the exemplar followed an understandable and acceptable pattern. We used custom scripts developed in the NCL programming language to check the range, continuity, and monotonicity of the time coordinate and the range of the data values. We then found the mean, maximum, and minimum of the data in each file over time and space and plotted the results. We also plotted transects of the data along each dimension. While it is difficult to know a priori what form data errors will take, these visualizations typically make any errors stand out plainly to the human eye, at which point further investigation can determine the nature of the error and what steps must be taken to correct it. In many cases, the error becomes obvious when the visualization script fails to plot the data correctly, or at all.

When we found correctable errors involving metadata, ancillary data (i.e., the coordinates or dimensions), or file structure, we were usually able to correct the files using the NCO toolkit, a collection of command-line utilities for manipulation of NetCDF files. In many cases, we were also able to resolve problems with incorrect units this way as well. When we found errors involving incorrect or missing data values, we would have to contact the modelers and request that they regenerate and resubmit the data files. In some cases, doing so proved impossible, and then we either excluded that data from the archive (if the problem rendered a large fraction of the data set unusable) or made note of the issue on the project website to inform users about known problems.

Other lessons learned from experience with these data sets include using programmatic naming conventions to ensure that all data set elements can be referenced uniquely and versioning files and data sets to handle changes over time. These standardization techniques should be applied as early as possible by publishing standards and guidelines for data generation and pushing them upstream in the data management pipeline to the data producers (McGinnis, 2014). All these efforts aim to reduce variety in the output data sets as much as possible, easing the use of these data sets in scientific and impacts analysis and enabling the use of smart tools such as GIS systems and the Climate Data Operators (CDO) and NetCDF Operators (NCO) toolkits.

Retaining Variety: Lessons From the Advanced Cooperative Arctic Data Information Service

The Advanced Cooperative Arctic Data Information Service (ACADIS) is a data archive for all arctic science projects funded by the National Science Foundation's Arctic System Science Program. Its purpose is to provide a common repository and portal for scientific data relating to the Arctic in support of research into the relationships between physical, chemical, biological, and human processes in the arctic system. It houses nearly 3500 distinct data sets comprising nearly half a million individual files totaling almost 5 TB. Because of their deeply interdisciplinary nature, these data are deeply heterogeneous and include images, audio recordings, text, spreadsheets, numeric data, and many other types of information (UCAR, 2015).

Because Arctic data is used as an input to the larger climate models that scientists use to understand and project current and future climate, it has become increasingly important to the field of climate modeling (Lawrence and Swenson, 2011). Climate models and weather forecasts have become routine inputs for hazard assessment and catastrophe modeling, making it important to both understand and appropriately integrate these data, with their intrinsic variability, into those models. The ACADIS archive also serves as an exemplar of the issues associated with integrating multidisciplinary data into hazard assessment analyses.

Variety in the ACADIS archive can only be reduced by a small amount using the techniques described in the previous section, so the strategy for maximizing the utility of the archive is to maximize the discoverability of the data it houses.

Essential to good discoverability is a basic "discovery metadata" profile, which is a reasonably short list of metadata elements that are common to everything in the archive. The ACADIS metadata profile was based on a profile evolved from the Dublin Core metadata element set for International Polar Year scientific activities; a newer metadata profile would likely make use of the ISO 19115 schema for geographic information. The ACADIS profile includes a controlled vocabulary for instruments and scientific keywords. Note that good metadata profiles need not be comprehensive; one that covers the bulk (say, 80%) of the data in the archive well will make a very large improvement in the discoverability of data. It is also essential to document the profile well.

The ACADIS Data Provider's Guide (UCAR, 2013) requires only 13 pieces of information, which comprise the common metadata profile. Free-form pieces of metadata include a title for the data set, a short (one-word) name, and a brief human-readable description of its contents. Names and contact information are recorded for the people or groups who created the data (the authors), who distribute the data, and who are responsible for the metadata record. Selections from controlled vocabularies are recorded for the locations, instruments, instrument platforms, science keywords (from the NASA Global Change Master Directory terms), ISO Topics, and file formats of the data. Finally, each data set is associated with the NSF Award that funded its collection.

The Data Provider's Guide also recommends a number of additional pieces of metadata that should be provided if possible, but which may not be meaningful for every data set submitted to the archive. These include information about the spatial and temporal coverage of the data; spatial sampling type, spatial resolution, and reporting frequency of the data; progress status of data collection; related resources (data, documentation, and/or publications); and the language used in the data set. If data sets have access restrictions or use constraints, they should be indicated in the metadata as well. Finally, because the archive collects metadata for data sets housed externally, links to external data access can be provided.

Collecting high-quality documentation and metadata beyond the basic profile for data in an archive is valuable but difficult. Data producers in the cross-disciplinary long tail are usually a very different group from data consumers. Producers know all the information that goes into good documentation, but do not usually need it, while consumers have the best motive to create the documentation, but do not know the information. This mismatch is, generally speaking, an unsolved problem. Attempts are often made to solve it by declaring that documentation is required, but this is rarely successful: unfunded mandates are hard to fulfill. The best solution is to provide producers with strong incentives to produce good documentation, although this can be difficult to achieve in practice. Financial incentives may be a good approach if the data consumers or archive developers have a budget for them. Ideas exist for crowd-sourcing data standardization and creation of discovery metadata, but this is an area fraught with peril regarding quality, trust, and competence issues.

Note also that data providers will share their data only if their own needs are met in doing so. Academics need to be given credit for their efforts in producing data, so a system that makes data citations and DOIs (digital object identifiers) readily accessible to data consumers can improve the incentives for academics to deposit data in the archive. The archive may also need to allow for an embargo period after the initial deposit to allow researchers to publish first using their own data. Nonacademic data providers such as private businesses will have an entirely different set of needs that must be met.

A big data set with a high variety will require a significant amount of curation, and human effort is still superior to automation for that purpose. This is reflective of a broader point that a successful data archive is a part of a data supply chain. It is a middle man who should add value to the transfer of data, not an unattended self-service dumping point. The value added to big data with retained variety is in providing systems that enable data consumers to find useful data despite the variety.

LESSONS FROM SOLID EARTH SCIENCE

Big data with high velocity and large volume presents a challenge in the context of the need to respond rapidly and effectively to natural disasters. Real-time and near real-time products generated from a wide variety of remote sensing images can provide invaluable information to emergency hazard responders. However, generating these products depends not only on handling their volume and velocity, but preserving value as well.

Seismic hazard maps have many practical and important applications and are widely used by policymakers to assign earthquake-resistant construction standards, by insurance companies to set insurance rates, by civil engineers to estimate structural stability, and by emergency hazard agencies to estimate hazard potential and the associated response. In addition, in an increasingly competitive market, insurance companies are seeking improved insights into short-term changes in hazard and risk. Improvements in seismic hazard and ground-shaking maps can provide better estimates of potential damage from a large event. Here we provide an example of how innovative streaming techniques can both automate and improve those hazard maps in near real-time, resulting in more rapid deployment and efficient allocation of hazard responders and associated resources.

Today, probabilistic seismic hazard analysis (PSHA) (Cornell, 1968; McGuire, 2004, 2008) is the most widely used technique for estimating seismic hazard, particularly seismic hazard map calculations. Seismic hazard maps provide the probability of a ground motion acceleration value occurring at any given site. This hazard is estimated by integrating conditional probabilities over all possible distance and magnitude values.

The spatial resolution, or grid density, of hazard maps is one of the most important technical issues for this field (Musson and Henni, 2001). If the grid is too coarse, there is a loss of detail for smaller source zones, which could result in an underestimation of the maximum ground motion level value (Musson and Henni, 2001). However, computation on a finer grid requires considerably more computational resources, particularly for a complex model. For example, calculation at more than 6.5 trillion locations is required to obtain a hazard map for eastern Canada with a resolution of 1 m. In addition, the Monte Carlo simulations needed to handle the temporal variations associated with aftershocks require significant computation (Musson, 1999). Recent technological breakthroughs in high performance computing (HPC) have improved access to the needed computational power, allowing complex problems to be partitioned into smaller tasks, programmed into pipelines, and deployed on large computer clusters to deal with big data velocity issues.

IBM InfoSphere Streams (IBM, 2014) is an advanced streaming computing platform designed for parallelism and deployment across computing clusters, enabling continuous analysis of big data with high speed and low latency. Here, PSHA mapping programs are run in parallel on large-scale HPC clusters, producing better and faster real-time seismic hazard analysis through distributed computing networks. The result is an innovative computational technique for PSHA mapping by integrating different input data sources and existing processing tools into a streaming and pipelined computing application. We produce high-resolution maps of the probability of exceeding certain ground motion levels across eastern Canada. We demonstrate that the use of the pipelining and streaming techniques provided makes possible the production of high-resolution hazard maps in near real time with virtually no resolution limits. This approach is flexible and can be used in any region where seismic sources and ground motion characteristics are available.

Monte Carlo Simulation for Probabilistic Seismic Hazard Analysis Mapping

One of the most straightforward and flexible PSHA methodologies uses Monte Carlo simulation. Synthetic earthquake catalogs are generated from the historical earthquake record in the region. This synthetic catalog is then used to estimate the distribution of ground motions at a number of sites using selected ground motion prediction equations (GMPEs) (Musson, 1999). Finally, the probability of exceeding a certain ground motion level at every point across a region is estimated. The resulting maps take into account uncertainties in earthquake location, size, and ground motions.

The Monte Carlo PSHA tool we use is the open-source EqHaz software suite of programs developed by Assatourians and Atkinson (2013). This suite consists of three programs. EqHaz1 creates synthetic earthquake catalogs with user-specified seismicity parameters. EqHaz2 estimates the ground motion, mean hazard probability curves, and mean hazard motions at specified return periods for a grid of points. EqHaz3 deaggregates the hazard, producing the relative contributions of the different earthquake sources as a function of distance and magnitude. The EqHaz suite has a number of limitations; most significantly for this application, it cannot be applied to hazard calculations with more than 1,000,000 records or more than 100,000 grid points in a synthetic catalog. These limits present barriers to the production of high-resolution hazard maps.

Implementation of Pipelined Probabilistic Seismic Hazard Analysis in Streams

IBM's Streams software package provides a runtime platform, programming model, and associated toolkits for streaming computation. Its programming language, Stream Processing Language (SPL) is used to build stream processing applications in a data-centric programming model (IBM, 2014).

The building blocks of Streams are data streams and operators. Data streams consist of user-defined packets of data. Processing routines are implemented as operators that are connected together into flexible, scalable processing pipelines. Operators and streams are assembled into applications via a data-flow graph that defines the connections between data sources (inputs), operators, and data sinks (outputs).

Streams deployment is supported by an underlying runtime system consisting of distributed processes. Single or multiple operators form a processing element (PE). The Streams compiler and runtime services determine where and how to best deploy PEs across the runtime system to meet the resource requirements (Bauer et al., 2010). The Streams programming interface also has a set of built-in operator toolkits that can be employed to speed up development work (IBM, 2014).

To implement PSHA mapping in Streams, the EqHaz1 and EqHaz2 source codes were modified and compiled into shared system object libraries and linked into Streams applications. Two primitive operators were implemented for this PSHA hazard map calculation: a Catalog operator, which generates a synthetic catalog using the EqHaz1 procedures, and a Map operator, which generates a map using the EqHaz2 procedures. The calculation pipeline is shown in Fig. 8.1 and is divided into two sections. The first (upper) section shows the process required to gather input parameters. The second (lower) section contains PEs performing PSHA procedures and outputting results (Kropivnitskaya et al., 2015). Connections between the two sections are indicated by filled circles.

This Streams hazard map application generates synthetic catalogs using the operator developed from the EqHaz1 program and produces GMPEs based on the EqHaz2 program operator. Subsequently, the JSON map file produces spectral acceleration for a given period of time at some level of probability at each point of the grid. These outputs are combined with the map coordinates for the grid by a specially designed JAVA script to produce a map that is displayed on the web page (Kropivnitskaya et al., 2015).

High-resolution PSHA maps require heavy processing in the catalog generation step because of the Monte Carlo simulations and in the map generation step because of calculations on a dense grid. Therefore, either the former or latter or both are bottlenecks to the entire process of generating high-resolution PSHA maps. To overcome these limitations and reduce the application execution time, the workload was decomposed so that these two components could be split into multiple parallel pipelines. As a result, each PE executes on a single core. This implementation increases the number of records in the synthetic catalog to 10 million and the number of sites in the hazard map to 2.5 million, allowing for the production of high-resolution maps while significantly decreasing the execution time (Kropivnitskaya et al., 2015).

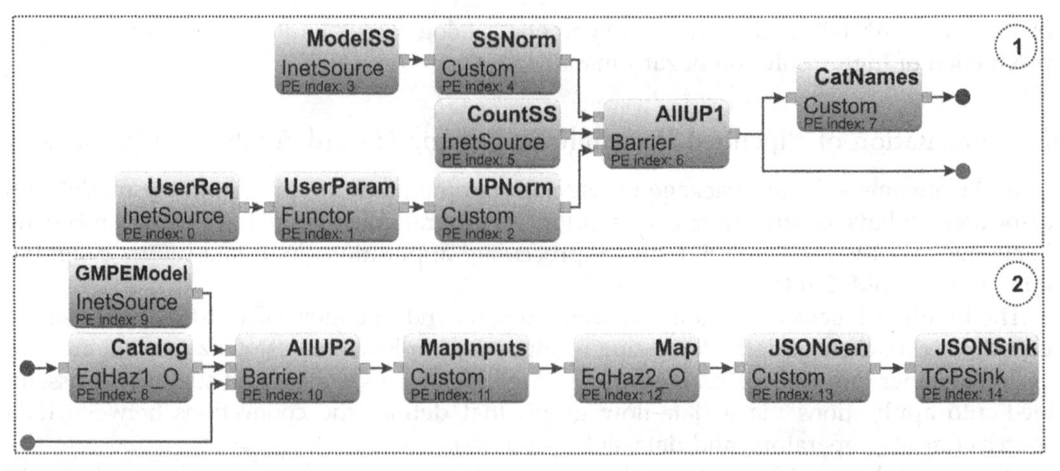

FIGURE 8.1 Pipelined probabilistic seismic hazard analysis mapping application graph (captured from Streams Studio) (Kropivnitskaya et al., 2015).

Experimentation

The experimental environment consists of a cluster of four machines running Linux, each with dual Xeon quad-core 2.4 GHz CPUs and 16 GB RAM. The cluster has 32 cores in total, allowing at least 32 processes to be run in parallel using InfoSphere Streams Version 3.2. To demonstrate the performance improvement by running multiple pipelines in parallel for PSHA mapping, we executed the same processing workload sequentially and in parallel and measured the execution time for both implementations. For the parallel experiment, 10 pipelines were employed at the catalog generation stage and 25 pipelines were employed for map generation. The number of records in the synthetic catalog was limited to 1 million and the number of grid points to 100,000 because those are the maximum sizes for the sequential program. The timing results are summarized in Table 8.1 (Kropivnitskaya et al., 2015).

The total speedup obtained with the pipelined implementation and parallel execution is a factor of 27.3. Note that the performance increase is obtained not by improving the execution of the primitive operators, but by simply subdividing the input streams and pipelining the primitive operators.

TABLE 8.1 Results of Parallel Pipelining in Probabilistic Seismic Hazard Analysis Mapping

	Sequential Processing (s)	Parallel Processing (s)	Speedup Factor
Catalog (1 M records)	5.68	0.43	13.2
Map (100 k grid points)	3780.3	138.2	27.4
Total	3785.98	138.63	27.3

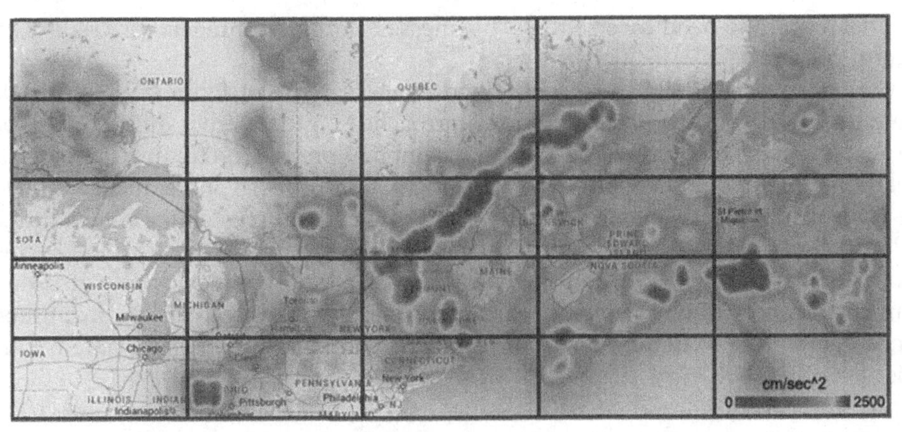

FIGURE 8.2 Mean hazard map for eastern Canada with a 2475-year return period for pseudoacceleration at T = 0.1 s period. The dynamic heatmap.js JavaScript library is used for visualization purposes (Wied, 2014; Kropivnitskaya et al., 2015).

An example of a final output map is shown in Fig. 8.2. Red lines show the spatial decomposition of the map grid for processing. Each red rectangle contains 100,000 points; the map contains 2.5 million points in total and a map resolution of 1 km.

The total execution time for the entire procedure is 5.6 min, a significant improvement in PSHA map generation. Assatourians and Atkinson (2013) produced hazard maps with a resolution of 3326 points using the EqHaz suite. The PSHA procedure execution time for that case was ~90 min on one dual quad-core 2.4 GHz CPU with 8 GB of RAM running Windows 7 Professional 64-bit (Kropivnitskaya et al., 2015).

Although the volume of this big data is not as large as some remote sensing data sets, the velocity requirements present a particular challenge. Here the value of these data has been enhanced by using parallelism to cope with velocity. In the future, additional data sets can be incorporated into this advanced computing environment, expanding the volume and variety of those sources and thereby increasing the value of the associated tools and products. This example demonstrates that scaling up calculations using advanced software and computational architecture to deal with big data velocity and volume can be relatively straightforward if the associated calculations can be segmented into independent operations that are easily parallelizable. This is often the case in practice, and many problems of significant real-world value are "embarrassingly parallel," requiring very little effort to separate into parallel tasks with little or no interdependency.

CONCLUSIONS

While big data presents important challenges today in the field of hazards modeling and response, innovative technological advancements can provide cost-effective solutions. Although volume often is considered to be the most important feature of big data, here we show that three other V's—velocity, variety, and value—are at least as significant when considering the need by scientists, businesses, and governments for credible, timely, and useful products.

We present two methods designed to handle one or more of those 4 V's of big data for climate and seismic hazard analysis. Each approaches big data from different perspectives and provides effective solutions based on a combination of structural, technological, and computational methodologies. Effective implementation of these and similar advancements will be essential to effective communication and interaction between the relevant organizations that collect and analyze big data in the natural hazards arena today. The resulting improvements in modeling of natural hazards will allow a wide variety of users, from government planners to emergency hazard response providers to financial institutions and insurance companies, to better assess and mitigate the risk and cost associated with these hazards.

Acknowledgments

The research of KFT and YK was made possible by a MITACS Accelerate grant and an NSERC Discovery Grant and is the result of collaborations between the Western University Computational Laboratory for Fault System Modeling, Analysis, and Data Assimilation and the consortium of Canadian academic institutions, a high performance computing network SHARCNET. We are grateful to Dr. K. Assatourians for his assistance with EqHaz. The system object libraries are compiled using the Intel Fortran Compiler, providing an opportunity to adapt the EqHaz FORTRAN code with no major modifications. Special thanks to Eric Nienhouse for sharing his experiences with the ACADIS program.

References

Assatourians, K., Atkinson, G.M., 2013. EqHaz: an open-source probabilistic seismic-hazard code based on the Monte Carlo simulation approach. Seismological Research Letters 84 (3), 516–524.

Bauer, M.A., Biem, A., McIntyre, S., Xie, Y., 2010. A pipelining implementation for parsing x-ray diffraction source data and removing the background noise. Journal of Physics Conference Series 256, 012017.

Biehn, N., 2013. The Missing V's in Big Data: Viability and Value. Wired (online). www.wired.com/insights/2013/05/the-missing-vs-in-big-data-viability-and-value/.

Cattell, J., Chilukuri, S., Levy, M., 2013. How Big Data Can Revolutionize Pharmaceutical R&D. Whitepaper, McKinsey & Co.

Cooper, M., Mell, P., 2012. Tackling Big Data (online). csrc.nist.gov/groups/SMA/forum/documents/june2012presentations/f%csm_june2012_cooper_mell.pdf.

Cornell, C.A., 1968. Engineering seismic risk analysis. Bulletin of the Seismological Society of America 58 (5), 1583–1606.

Dilling, L., Lemos, M.C., 2011. Creating usable science: opportunities and constraints for climate knowledge use and their implications for science policy. Global Environmental Change 21, 680–689.

Dlugolecki, A., Bermingham, K., Crerar, C., 2009. Coping with Climate Change: Risks and Opportunities for Insurers. Chartered Insurance Institute, London.

Gamba, P., Du, P., Juergens, C., Maktav, D., 2011. Foreword to the special issue on "Human Settlements: A Global Remote Sensing Challenge". IEEE Selected Topics in Applied Earth Observations and Remote Sensing 4 (1).

Gantz, J., Reinsel, D., 2011. Extracting value from chaos. IDC IView 1–12.

Gregory, J., 2003. The CF metadata standard. CLIVAR Exchanges 8 (4).

Giorgi, F., Jones, C., Asrar, G.R., 2009. Addressing climate information needs at the regional level: the CORDEX framework. World Meteorological Organization Bulletin 58 (3), 175.

Guston, D., 2001. Boundary organizations in environmental policy and science: an introduction. Science, Technology, and Human Values 26 (4), 399–408.

Gutowski, W., McGinnis, S., 2007. Requirements for Standard Output Contributed to the NARCCAP Archive. North American Regional Climate Change Assessment Program (online), (updated 2010). narccap.ucar.edu/data/output_requirements.html and narccap.ucar.edu/data/output_archive.html.

Hilbert, M., López, P., 2011. The World's technological capacity to store, communicate, and compute information. Science 332.

Hu, H., Wen, Y., Chua, T.-S., Li, X., 2014. Toward scalable systems for big data analytics: a technology tutorial. IEEE Access 2.

IBM, (2014). IBM Knowledge Center (online). (www-01.ibm.com/support/knowledgecenter/).

Karsten, H., Lyytinen, K., Hurskainen, M., Koskelainen, T., 2001. Crossing boundaries and conscripting participation: representing and integrating knowledge in a paper machinery project. European Journal of Information Systems 10 (2), 89–98.

Khan, M.A., Fahim Uddin, M., Gupta, N., 2014. Seven V's of big data, understanding big data to extract value. In: Proceedings of 2014 Zone 1 Conference of the American Society for Engineering Education (ASEE Zone 1).

Kouzes, R.T., Anderson, G.A., Elbert, S.T., Gorton, I., Gracio, D.K., 2009. The changing paradigm of data-intensive computing. Computer 42 (1), 26–34.

Kropivnitskaya, Y., Qin, J., Tiampo, K.F., Bauer, M.A., 2015. A pipelining implementation for high resolution seismic hazard maps production. Procedia Computer Science 51. http://dx.doi.org/10.1016/j.procs.2015.05.337.

Laney, D., 2001. 3-D Data Management: Controlling Data Volume, Velocity and Variety. Application Delivery Strategies (online). blogs.gartner.com/doug-laney/files/2012/01/ad949–3D-Data-Management-Controlling-Data-Volume-Velocity-and-Variety.pdf.

Lawrence, D.M., Swenson, S., 2011. Permafrost response to increasing Arctic shrub abundance depends on the relative influence of shrubs on local soil cooling versus large-scale climate warming. Environmental Research Letters 6, 045504. http://dx.doi.org/10.1088/1748-9326/6/4/045504.

Lynch, C., 2008. How do your data grow? Nature Commentary 455, 28–29.

Ma, Y., Wang, L., Liu, P., Ranjan, R., 2015. Towards building a data-intensive index for big data computing - a case study of remote sensing data processing. Information Sciences 319, 171–188.

Manyika, J., Chui, M., Brown, B., Bughin, J., Dobbs, R., Roxburgh, C., Byers, A.H., 2011. Big Data: The Next Frontier for Innovation, Competition, and Productivity. McKinsey Global Institute, San Francisco, CA, USA, pp. 1–137.

Marx, V., 2013. Biology: the big challenges of big data. Nature 498, 255–260.

McAfee, A., Brynjolfsson, E., October 2012. Big data: the management revolution. Harvard Business Review 90 (10), 60–128.

McGinnis, S., 2014. Lessons Learned from NARCCAP about Data Archiving and Quality Control. North American Regional Climate Change Assessment Program (online). narccap.ucar.edu/contrib/lessons-learned.html.

McGuire, R., 2004. Seismic Hazard and Risk Analysis. Earthquake Engineering Research Institute, Oakland, California.

McGuire, R., 2008. Probabilistic seismic hazard analysis: early history. Earthquake Engineering and Structural Dynamics 37, 329–338.

McNulty, E., 2014. Understanding Big Data: The Seven V's. Dataconomy (online). dataconomy.com/seven-vs-big-data/.

Mearns, L.O., Gutowski, W.J., Jones, R., Leung, L.-Y., McGinnis, S., Nunes, A.M.B., Qian, Y., 2009. A regional climate change assessment program for North America. EOS 90 (36), 311–312.

Meijer, E., 2011. The world according to LINQ. Communications of the ACM 54 (10), 45–51.

Miller, S.D., Kleese van Dam, K., Li, D., 2011. Challenges in Data Intensive Analysis at Scientific Experimental User Facilities. Springer, Berlin, pp. 249–284.

Moore, R., Prince, T.A., Ellisman, M., 1998. Data-intensive computing and digital libraries. Communications of the ACM 41 (11), 56–62.

Musson, R.M.W., 1999. Determination of design earthquakes in seismic hazard analysis through Monte Carlo simulation. Journal of Earthquake Engineering 3, 463–474.

Musson, R.M.W., Henni, P.H.O., 2001. Methodological considerations of probabilistic seismic hazard mapping. Soil Dynamics and Earthquake Engineering 21, 385–403.

NASA, 2015. An Overview of EOSDIS. Earthdata (online). earthdata.nasa.gov.

Rew, R., Davis, G., 1990. NetCDF: an interface for scientific data access. Computer Graphics and Applications, IEEE 10 (4), 76–82.

Schadt, E.E., Linderman, M.D., Sorenson, J., Lee, L., Nolan, G.P., 2010. Computational solutions to large-scale data management and analysis. Nature Reviews 11, 647–657.
Star, S.L., Griesemer, J.R., 1989. Institutional ecology, 'translations' and boundary objects: amateurs and professionals in Berkeley's Museum of Vertebrate Zoology, 1907–39. Social Studies of Science 19, 387–420.
Taylor, K., 2005. Requirements for IPCC Standard Output Contributed to the PCMDI Archive (online). http://rainbow.llnl.gov/ipcc/IPCC_output_requirements.htm.
Taylor, K., Stouffer, R., Meehl, G., 2013. An overview of CMIP5 and the experiment design. Bulletin of the American Meteorological Society 93, 485–498.
UCAR, 2013. ACADIS Data Providers Guide 5.1 (online). www.aoncadis.org/media/ProvidersGuide.pdf.
UCAR, ACADIS website (online). (www.aoncadis.org).
U.S. Department of Energy, 2004. The Office of Science Data-management Challenge. Technical Report, pp. 1–2.
U.S. Department of Energy, 2013. Synergistic Challenges in Data-Intensive Science and Exascale Computing. Summary Report of the Advanced Scientific Computing Advisory Committee (ASCAC) Subcommittee.
Vorhies, B., 2013. How Many "V"s in Big Data – The Characteristics that Define Big Data. Data-Magnum (online). data-magnum.com/how-many-vs-in-big-data-the-characteristics-that-define-big-data.
Wied, P., 2014. Dynamic Heatmaps for the Web (online). www.patrick-wied.at/static/heatmapjs/.
Zikopoulos, P., Eaton, C., 2011. Understanding Big Data: Analytics for Enterprise Class Hadoop and Streaming Data. McGraw-Hill, New York, NY, USA.

CHAPTER

9

Progress Toward Hyperresolution Models of Global Flood Hazard

Paul D. Bates[1,2], *Jeff Neal*[1,2], *Chris Sampson*[1,2], *Andy Smith*[1,2], *Mark Trigg*[1]

[1]School of Geographical Sciences, University of Bristol, Bristol, UK; [2]SSBN Flood Risk Solutions, Cardiff, UK

INTRODUCTION

Floods are the costliest and most deadly class of natural disaster each year. According to the UN's EM-DAT database (Guha-Sapir et al., 2015), out of 324 reported major natural disasters in 2014, floods and landslides caused by hydrological events affected 42.3 million people, killed ~4600 people (~58% of the annual total), and caused ~$37 Bn of damages (~38% of the total). A single flood event in June 2014 in China affected 15 million people, and the largest economic loss due to flooding occurred in the Jammu and Kashmir region of India and resulted in $16 Bn of damages. Although these figures are shocking, it is worth noting that 2014 had the third lowest number of reported disasters out of the preceding decade, and the total number of disasters was below the annual average observed from 2004 to 2013. The patterns observed in 2014 are repeated annually, and high-profile flood events (e.g., Australia and Thailand in 2011; Central Europe in 2013; India and Pakistan in 2014) are an all too regular occurrence. As a result there has been, and continues to be, sustained public, commercial, political, and scientific interest in flood risk.

Flood risk (defined here as the product of event probability and consequences) is really only well understood in a very small number of territories worldwide (the Netherlands, the United Kingdom, Germany, Japan, the United States, and a handful of others) and even here while data and models that can produce accurate local hazard estimates exist (e.g., Neal et al., 2009), there can often be a significant mismatch between national scale model estimates of flood risk and long-term loss observations (Penning-Rowsell, 2015). Despite this, sufficiently mature models and detailed loss data mostly exist to support financially efficient insurance markets that are capable of absorbing the impact of large events. As a result, these

Risk Modeling for Hazards and Disasters
http://dx.doi.org/10.1016/B978-0-12-804071-3.00009-4

territories are (relatively) well provided with insurance cover and capital; however, the prospects for significant profit growth are constrained as a consequence.

National scale flood risk for the rest of the planet has until quite recently been largely unmodeled, and for the insurance and reinsurance industries, this is of real concern due to the rapid expansion of insurance services into emerging markets in South America and Southeast Asia. The EM-DAT figures quoted above are particularly worrying in this respect because it is clear that large floods and significant losses occur in many of these currently unmodeled territories where our ability to price flood risk (even approximately) is extremely limited. Lack of skill in global flood models to date has been due to poor or nonexistent monitoring of river networks in many regions (Fekete et al., 2002), the low suitability of global terrain data sets for flood modeling (Sanders, 2007), the restricted spatiotemporal coverage of suitable remotely sensed data sets used to map flood extent (Bates et al., 2014; Prigent et al., 2007), and the limited spatial resolution and process representation that can be achieved in global flood models because of computational constraints (Bell et al., 2007; Ngo-Duc et al., 2007). Like politics[1], all flooding is local and, in addition, flood extent is determined by complex and nonlinear processes. For example, correct representation of the physics of flood waves (i.e., hydrodynamics) is often required to determine whether a flood defense overtops or not, and this can result in huge differences in predicted flooded area. As a result, low-resolution and nonhydrodynamic models may significantly misestimate exceedance probability curves because they do not have enough local skill to produce sufficiently correct results, even when aggregated to coarser spatial scales.

Looking to the future, the continued expansion of cities located on river floodplains and coastal deltas due to population growth and migration will inevitably produce a significant increase in flood exposure (Hallegatte et al., 2013; Hirabayashi et al., 2013; Jongman et al., 2012). Economic losses may also increase as people are lifted out of poverty, living standards rise, and a global middle class with western consumption patterns emerges (however, see Mechler and Bouwer, 2015; Jongman et al., 2015 for a more nuanced discussion of such effects). If current trends continue, then populations will grow, age, become more affluent, and migrate to zones of higher flood risk. It is also very likely that anthropogenic warming will lead to changes in flood hazard (e.g., Merz et al., 2014). Recent modeling studies (e.g., Chapter 2 in Mace et al., 2014) have shown that exposure to flooding and climate change over the forthcoming decades will increase most in exactly those regions (Africa, less developed parts of Southeast Asia) where we currently have the most limited risk information. At the same time there will be strong economic drivers for the insurance industry to develop these new markets to continue, or even increase, profit growth.

There are therefore very strong economic and humanitarian reasons to improve our understanding of flood risk for the ~90% of the globe that is currently unmodeled. As a result the last 2–3 years has seen a significant push to develop and test the first global flood hazard and risk models. This work is still very much in its infancy, but there is already evidence for levels of model skill that would have been considered surprising (if not actually impossible) only 5 years ago. These new global models build on a rich heritage of algorithm and data set development in hydraulic modeling over the last 20 years, and conceptually the challenges

[1]"All politics is local" is a maxim widely attributed to the former Speaker of the US House of Representatives Tip O'Neil.

of global flood modeling are the same as those faced at local and reach scales. Given that flooding is locally determined and the processes involved are strongly nonlinear, even for global simulations, there is a clear need to develop true hydrodynamic models at very high (subkilometer) resolution for the entire terrestrial land surface despite the huge computational cost of such an endeavor. To put this in context, even the very best operational global climate models only have resolutions of ~100 km (Palmer, 2014), and climate scientists, unlike hydrologists, have access to some of the world's best supercomputers. For hydrodynamic models using explicit numerical solutions a halving of the horizontal grid size results in an order of magnitude increase in computational cost, and so moving from 100 to 1 km requires between 10^6 and 10^7 more calculations. In the field of hydrology, global models with approximate kilometer grid scales have been termed "hyper-resolution" by Wood et al. (2011) and are seen as a "Grand Challenge" for the forthcoming decade. In hydrodynamics the use of true physically based equations derived strictly from Newtonian physics means that this is potentially even more of a test, yet this is exactly what addressing the global flood risk problem ultimately requires.

In this chapter we therefore review recent progress in global flood modeling toward hyperresolution simulations using true hydrodynamic equations. In Section Progress in Global Flood Modeling: The Drive to Hyperresolution we discuss the developments in hydrodynamic modeling that have enabled a rapid shift from local and reach scale models to schemes that can be globally applied. An example of one of the few such models yet to be developed, SSBN*flow*, is discussed in Section Case Study: Hyperresolution Global Flood Hazard Modeling Using SSBN*flow*, with a particular emphasis on the validation and testing of the scheme at various aggregation levels. In Section Integrating Global Exposure Data in Global Flood Risk Calculations we discuss the challenges in integrating available global exposure data with hyperresolution flood hazard calculations, and in Section Understanding Model Deficiencies we address the current limitations of global flood models with a specific focus on their use in (re)insurance applications. Conclusions are drawn in Section Conclusions.

PROGRESS IN GLOBAL FLOOD MODELING: THE DRIVE TO HYPERRESOLUTION

Before 2007 all hydrodynamic simulations in 2D and above were restricted to local models covering a few square kilometers or to reach scale models covering channel lengths of a few tens of kilometers and domain areas of tens to hundreds of kilometers squared. Regional scale one-dimensional models with floodplain storage cells had been in existence since the late 1960s, for example, the Zanobetti et al. (1970) model of the Mekong Delta, Cambodia, which covered an area of 50,000 km^2 and was used for the simulation and prediction of seasonal flooding episodes. However, the technical challenge of developing one-dimensional models, which need to specify flow paths a priori at this scale, was (and is) a significant limitation to their widespread adoption. Floodplain inundation is very clearly at least a two-dimensional process where the inundation front propagates according to the local terrain and dynamic water surface gradients that evolve rapidly in space and time, and this requires models of commensurate dimensionality. Unfortunately, two-dimensional hydrodynamic models are significantly more expensive to run than one-dimensional schemes, and until about the mid-2000s this fundamentally limited the size of domain to which they could be applied at reasonable resolutions (grid sizes of, say, 20–100m). Moreover, before this point some of the key data sets required by large-scale hydrodynamic models did not yet exist (e.g., global return period flow estimates), or if they did (e.g., global bare earth digital elevation data) then they had not yet been quality checked, cleaned up, and made fit for routine use in flood inundation modeling.

Three developments in particular began to change this situation. First, a new generation of computationally efficient two-dimensional flood inundation models had started to be developed for local and reach scale application to take advantage of the wealth of modeling possibilities afforded by new laser terrain mapping (LiDAR) technologies. Numerical experiments by Bates and De Roo (2000) showed that model skill improved more quickly if higher resolution and accuracy terrain data were used than if the model physical basis was upgraded. This led to the realization that many existing two-dimensional models were over-specified in terms of their physics and created an impetus to develop flood inundation models that were as simple and as efficient as possible because these could then be applied at the highest feasible grid resolution. Although not developed specifically with regional to global scale applications in mind, such schemes were for the first time capable of being put to such a use. Second, Moore's Law (and variants thereof) continued to hold and meant that increasing computer power was available for simulations. Together these two developments allowed the first national scale flood risk models to be developed, both by regulators and catastrophe modeling companies, for territories with high-quality Digital Elevation Model (DEM) mapping. The United Kingdom was one of the first such territories to be modeled in this way, initially using simple nonmodel approaches to the hydraulics because of the computational cost (Hall et al., 2003), but later with two-dimensional diffusive wave hydraulic schemes (Bradbrook et al., 2004, 2005). Extending such approaches to territories with poor national mapping required one further advance: the development of bespoke versions of global terrain data sets (principally the Shuttle Radar Topography Mission data set, or SRTM: Farr et al., 2007) expressly suitable for flood inundation modeling.

These three developments opened the way to the first regional scale two-dimensional flood inundation models (i.e., models covering domain areas, not of a few tens of kilometers squared, but tens of thousands). One of the first such applications was that of Wilson et al. (2007) using the LISFLOOD-FP scheme to model a 30,000 km^2 region of the central Amazon floodplain surrounding the confluence of the Solimões and Purus rivers 250 km upstream of Manaus, Brazil, using a grid resolution of 270 m. The model was constructed using topographic data from the Shuttle Radar Topography Mission, bathymetric survey data for the river geometry, and as boundary conditions used hydrographs from upstream and downstream gauging stations. Wilson et al. were the first to realize that new developments in high-resolution reach scale flood inundation modeling could also allow coarser (yet still comparatively well resolved) wide-area models to be developed and that, conceptually, the challenges to be overcome were no different from those that pertained for reach scale applications. The earlier study of Bates and De Roo (2000) had already showed that the importance of good terrain data for flood modeling, and previous studies with LiDAR data had given experience of methods for removing surface artifacts from digital surface models to recover the bare earth surface (e.g., Mason et al., 2003). To correct the SRTM, Wilson et al. simply conducted field surveys of vegetation height in different ecological regions of the floodplain and then used these along with a vegetation distribution map and assumptions about the penetration depth of the SRTM X band radar into the vegetation canopy to determine the necessary correction. The result was a "first-order accurate" bare earth surface that could then be used to model flood evolution in space. Not only did this model cover a domain area of orders of magnitudes larger than previous two-dimensional model applications, it was also used to simulate multiannual flood sequences rather than events lasting just a few days or hours. The model was calibrated against water elevation data from satellite altimetry (Birkett et al., 2002; Frappart et al., 2006) and flood extent Synthetic Aperture Radar (SAR: Hess et al., 2003). Simulation results were most accurate at high water, with 73% flood extent agreement and altimetry RMSE ~2 m, but this declined at low water to 23% and RMSE of 4–5 m. Although the LISFLOOD-FP variant in use in 2007 was not a true hydrodynamic model as it lacked a local inertial term (just like the previous Bradbrook et al., 2005 national scale model for the UK), the study did demonstrate that both the data and algorithms now existed for regional scale models to be developed anywhere on the globe.

Subsequently, Biancamaria et al. (2009) used a similar approach to create a LISFLOOD-FP model of ~900 km of the Ob River in Siberia, and subsequent developments to both the data and numerical code used in Wilson's Amazon model allowed substantial improvements in performance. First the diffusive hydraulics of flow within the main river was accounted for through the development of a 1D diffusion wave numerical scheme (Trigg et al., 2009), whereas the computational efficiency of floodplain flow calculations was upgraded to a proper hydrodynamic model solving the local inertial version of the shallow water equations (Bates et al., 2010). Vegetation artifact errors within the SRTM were further reduced using a new technique that incorporated remotely sensed vegetation heights (Baugh et al., 2013), and ICESat altimetry data were used to geodetically level the gauging stations on the river channel to improve the validation of simulated water surface elevations (Hall et al., 2012). As a result, the latest version of this model of the Solimões-Purus confluence plain model now has flood extent prediction accuracy of 90% at high water and 30% at low water, with altimetry RMSE improving to 1.5 m throughout the entire flood cycle. This is within the vertical error in the SRTM DEM at 270 m spatial resolution. For this 30,000 km^2 area, the model is now able to inundate and drain a similar volume of water (76 km^3) to that estimated from GRACE gravimetric anomalies (60–80 km^3) through the same hydraulic processes as observed in field and remote sensing data sets (Baugh et al., 2013).

By 2010 it was therefore clear that regional scale flood inundation modeling at reasonable grid resolutions (10^2–10^3 m) was now a realistic proposition, and that the data sets now existed to allow such modeling to be rolled out for any place on the planet. As a result, a number of detailed hydraulic models have since been constructed for large river reaches in data sparse regions including the Amazon (de Paiva et al., 2013; da Paz et al., 2011), the Niger (Neal et al., 2012), the Congo (Jung et al., 2010), the Mekong delta (Yamazaki et al., 2014b), and the Zambezi (Schumann et al., 2013). These approaches vary in complexity from coupled 1D/floodplain storage models (de Paiva et al., 2013), through coupled 1D channel/simplified 2D floodplain models (Biancamaria et al., 2009; da Paz et al., 2011), to 2D models with subgrid representations of minor channels (Neal et al., 2012; Schumann et al., 2013). However, continental scale studies that employ detailed hydraulic models are scarce because the difficulties of data availability and computational expense. Feyen et al. (2012) produced estimates of fluvial flood risk in Europe at local scales (~100 m) using a model cascade that converted estimated river water depths into inundation extents using a simple nonmodel planar approximation for inundation extent. A more recent study produced by the European Commission's Joint Research Center used a combination of distributed hydrological and hydraulic models, driven using observed meteorological data, to derive a 100-year return period pan-European flood map in which inundation extents were simulated using over 37,000 two-dimensional hydrodynamic models, each representing a ~5–10 km reach of main river channel (Alfieri et al., 2013).

Given this success, an obvious next step was therefore to begin the development of global flood models. For this task Sampson et al. (2015) identified six key challenges that need to be solved:

- Development of global bare-earth DEMs of a resolution and accuracy suitable for flood modeling.
- Development of suitable methods to determine return period discharge anywhere on the global river network.
- High-resolution global scale databases for both river network typology and geometry.
- Global data sets detailing flood defense standards and locations.
- A highly computationally efficient hydraulic engine.
- A software framework to implement and automate this complex calculation chain.

Each of these tasks is a substantial research project in its own right, yet over the last decade enormous progress has been made to the extent that the first global flood inundation simulations are now possible (see for example, Hirabayashi et al., 2013; Winsemius et al., 2015). Researchers have developed near-automated methods to correct SRTM terrain data to recover bare earth (e.g., Baugh et al., 2013), applied atmospheric reanalysis data and land surface models (Pappenberger et al., 2012; Winsemius et al., 2013), or regional flood frequency analysis (Smith et al., 2015) to estimate global return period flows, published global data of river width (Yamazaki et al., 2014a), developed new numerical analysis methods to allow highly efficient solution of true hydrodynamic shallow water equations result the first global models have been created and the process of testing these has begun.

As with the national scale models discussed earlier, the development path to global scale simulations was first to employ nonphysics approaches for the hydraulics that used simple planar water surface approximations or volume spreading algorithms to route water in excess of channel bankfull capacity across the floodplain (e.g., GLOFRIS, Ward et al., 2013), before proceeding to more complex schemes that solve true hydrodynamic versions of the shallow water equations (e.g., Sampson et al., 2015). A small but growing number of global models now exist, and these vary significantly in terms of physical complexity, how boundary conditions (i.e., return period flows) are generated, and the type and quantity of validation testing that has been conducted. There are also significant differences in the spatial resolution of the model calculation, with final output ranging from 1/120 degrees or ~900 m at the equator to 1/1200 degrees or ~90 m. All models to date are based on processed versions of the Shuttle Radar Topography Mission Digital Elevation Model (SRTM DEM) and Hydrosheds river network to provide near global coverage.

This diversity is unsurprising, given the relatively early developmental stage of such schemes and the fact that a consensus around the strengths and weaknesses of the different approaches has yet to emerge. An important part of generating the now required comparative evidence base (and hence promoting a rapid evolution of the models) is the use of consistent, rigorous and nontrivial validation, and benchmarking tests. As an example of such testing the next section reviews recent analysis and validation of the SSBN*flow* global hydrodynamic model to determine the range of problems that such models can (and perhaps more importantly, cannot) be applied to.

CASE STUDY: HYPERRESOLUTION GLOBAL FLOOD HAZARD MODELING USING SSBN*FLOW*

SSBN*flow* is a 1/1200 arcsecond spatial resolution (~90 m at the equator) global flood inundation model that solves the local inertial (i.e., a true hydrodynamic) form of the shallow water equations. The hydraulic engine is a clone of the LISFLOOD-FP model, the blueprint for which is described in detail by Bates et al. (2010), and uses the subgrid scale approach to river channels outlined by Neal et al. (2012). Boundary conditions for the flood extent calculations are derived from the regional flood frequency approach for global applications developed by Smith et al. (2015), which gives the magnitude of extreme return period river flows and rainfall anywhere on the globe. Digital elevation data for the model come from a bespoke version of the SRTM data set specifically adapted for flood inundation using some of the techniques outlined by Baugh et al. (2013), but extended to also identify and remove building artifacts using night-time light intensity data to improve the representation of the ground surface in built-up areas. River widths are estimated using a hybrid geomorphological/web-survey technique in which river widths along major rivers within a domain are measured and recorded along with their corresponding upstream accumulating areas. By assuming a bankfull discharge return period of approximately 1 in 2 years (Andreadis et al., 2013; Harman et al., 2008; Leopold, 1994), the flow generation algorithm of Smith et al. (2015) is used to generate an estimate of bankfull discharge. By combining bankfull discharge, channel width, and an estimate of valley slope calculated from the DEM, it is then possible to produce an estimate of channel depth using Manning's equation. Linking channel geometry to discharge return period in this manner ensures that the channels are appropriately sized for the flows being simulated, mitigating against the problem of gross mismatches between discharge and channel conveyance. Flood defenses are accounted for by modifying local channel conveyance capacity using an empirical relationship between socioeconomic factors, a spatial measure of urbanization and a database of known defense standards. Inundation dynamics on large rivers are simulated at 30-arcsecond resolution rather than 3-arcsecond resolution because DEM noise is reduced on the coarser grid, enabling a more stable estimate of water surface elevation to be produced over large floodplains. This has the additional advantage of reducing model runtimes by more than two orders of magnitude, allowing the long inundation events associated with seasonal floods on major rivers to be simulated fully. A smooth water surface is then calculated by interpolating between points at the center of each 30-arcsecond cell, enabling water heights to be reprojected onto the 3-arcsecond DEM to obtain simulated water depths at the higher resolution. For smaller channels and surface water inundation, where flows are driven by precipitation rather than river flow, simulations are undertaken directly on the 3-arcsecond

DEM. Global simulations are undertaken for 10 degrees by 10-degree tiles for 10 specified return period flows from 1 in 2 years to 1 in 1000 years over the entire area covered by SRTM. A 1-degree overlap between tiles is used, across which a weighted blend is calculated to reduce the impact of anomalous boundary effects at the edge of each tile. The calculations are performed on a high-throughput computing system using an automation framework consisting of some 40,000 lines of Mathworks Matlab code, augmented with python-based GIS routines. The total runtime for a typical 10 degrees by 10-degree tile equates to approximately 2000 h on a single core of a 3.2 GHz Intel i7 processor, although the exact value for each tile varies according to several factors, including percentage land surface, climate (drier arid tiles are typically faster to simulate than wetter tropical tiles) and hydrograph durations on the largest rivers in the tile. For full details of the SSBN*flow* approach the reader is referred to Sampson et al. (2015); a schematic diagram that summarizes the SSBN*flow* approach is Fig. 9.1.

Sampson et al. (2015) report wide area validation tests of the SSBN*flow* approach. Validation of flood hazard maps is challenging because they do not attempt to recreate the characteristics of a single observable event. To overcome this issue, Sampson et al. compare global results to analogous regional flood hazard data sets constructed using high-resolution models and detailed local data for regions of Canada and the United Kingdom. No local data (for example, enhanced terrain data or local flow records) were used to calibrate or otherwise modify the global model. The performance measures used were spatial variants of the commonly employed hit rate (HR), false alarm ratio (FAR), and critical success index (CSI). Performance measures were applied at the scale of individual model grid cells (3 arcseconds, ~90 m resolution). In addition, for the catchment scale applications in the United Kingdom the percentage error in flooded fraction between model and observed data aggregated to 30 arcsecond or ~1 km scale could also be meaningfully calculated.

The Canadian benchmark data covered the Calgary, Edmonton, and Red Deer urban regions of Alberta State and were supplied by the Informatics Branch of the Alberta State Government. It consisted of vector-based binary wet/dry flood hazard maps produced as part of the Alberta Flood Hazard Identification Programme by simulating 1 in 100-year design floods using steady-state 1D Hydrologic Engineering Center River Analysis System (HEC-RAS) models. The 1D water levels were reprojected onto the DEM to create the 2D flood outline vector. The models were built using ground-surveyed channel cross sections (including defenses) in conjunction with 1:5000 scale, 1 m vertical resolution contour maps; design flows were derived from federal flow records from the Water Survey of Canada. Fig. 9.2 shows the 1 in 100-year flood outline from the local HEC-RAS models superimposed on the output for the Sampson et al. (2015) global model. The global model covers watercourses down to basins areas of about 10 km^2, whereas the Alberta State model is only built for specific large river reaches.

The data from the United Kingdom comprised catchment-wide 1 in 100-year return period flood outlines for the Severn and Thames catchments (both around 10,000 km^2 in area) and were supplied by the Environment Agency of England and Wales as shapefiles. The data are multisource and are comprised, in descending order of preference, of: (1) flood extent observations of events with an estimated return period of ~100 years; (2) detailed 1D or 1D/2D hydraulic models build using airborne LIDAR data with a spatial resolution of ~2 m and a vertical accuracy of <10 cm; (3) a 2D hydraulic model constructed using airborne interferometric Synthetic Aperture Radar (IfSAR) terrain data with a spatial resolution of

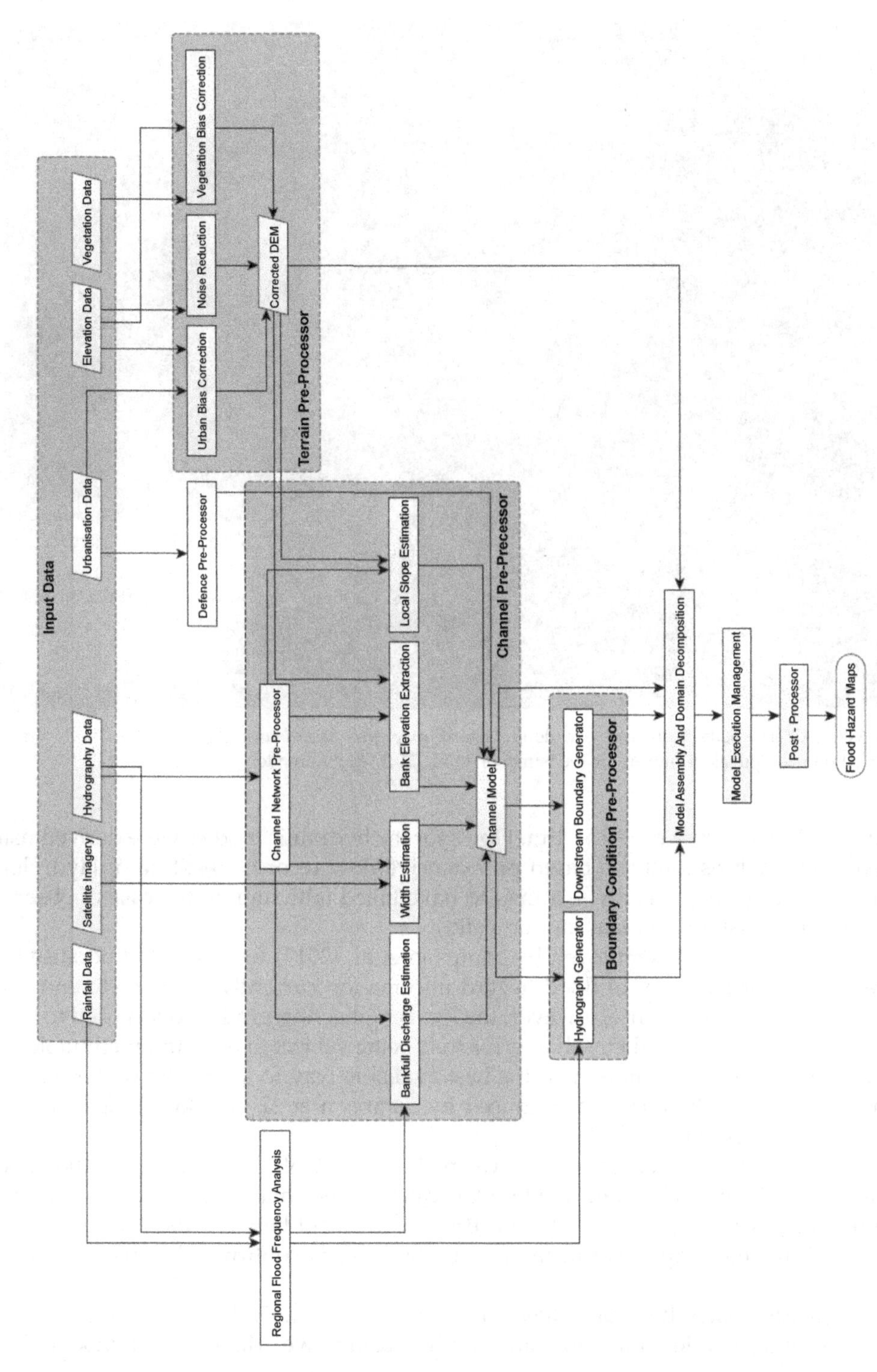

FIGURE 9.1 Conceptual flowchart of the SSBN*flow* global flood hazard model framework.

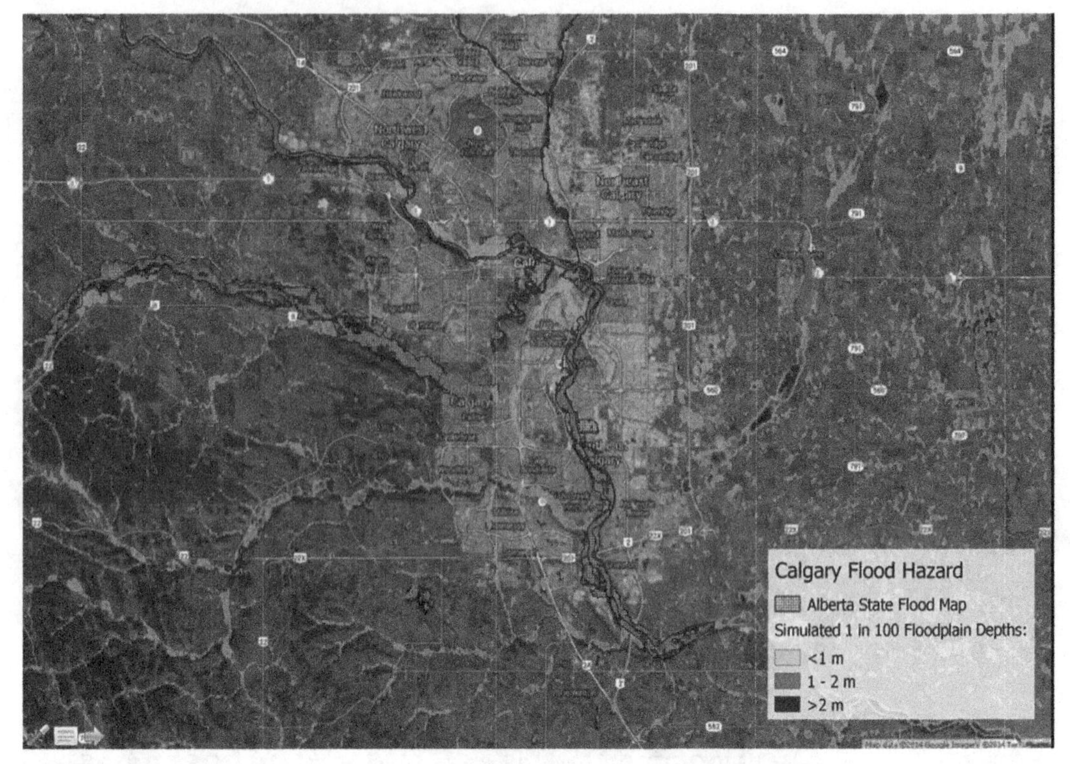

FIGURE 9.2 Alberta State 1 in 100-year flood hazard map [*red hashed area* (*light gray in print versions*)] superimposed on the equivalent output for the Sampson et al. (2015) global model.

~5 m and vertical accuracy of ~0.5–1 m. Flows for the hydraulic models were derived using flood frequency analysis methods based on extensive flow records (Institute of Hydrology, 1999) and defenses were generally assumed to have failed (although in the case of observed flood outlines this assumption may not be valid).

The benchmark data sets employed by Sampson et al. (2015) were selected because they are representative of the types of flood hazard information currently available to end users in some more developed regions. However, the local engineering-grade models used to create the benchmark data were not themselves error free, being subject to data and methodological limitations. Although such data remain the best available way to assess the performance of global hazard models, the analyses produced by Sampson et al. should be viewed in the context of these limitations.

Figs. 9.3–9.6 show the comparison between the output of the benchmark and global models for two sites in Canada and the two UK catchments. Quantitative metrics summarizing the performance are given in Table 9.1. Broad agreement between the benchmark data and the global models is apparent in the figures, and this is confirmed by the quantitative analysis.

For the Canadian sites the CSI values vary from 0.59 to 0.65. To put these scores into context, engineering-grade local modeling studies using two–dimensional models have

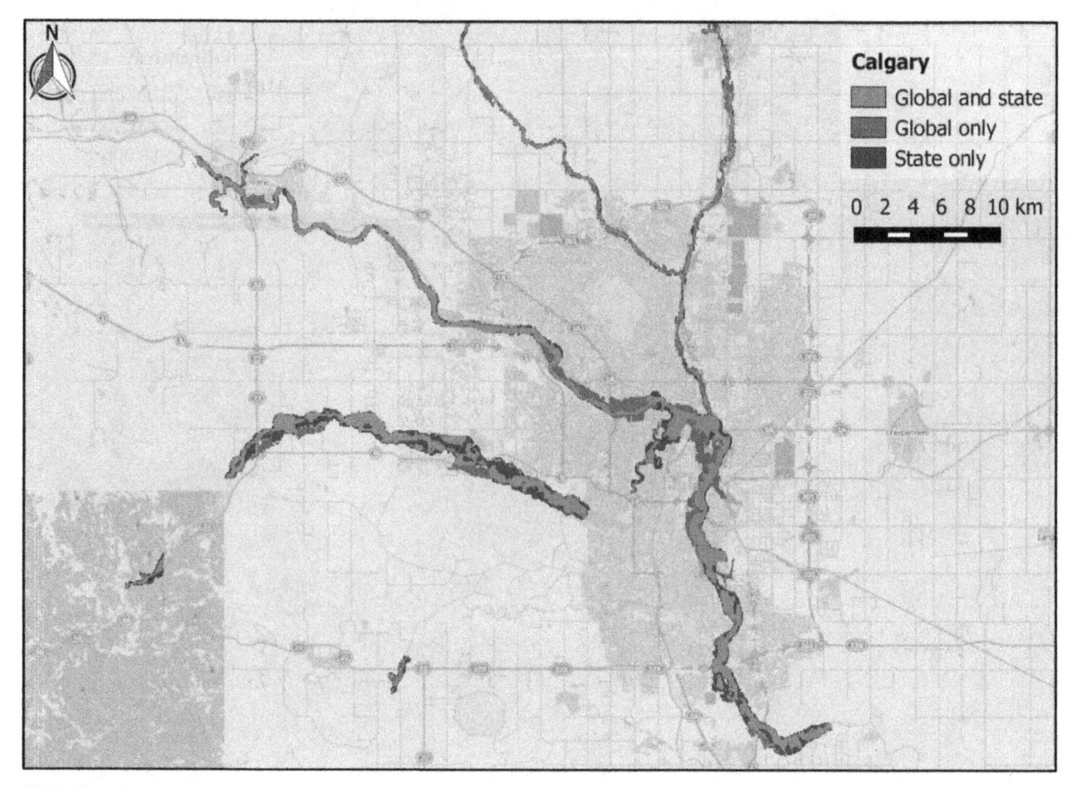

FIGURE 9.3 Map showing fit between global model and Canadian benchmark data for Calgary, Alberta.

achieved maximum CSI scores against observed extents of between 0.65 and 0.89 (Aronica et al., 2002; Horritt and Bates, 2002; Bates et al., 2006). Given that these local models employed higher quality topographic data (photogrammetry and LIDAR), gauged flow data, surveyed bathymetry and are calibrated to optimize friction parameters for local conditions, the relative skill of the global model is encouraging. For the Canadian sites there is a tendency to overpredict relative to the benchmark data that is particularly evident in the blue areas visible around the center of Calgary in Fig. 9.3. This behavior may be explained by the global model's reduced ability to resolve objects on the floodplain that restrict flow, such as high-density urban developments in city centers, but it may also represent a limitation of the one-dimensional hydraulic model used to produce the benchmark data. Such a model would have been limited by the width of the cross sections used to construct the model domain, and it is therefore not possible to ascertain the cause of this discrepancy between the global and benchmark data. The relatively small amount of underprediction that occurs within urban areas across the three test cases indicates that the filter employed to reduce the positive elevation bias typically present in SRTM data is functioning well.

For the UK sites visual inspection of Figs. 9.5 and 9.6 also shows a reasonably good match to the benchmark data. In addition to performance metrics for the whole catchment, Table 9.1 also gives the model performance for larger (>500 km^2) catchments only. It is apparent that

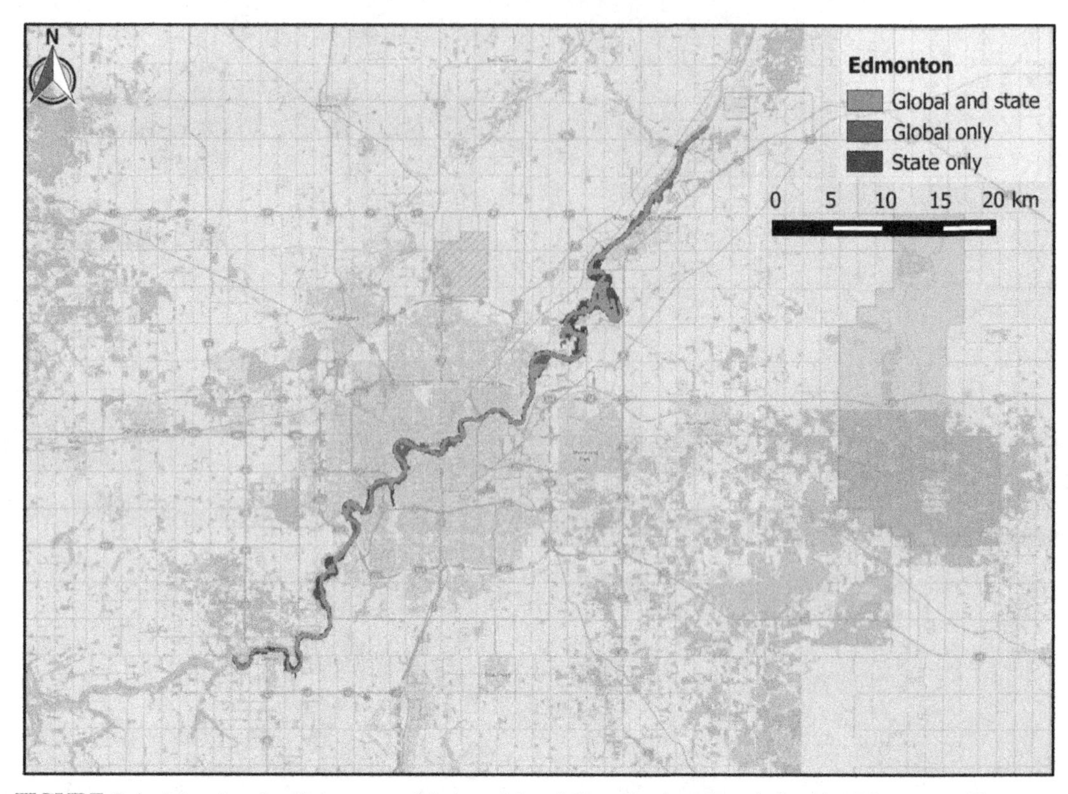

FIGURE 9.4 Map showing fit between global model and Canadian benchmark data for Edmonton, Alberta.

model performance is consistently better on the Severn than on the Thames, and that model performance declines significantly for smaller channels compared with larger channels. The difference in skill between the Thames and Severn has several explanations. The Severn is a relatively simple catchment in terms of topography, with the majority of its larger floodplains consisting of rural pasture. In comparison, the Thames catchment has complex chalk hydrology and is heavily urbanized, flowing through a number of sprawling towns upstream of London before continuing through the center of the capital itself. As a result, the SRTM terrain model is more closely matched to the high-quality terrain data used for the benchmark hazard maps over the Severn catchment than it is over the Thames. Furthermore, flooding along the Thames is prevented or reduced by a number of substantial flood defense and alleviation schemes that are more extensive than any modifications made to the Severn and its tributaries. Visual inspection of Fig. 9.6 shows the global model to underestimate flood hazard in the eastern part of the Thames catchment. This is due to an assumed sea level of 0 m (i.e., mean sea level) in the global model, whereas the Environment Agency flood hazard map assumes a 1 in 200-year coastal flood along with failure of the Thames tidal barrier, thus exposing a large area of central London to tidal surge flooding. This is not represented in the global model, thus explaining the underprediction in this location.

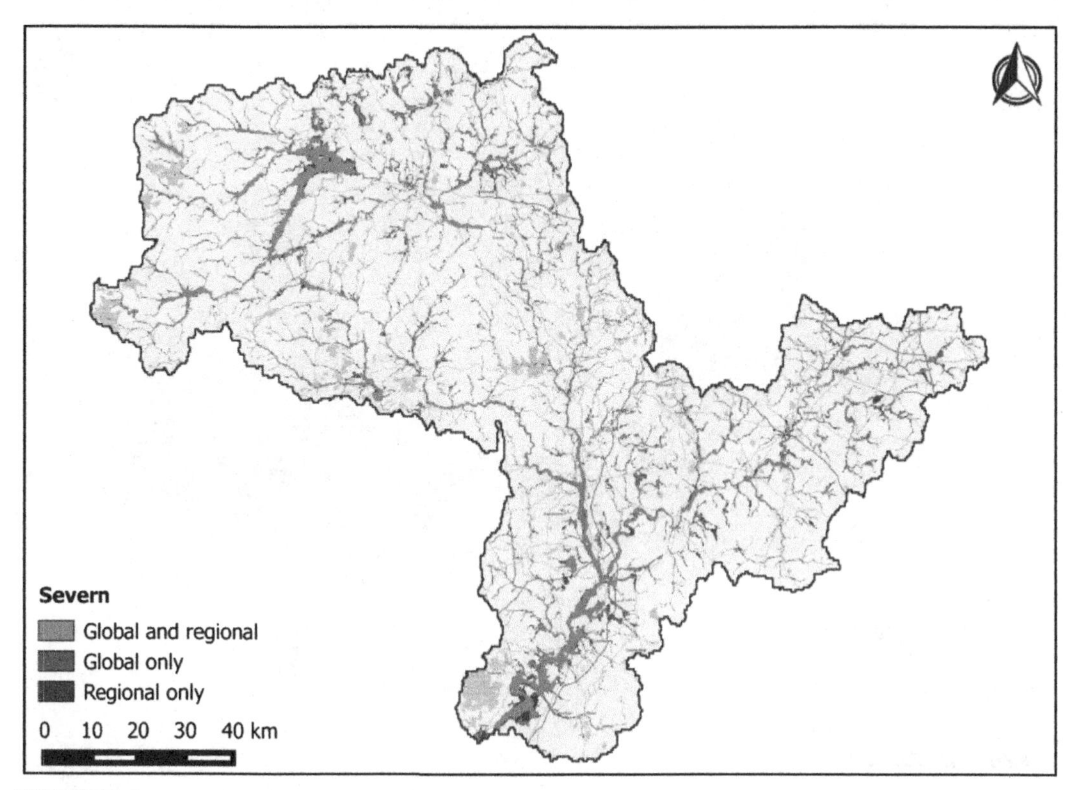

FIGURE 9.5 Map showing fit between global model and UK Environment Agency benchmark data for the Severn catchment.

The reduction in model performance over smaller catchments is unsurprising, given the ~90 m resolution and poor vertical accuracy (relative to typical flood wave amplitudes) of the SRTM data used in global study. The benchmark data are constructed using DEMs of (at worst) 5 m resolution with ~1 m vertical accuracy, enabling them to capture small-scale topographic features that are smaller than the individual pixels of the global model. The result of this is that topography is smoothed in the global model, often leading to an overprediction of flood extents. Other errors relating to the limited vertical precision of the SRTM data are also more pronounced for smaller channels, such as channel location errors. Noise in the DEM may cause small channels to be located incorrectly, as can pixilation effects owing to the limited resolution; this can adversely affect model performance statistics when compared with a higher resolution benchmark data set. However, despite the significant fall in CSI, HRs remain quite high indicating that the model is indicating risk in broadly the correct areas.

Model performance for the UK catchments when aggregated to a ~1-km grid is strong with errors in flooded fraction typically around 5%. Moreover, the difference between small and large catchments is not evident at this scale; this is because for minor channels, the area of flooding relative to the total pixel area is small, leading to a small aggregate error. Furthermore,

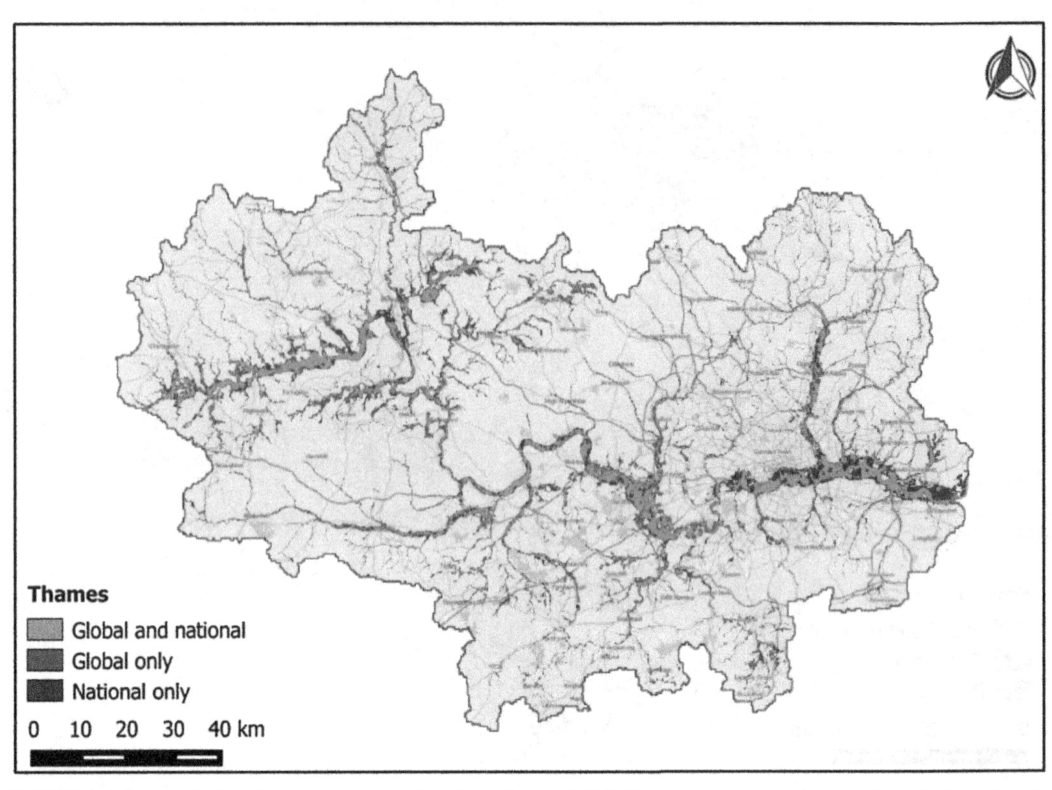

FIGURE 9.6 Map showing fit between global model and UK Environment Agency benchmark data for the Thames catchment.

TABLE 9.1 Quantitative Model Performance

Model	Hit Rate	False Alarm Ratio	Critical Success Index	Aggregate Error at 30 arcseconds (%)
Calgary	0.75	0.26	0.6	–
Edmonton	0.67	0.18	0.59	–
Red Hat	0.81	0.22	0.65	–
Thames	0.65	0.45	0.43	0.04
Severn	0.74	0.49	0.43	0.02
Thames (catchments > 500 km^2)	0.73	0.3	0.56	0.02
Severn (catchments > 500 km^2)	0.83	0.23	0.67	0.01

where the pixel-by-pixel comparison penalizes imperfect alignment in flooded areas caused by errors in the SRTM terrain data, the aggregate comparison only penalizes if the offset is large (i.e., if there is a bias).

The study by Sampson et al. (2015) therefore gives a very clear picture of current capability in global flood hazard modeling. At 3 arcsecond scale the SSBN*flow* model shows pleasing skill, although the CSI values at this scale are not yet high enough to confidently determine hazard at the scale of individual buildings.

INTEGRATING GLOBAL EXPOSURE DATA IN GLOBAL FLOOD RISK CALCULATIONS

Floods pose a distinct challenge for risk modelers because of the dominant control of relatively small-scale topographic features on the flow of water and the resulting water depths and velocities. Other hazards, such as tropical cyclones or earthquakes, typically interact with all features within or along their large and spatially continuous footprints. Such large footprints allow a degree of uncertainty within the geocoding of asset locations to be tolerated; for example, if the location of a building is only known to postcode level, this is still likely to be sufficiently precise when intersecting it with the footprint of a large hurricane. However, when considering floods, the importance of precise and accurate exposure data is increased because a small difference in location can determine whether or not an asset is exposed.

Although exposure data in developed nations are often sufficient to allow end users to confidently assess whether or not an asset lies within a modeled flood footprint, the same cannot be claimed for exposure data across much of the developing world. The lack of high-quality exposure data therefore poses a large challenge for operators in these markets, and for those interested in flood risk at the global scale. Some general exposure data sets do currently exist, for example, the Landscan and GRUMP population data sets, global land use data sets (e.g., Klein Goldewijk et al., 2010, 2011; Seto et al., 2012; Güneralp and Seto, 2013), and satellite nightlight data (e.g., Ceola et al., 2014); however, insurance portfolio data can be limited. Unfortunately, there is no easy fix for this situation, and the problem will likely only be fully solved when primary insurers or regulators consistently insist upon the collection of precise geolocation data for insured assets in these regions.

The impact of imprecise geocoding on flood risk analysis can, however, be somewhat alleviated using methods that integrate the hazard layer with global or regional "proxy" exposure data sets. The idea behind such methods is to use other indicators of development to estimate where an asset is most likely to be located when only coarse resolution exposure data are available. For example, if the location of an asset in a portfolio is only known to postcode level, but the hazard data are available at 90 m resolution, then the postcode area is likely to contain many hazard pixels. Rather than just taking a central measure (mean or median) of all values, it is possible to calculate a weighted mean by assigning each hazard pixel a weight according to how likely it is that the asset is located in that pixel. One data set commonly employed to supply such weights is satellite luminosity data (or "night-time light" data). In the luminosity data, pixels with high values are those that appear bright at night; consistent brightness suggests the presence of electrical lighting, and thus it can be

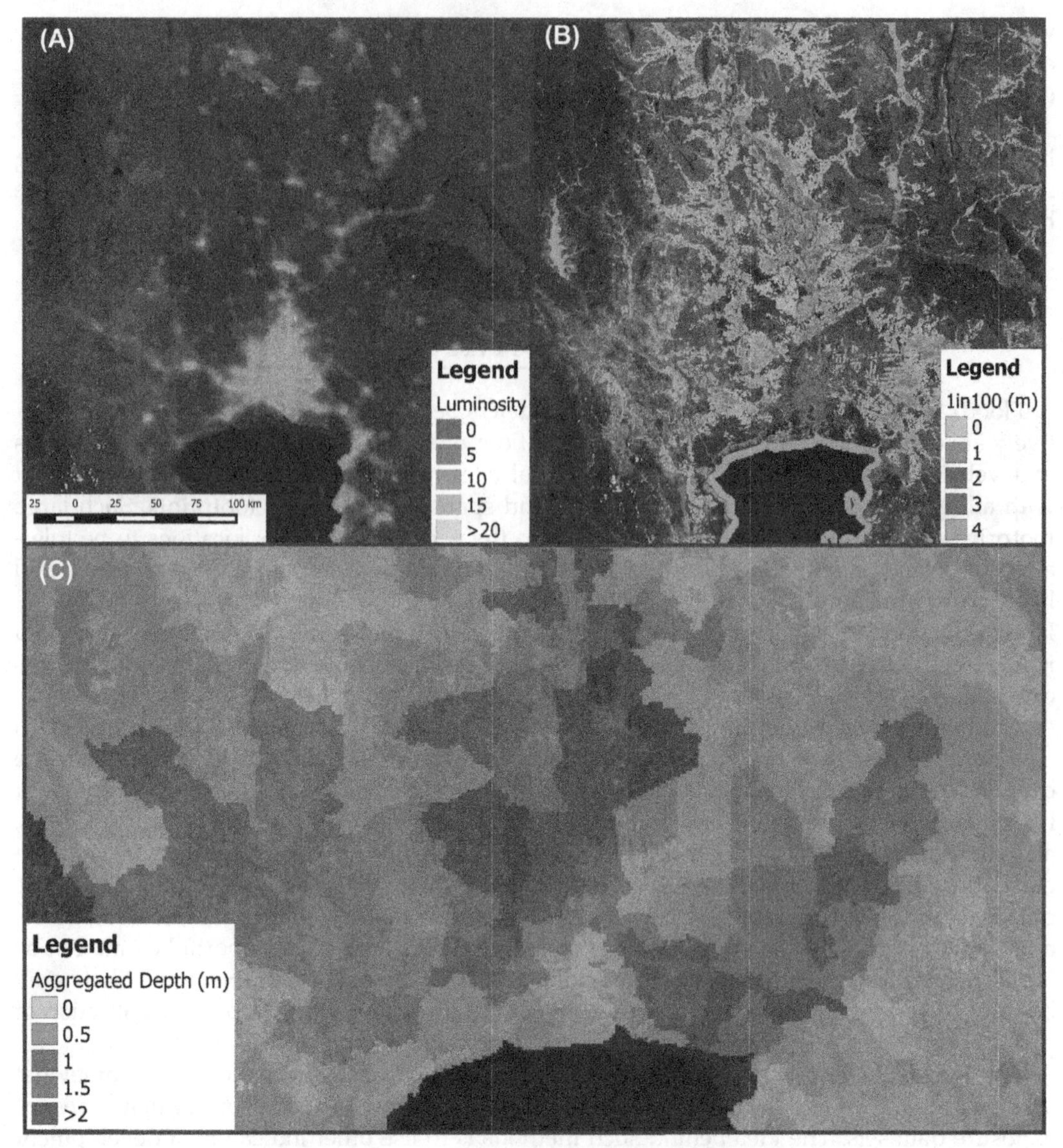

FIGURE 9.7 Flood hazard data aggregated to district level, using night-time light data. (A) Luminosity or night-time light data over Bangkok, Thailand, (B) SSBNflow fluvial flood hazard data, (C) Luminosity and flood hazard data integrated and aggregated to district level.

inferred that buildings are more likely to be situated in bright pixels than in dark ones. Using the method described above it is possible to use this proxy exposure data set to aggregate data to any administrative level; in Fig. 9.7 an example of this has been carried out for Southern Thailand, with SSBN*flow* flood hazard layers being aggregated to district level.

UNDERSTANDING MODEL DEFICIENCIES

Section Case Study: Hyperresolution Global Flood Hazard Modeling Using SSBN*flow* has highlighted that global models have sufficient skill for a number of practical applications, and indeed they are now being widely employed across a number of industries, including (re)insurance. However, before such models can be used appropriately, it is important that practitioners have an adequate understanding of model deficiencies. All models will have relative strengths and weaknesses, scenarios, or physical processes that are better represented than others, and this is particularly true of global models. Gaining an understanding and an appreciation of these uncertainties is crucial if such models are to be used effectively. An excellent overview of this topic is given by Ward et al. (2015), and here we highlight aspects with specific relevance to the insurance and reinsurance industries.

As mentioned previously, global models can be expected to be far more reliable for larger catchments, with large well-defined floodplains. This is due to the accuracy of global DEMs; with ~90 m SRTM being the best available data set for hydrodynamic modeling, small topographical features that govern local flow paths are likely to be poorly represented. For larger floodplains, however, the spatial resolution of SRTM is more likely to be sufficient to adequately capture floodplain dynamics. Also, the location of river channels in a global model needs to be automatically derived from the underlying DEM, and this in turn results in model errors. As the upstream catchment area of a river decreases, the more likely it is that the derived channel network will be erroneous as small errors in the DEM will begin to influence the derived channel location. Channel location errors may also arise in large flat floodplains where the drainage algorithm ends up with multiple parallel paths across the area and for areas in which anthropogenic influences have significantly altered river flow paths. The manipulation of river channels typically occurs in urbanized areas, with channels being diverted for numerous reasons and often over long time periods. Capturing these influences in a global DEM such as SRTM is inherently difficult, because the data lacks the spatial resolution to capture these engineered structures. Instead, in global models the derived river channel will likely be placed in its natural location, as defined by the local topography, and this may be very different from where channels are in reality. Lastly, current global flood hazard maps show flood inundation extent and depth. However, they generally do not contain information on flood durations, whereas recent flood disasters, such as those in England and Wales in 2014, Mississippi in 2012, Thailand in 2011, Queensland (Australia) in 2010–11, and Pakistan in 2010, have shown that the duration of flooding is also an important parameter in determining the impacts of flooding. Indeed, several risk papers have shown duration to be very important for damage estimation (e.g., Lekuthai and Vongvisesisomjai, 2001; Elmer et al., 2010; Dang et al., 2011). True hydrodynamic models are theoretically capable of predicting flood duration, but their simulation ability in this respect has yet to be adequately investigated.

Output from global models should also be treated differently across different climate zones, with model performance varying depending on climatic conditions. Firstly, errors relating to the location of rivers are likely to increase in dry, arid regions. A lack of rainfall, and subsequently river discharge, means that well-defined channels are less likely to form. Moreover, errors in the estimation of return period flows can be expected to be far larger in arid regions, regardless of the underlying methodology used to define them. For example, when applying a regionalized flood frequency analysis at the global scale, Smith et al. (2015) found that errors in return period flow estimates more than doubled in arid regions, when compared with those in stations located in other climates. The capacity for carrying out rainfall–runoff modeling in arid regions is also limited, owing to the poor representation of rainfall dynamics in climate models and the sparse rainfall gauging networks that exist in many of these regions (El Kenawy and McCabe, 2015).

Global flood models may be utilized by the (re)insurance industry in a number of ways. Output in the form of flood hazard footprints can be used directly for underwriting purposes or, when combined with stochastic, vulnerability, and loss components, a catastrophe (CAT) model can be constructed. However, that the output of global flood models can be integrated into a CAT modeling framework, flood depths must be converted into loss or damage. This is achieved using a vulnerability function in which the modeled flood depths are used to infer damage based on empirically derived depth–damage relationships (e.g., Sampson et al., 2014). However, as Sampson et al. (2015) have shown, local (i.e., 3 arcsecond scale) flood depths predicted by SSBN*flow* have significantly higher uncertainty than spatially aggregated data and the coarse spatial resolution of global DEMs renders them unsuitable for defining local/building scale water depths. Moreover, the significant pixel-to-pixel vertical noise that exists within global DEM products such as SRTM (~6 m at 3 arcseconds, Rodriguez et al., 2006) adds an additional uncertainty component. Before the output of global flood models can be used as part of a larger CAT modeling framework, modelers should be aware that the absolute depths are highly uncertain, compounding the uncertainty inherent within depth–damage relationships. It is critical that end users are aware of such limitations, particularly if the model is to be used for site-specific analysis.

Although global models produce flood hazard footprints at a variety of recurrence intervals, it is important that end users also understand that estimating the magnitude of return period flows is highly uncertain, particularly for high-magnitude, low-probability events. Indeed, even in well-gauged areas with a 1-m spatial resolution LiDAR based DEM, simulating the flood extent of a given return period event is very difficult, and subject to large uncertainty because of errors in flood frequency relationships (Neal et al., 2013). Given the typical scenario within a global model of no local information being incorporated, these uncertainties are amplified further. End users should therefore be aware that although differing return period hazard layers are provided, they are inherently uncertain data, and as with all model output should be viewed cautiously.

For portfolio analysis to be conducted, end users will likely want to convert flood hazard layers into a realistic set of scenario or event footprints. This is required because a return period hazard layer is not representative of a realistic event, i.e., the 100-year event would not occur everywhere at the same time. For these scenarios to be derived, some form of statistics defining the correlation between catchments is required (e.g., Jongman et al., 2014). In developed regions, such as the United Kingdom, these data are available via a dense network of river gauging stations. Without a gauging network that has a sufficiently long history and spatial coverage, defining correlation relationships between catchments becomes very difficult, and modeled or remotely sensed precipitation data must be used. However, global precipitation data sets poorly represent extreme precipitation (Sampson et al., 2014). Moreover, flood response is not always consistent between neighboring catchments, as basin area, land use, topography etc. will all have a significant impact on catchment storm response. Therefore when conducting portfolio analysis outside data-rich, well-gauged regions, it should be noted that any derived scenarios will again be subject to significant uncertainty and should be treated with appropriate caution.

CONCLUSIONS

This chapter has reviewed the recent development of global flood models and attempted to describe our current best understanding of their strengths and weaknesses. Flood modeling at global scales represents a revolution in hydraulic science (Hall, 2014) and has the potential to transform decision-making and risk management in a wide variety of fields. Such modeling draws on a rich heritage of algorithm and data set development in hydraulic modeling over the last 20 years, but conceptually the challenges of global flood modeling are the same as those faced at local and reach scales. Given the current rates of progress, this is likely to be a rapidly changing field, and significant work remains to be done. Obvious targets for future work include a better global database of flood defense standards and hydraulic control infrastructure, better estimates of global precipitation and river flow extremes, a global database of river bathymetry, construction of better model validation data sets, and the setting up of rigorous model intercomparison studies. The forthcoming SWOT satellite mission (see http://swot.jpl.nasa.gov/), which will, approximately every 10 days, measure water surface elevation and slope to high accuracy in all rivers greater than 100 m wide will significantly help with many of these studies. However, the key constraint on the performance of all flood models is the quality of the DEM data that are available. In this respect current global DEMs do not resolve adequately the detail of terrain features that control flooding (Schumann et al., 2014), and a new global topographic mapping mission with sub 30 m spatial resolution and sub 1 m vertical accuracy is desperately required. Such a global DEM would likely be the biggest single thing we could do to improve disaster risk reduction, disaster forecasting, risk finance, and humanitarian relief efforts.

References

Alfieri, L., Salamon, P., Bianchi, A., Neal, J., Bates, P., Feyen, L., 2013. Advances in pan-European flood hazard mapping. Hydrological Processes 28, 4067–4077.

Andreadis, K.M., Schumann, G.J.-P., Pavelsky, T., 2013. A simple global river bankfull width and depth database. Water Resources Research 49, 7164–7168.

Aronica, G., Bates, P.D., Horritt, M.S., 2002. Assessing the uncertainty in distributed model predictions using observed binary pattern information within GLUE. Hydrological Processes 16, 2001–2016.

Bates, P.D., De Roo, A.P.J., 2000. A simple raster-based model for flood inundation simulation. Journal of Hydrology 236, 54–77.

Bates, P.D., Wilson, M.D., Horritt, M.S., Mason, D.C., Holden, N., Currie, A., 2006. Reach scale floodplain inundation dynamics observed using airborne synthetic aperture radar imagery: data analysis and modelling. Journal of Hydrology 328, 306–318.

Bates, P.D., Horritt, M.S., Fewtrell, T.J., 2010. A simple inertial formulation of the shallow water equations for efficient two dimensional flood inundation modelling. Journal of Hydrology 387, 33–45.

Bates, P.D., Neal, J.C., Alsdorf, D., Schumann, G.J.-P., 2014. Observing global surface water flood dynamics. Surveys in Geophysics 35, 839–852.

Baugh, C.A., Bates, P.D., Schumann, G., Trigg, M.A., 2013. SRTM vegetation removal and hydrodynamic modeling accuracy. Water Resources Research 49 (9), 5276–5289.
Bell, V.A., Kay, A.L., Jones, R.G., Moore, R.J., 2007. Development of a high resolution grid-based river flow model for use with regional climate model output. Hydrology and Earth System Science 11, 532–549.
Biancamaria, S., Bates, P.D., Boone, A., Mognard, N.M., 2009. Large-scale coupled hydrologic and hydraulic modelling of an arctic river: the Ob river in Siberia. Journal of Hydrology 379, 136–150.
Birkett, C.M., Mertes, L.A.K., Dunne, T., Costa, M.H., Jasinski, M.J., 2002. Surface water dynamics in the Amazon Basin: application of satellite radar altimetry. Journal of Geophysical Research-Atmospheres 107 (D20), 8059.
Bradbrook, K.F., Lane, S.N., Waller, S.G., Bates, P.D., 2004. Two-dimensional diffusion wave modelling of flood inundation using a simplified channel representation. International Journal of River Basin Management 2, 211–223.
Bradbrook, K., Waller, S., Morris, D., 2005. National floodplain mapping: datasets and methods - 160,000 km in 12 months. Natural Hazards 36 (1–2), 103–123.
Ceola, S., Laio, F., Montanari, A., 2014. Satellite nighttime lights reveal increasing human exposure to floods worldwide. Geophysical Research Letters 41, 7184–7190.
Dang, N.M., Babel, M.S., Luong, H.T., 2011. Evaluation of food risk parameters in the day river flood diversion area, red river delta, Vietnam. Natural Hazards 56, 169–194.
de Paiva, R.C.D., Buarque, D.C., Collischonn, W., Bonnet, M.-P., Frappart, F., Calmant, S., Bulhões Mendes, C.A., 2013. Large-scale hydrologic and hydrodynamic modeling of the Amazon River basin. Water Resources Research 49, 1226–1243.
Elmer, F., Thieken, A.H., Pech, I., Kreibich, H., 2010. Influence of flood frequency on residential building losses. Natural Hazards and Earth System Sciences 10, 2145–2159.
Farr, T.G., Rosen, P.A., Caro, E., Crippen, R., Duren, R., Hensley, S., Kobrick, M., Paller, M., Rodriguez, E., Roth, L., Seal, D., Shaffer, S., Shimada, J., Umland, J., Werner, M., Oskin, M., Burbank, D., Alsdorf, D., 2007. The Shuttle radar topography mission. Reviews of Geophysics 45, RG2004.
Fekete, B.M., Vörösmarty, C.J., Grabs, W., 2002. High-resolution fields of global runoff combining observed river discharge and simulated water balances. Global Biogeochemical Cycles 16, 15–21.
Feyen, L., Dankers, R., Bódis, K., Salamon, P., Barredo, J.I., 2012. Fluvial flood risk in Europe in present and future climates. Climatic Change 112, 47–62.
Frappart, F., Calmant, S., Cauhopé, M., Seyler, F., Cazenave, A., 2006. Preliminary results of ENVISAT RA-2-derived water levels validation over the Amazon basin. Remote Sensing of Environment 100 (2), 252–264.
Guha-Sapir, D., Hoyois, P., Below, R., 2015. Annual Disaster Statistical Review 2014: The Numbers and Trends. Centre for Research on the Epidemiology of Disasters, Université Catholique de Louvain, Brussels, Belgium, p. 46. Available from: http://www.emdat.be/.
Güneralp, B., Seto, K.C., 2013. Futures of global urban expansion: uncertainties and implications for biodiversity conservation. Environmental Research Letters 8, 014025.
Hall, J.W., 2014. Editorial: steps towards global flood risk modelling. Journal of Flood Risk Management 7 (3), 193–194.
Hall, J.W., Dawson, R.J., Sayers, P.B., Rosu, C., Chatterton, J.B., Deakin, R., 2003. A Methodology for National-Scale Flood Risk Assessment. In: Proceedings of the Institution of Civil Engineers, Water and Maritime Engineering, vol. 156, pp. 235–247.
Hall, A.C., Schumann, G.J.-P., Bamber, J.L., Bates, P.D., Trigg, M.A., 2012. Geodetic corrections to Amazon River water level gauges using ICESat altimetry. Water Resources Research 48 paper W06602.
Hallegatte, S., Green, C., Nicholls, R.J., Corfee-Morlot, J., 2013. Future flood losses in major coastal cities. Nature Climate Change 3, 802–806.
Harman, C., Stewardson, M., DeRose, R., 2008. Variability and uncertainty in reach bankfull hydraulic geometry. Journal of Hydrology 351, 13–25.
Hess, L.L., Melack, J.M., Novo, E.M.L.M., Barbosa, C.C.F., Gastil, M., 2003. Dual-season mapping of wetland inundation and vegetation for the central Amazon basin. Remote Sensing of Environment 87, 404–428.
Hirabayashi, Y., Mahendran, R., Koirala, S., Konoshima, L., Yamazaki, D., Watanabe, S., Kim, H., Kanae, S., 2013. Global flood risk under climate change. Nature Climate Change 3, 816–821.
Horritt, M.S., Bates, P.D., 2002. Evaluation of 1D and 2D numerical models for predicting river flood inundation. Journal of Hydrology 268, 87–99.
Institute of Hydrology, 1999. Flood Estimation Handbook. Institute of Hydrology, Wallingford, UK.

Jongman, B., Ward, P.J., Aerts, J.C.J.H., 2012. Global exposure to river and coastal flooding: long term trends and changes. Global Environmental Change 22, 823–835.
Jongman, B., Hochrainer-Stigler, S., Feyen, L., Aerts, J.C.J.H., Mechler, R., Botzen, W.J.W., Bouwer, L.M., Pflug, G., Rojas, R., Ward, P.J., 2014. Increasing stress on disaster-risk finance due to large floods. Nature Climate Change 4, 264–268.
Jongman, B., Winsemius, H.C., Aerts, J.C.J.H., Coughlan de Perez, E., Van Aalst, M.K., Kron, W., Ward, P.J., 2015. Declining vulnerability to river floods and the global benefits of adaptation. Proceedings of the National Academy of Sciences of the United States of America E2271–E2280.
Jung, H.C., Hamski, J., Durand, M., Alsdorf, D., Hossain, F., Lee, H., Hossain, A.K.M.A., Hasan, K., Khan, A.S., Hoque, A.K.M.Z., 2010. Characterization of complex fluvial systems using remote sensing of spatial and temporal water level variations in the Amazon, Congo, and Brahmaputra Rivers. Earth Surface Processes and Landforms 35, 294–304.
El Kenawy, A.M., McCabe, M.F., 2015. A multi-decadal assessment of the performance of gauge- and model-based rainfall products over Saudi Arabia: climatology, anomalies and trends. International Journal of Climatology. http://dx.doi.org/10.1002/joc.4374.
Klein Goldewijk, K., Beusen, A., Janssen, P., 2010. Long term dynamic modeling of global population and built-up area in a spatially explicit way: HYDE 3.1. Holocene 20, 565–573.
Klein Goldewijk, K., Beusen, A., van Drecht, G., de Vos, M., 2011. The HYDE 3.1 spatially explicit database of human induced land use change over the past 12,000 years. Global Ecology and Biogeography 20, 73–86.
Lekuthai, A., Vongvisessomjai, S., 2001. Intangible flood damage quantification. Water Resources Management 15, 343–362.
Leopold, L.B., 1994. A View of the River. Harvard University Press.
Mace, G., Balmford, A., Bates, P., Brown, K., Cox, P., Douglas, R., Godfray, C., Grimm, N., Head, P., Nicholls, R., Sokona, Y., Toulmin, C., Turner, K., Vira, B., Viana, V., Watson, B., 2014. Resilience to Extreme Weather. The Royal Society, London, UK, p. 124. Available at: https://royalsociety.org/~/media/policy/projects/resilience-climate-change/resilience-full-report.pdf.
Mason, D., Cobby, D.M., Horritt, M.S., Bates, P.D., 2003. Floodplain friction parameterization in two-dimensional river flood models using vegetation heights derived from airborne scanning laser altimetry. Hydrological Processes 17, 1711–1732.
Mechler, R., Bouwer, L.M., 2015. Understanding trends and projections of disaster losses and climate change: is vulnerability the missing link? Climatic Change 133 (1), 23–35.
Merz, B., Aerts, J., Arnbjerg-Nielsen, K., Baldi, M., Becker, A., Bichet, A., Blöschl, G., Bouwer, L.M., Brauer, A., Cioffi, F., Delgado, J.M., Gocht, M., Guzzetti, F., Harrigan, S., Hirschboeck, K., Kilsby, C., Kron, W., Kwon, H.–H., Lall, U., Merz, R., Nissen, K., Salvatti, P., Swierczynski, T., Ulbrich, U., Viglione, A., Ward, P.J., Weiler, M., Wilhelm, B., Nied, M., 2014. Floods and climate: emerging perspectives for flood risk assessment and management. Natural Hazards and Earth System Sciences 14, 1921–1942.
Neal, J., Bates, P.D., Fewtrell, T., Hunter, N., Wilson, M., Horritt, M., 2009. Distributed whole city water level measurements from the Carlisle 2005 urban flood event and comparison with hydraulic model simulations. Journal of Hydrology 368, 42–55.
Neal, J., Schumann, G., Bates, P., 2012. A subgrid channel model for simulating river hydraulics and floodplain inundation over large and data sparse areas. Water Resources Research 48. Paper no. W11506.
Neal, J.C., Keef, C., Bates, P.D., Beven, K., Leedal, D., 2013. Probabilistic flood risk mapping including spatial dependence. Hydrological Processes 27 (9), 1349–1363.
Ngo-Duc, T., Oki, T., Kanae, S., 2007. A variable streamflow velocity method for global river routing model: model description and preliminary results. Hydrology and Earth System Science Discussions 4, 4389–4414.
Palmer, T., 2014. Climate forecasting: build high-resolution global climate models. Nature 515, 338–339.
Pappenberger, F., Dutra, E., Wetterhall, F., Cloke, H.L., 2012. Deriving global flood hazard maps of fluvial floods through a physical model cascade. Hydrology and Earth System Science 16 (11), 4143–4156.
da Paz, A.R., Collischonn, W., Tucci, C.E.M., Padovani, C.R., 2011. Large-scale modelling of channel flow and floodplain inundation dynamics and its application to the Pantanal (Brazil). Hydrological Processes 25, 1498–1516.
Penning-Rowsell, E.C., 2015. A realistic assessment of fluvial and coastal flood risk in England and Wales. Transactions of the Institute of British Geographers 40, 44–61.

Prigent, C., Papa, F., Aires, F., Rossow, W.B., Matthews, E., 2007. Global inundation dynamics inferred from multiple satellite observations, 1993–2000. Journal of Geophysical Research: Atmospheres 112, D12107.

Rodriguez, E., Morris, C.S., Belz, J.E., 2006. A global assessment of the SRTM performance. Photogrammetric Engineering and Remote Sensing 72, 249–260.

Sampson, C.S., Fewtrell, T., O'Loughlin, F., Pappenberger, F., Bates, P.D., Freer, J.E., Cloke, H., 2014. The impact of uncertain precipitation data on insurance loss estimates using a flood catastrophe model. Hydrology and Earth System Science 18, 2305–2324.

Sampson, C.C., Smith, A.M., Bates, P.D., Neal, J.C., Alfieri, L., Freer, J.E., 2015. A high resolution global flood hazard model. Water Resources Research 51 (9), 7358–7381.

Sanders, B.F., 2007. Evaluation of on-line DEMs for flood inundation modeling. Advances in Water Resources 30, 1831–1843.

Schumann, G.J.-P., Neal, J.C., Voisin, N., Andreadis, K.M., Pappenberger, F., Phanthuwongpakdee, N., Hall, A.C., Bates, P.D., 2013. A first large scale flood inundation forecasting model. Water Resources Research 49 (10), 6248–6257.

Schumann, G.J.-P., Bates, P., Neal, J., Andreadis, K., 2014. Fight floods on a global scale. Nature 507, 169.

Seto, K.C., Güneralp, B., Hutyra, L.R., 2012. Global forecasts of urban expansion to 2030 and direct impacts on biodiversity and carbon pools. Proceedings of the National Academy of Sciences of the United States of America 109, 16083–16088.

Smith, A., Sampson, C., Bates, P., 2015. Regional flood frequency analysis at the global scale. Water Resources Research 51 (1), 539–553.

Trigg, M.A., Wilson, M.D., Bates, P.D., Horritt, M.S., Alsdorf, D.E., Forsberg, B.R., Vega, M.C., 2009. Amazon flood wave hydraulics. Journal of Hydrology 374, 92–105.

Ward, P.J., Jongman, B., Weiland, F.S., Bouwman, A., van Beek, R., Bierkens, M.F.P., Ligtvoet, W., Winsemius, H.C., 2013. Assessing flood risk at the global scale: model setup, results, and sensitivity. Environmental Research Letters 8 (4) paper 044019.

Ward, P.J., Jongman, B., Salamon, P., Simpson, A., Bates, P., De Groeve, T., Muis, S., Coughlan de Perez, E., Rudari, R., Trigg, M.A., Winsemius, H.C., 2015. Usefulness and limitations of global flood risk models. Nature Climate Change 5, 712–715.

Wilson, M.D., Bates, P.D., Alsdorf, D., Forsberg, B., Horritt, M., Melack, J., Frappart, F., Famiglietti, J., 2007. Modeling large-scale inundation of Amazonian seasonally flooded wetlands. Geophysical Research Letters 34. Paper no. L15404.

Winsemius, H.C., Van Beek, L.P.H., Jongman, B., Ward, P.J., Bouwman, A., 2013. A framework for global river flood risk assessments. Hydrology and Earth System Science 17 (5), 1871–1892.

Winsemius, H.C., Aerts, J.C.J.H., Van Beek, L.P.H., Bierkens, M.F.P., Bouwman, A., Jongman, B., Kwadijk, J., Ligtvoet, W., Lucas, P.L., Van Vuuren, D.P., Ward, P.J., 2015. Global drivers of future river flood risk. Nature Climate Change. http://dx.doi.org/10.1038/NCLIMATE2893 online first.

Wood, E.F., Roundy, J.K., Troy, T.J., van beek, L.P.H., Bierkens, M.F.-P., de Roo, A., Döll, P., Ek, M., Famiglietti, J., Gochis, D., van de Giesen, N., Houser, P., Jaffé, P.R., Kollet, S., Lehner, B., Lettenmaier, D.P., Peters-Lidard, C., Sivapalan, M., Sheffield, J., Wade, A., Whitehead, P., 2011. Hyper-resolution global land surface modeling: meeting a grand challenge for monitoring Earth's terrestrial water. Water Resources Research 47, 10. W05301.

Yamazaki, D., O'Loughlin, F., Trigg, M.A., Miller, Z.F., Pavelsky, T.M., Bates, P.D., 2014a. Development of the global width database for large rivers. Water Resources Research 50 (4), 3467–3480.

Yamazaki, D., Sato, T., Kanae, S., Hirabayashi, Y., Bates, P.D., 2014b. Regional flood dynamics in a bifurcating mega delta simulated in a global river model. Geophysical Research Letters 41 (9), 3127–3135.

Zanobetti, D., Longeré, H., Preissmann, A., Cunge, J.A., 1970. Mekong delta mathematical model program construction. American Society of Civil Engineers, Journal of the Waterways and Harbors Division 96 (WW2), 181–199.

SECTION III

MODEL INSURANCE USE-CASES

CHAPTER

10

Insurance Pricing and Portfolio Management Using Catastrophe Models

Helena Bickley[1], *Gero Michel*[2,3]

[1]Sompo International; [2]Western University, London, Ontario, Canada; [3]Chaucer Underwriting A/S, Copenhagen, Denmark

INTRODUCTION

This chapter explains how catastrophe models are used in the insurance industry. Although subject to company specific differences, the basic needs and processes are similar across the market: obtaining data from clients and brokers, running models, using model results for risk selection, pricing, portfolio management, and reporting, and so on. This chapter is aimed at readers who are interested in (1) learning how insurance companies use models, (2) how much they rely on model results, and (3) how much modern risk takers depend on the use of sophisticated modeling engines, which are paramount for both day-to-day underwritings and company strategy. The chapter reflects the everyday work of a new generation of actuarial analysts, managers and statisticians, typically referred to as reinsurance "modelers."

A reinsurer is a company that helps insurers to minimize capital cost i.e., a reinsurer insures insurers. Reinsurers are sophisticated users of risk models because they must typically manage both local and global risks and analyze various lines of business and territories. Their business is often heavily structured and complex. Reinsurance is largely about optimizing capital and capital cost, and any risk a reinsurer takes hence has a more or less direct relationship to capital.

Catastrophe insurance underwriting is subject to perils (e.g., hurricanes or earthquakes) for which the past does not adequately explain the future. This is because catastrophe events are rare and unlikely to repeat themselves in the foreseeable future. For example, the last significant earthquake in California occurred in Northridge in 1994, and the last major (Cat 3+) landfalling US Hurricane occurred in Wilma in 2005. In addition, the financial exposure to

Risk Modeling for Hazards and Disasters
http://dx.doi.org/10.1016/B978-0-12-804071-3.00010-0

perils such as earthquakes, windstorms, floods, or terror events changes over time because of factors including inflation, growth, risk mitigation, changes in building construction, policy coverage, and conditions. This means that a repeat of the 1994 Northridge earthquake hazard footprint would cause a vastly different loss today. In addition, insurers might focus on very specific parts of a loss with complex insurance or reinsurance policies, for which it would be unlikely that the historical data would be of sufficient detail to explain the risk as of today.

Aggregating risk over various policies or portfolios and often sophisticated hedging of the tail (severity) risk of a (re)insurance company requires specific means to process risk for a company. This raises several key questions. How can we process decision-making at often-large uncertainty across a company? How can we make sure that all underwriters view risk in the same way and that each underwritten risk is both reported, correlated, and aggregated in the same way in a (re)insurance portfolio? How can we make sure that our *own* company-wide *view of risk*, the underwriting process, and expense loadings are prudent and can help us make money in the long term or confirm that we have sufficient capital such that we can pay our customers in the event of extreme catastrophes?

The answers to these questions deal with the practical use of catastrophe models in a reinsurance company. The processes involved include much more than technical pricing and range from analyzing and reporting risk submissions through to thoroughly processing risk across a company. Individual tasks include accumulating risk, assessing, monitoring, and communicating risk for various submissions, territories and perils, allocating and optimizing capital, and setting a long-term risk strategy for the future of a company. Catastrophe models, hence, (1) act as pricing and trading platforms, (2) provide evidence that the underwriting process is prudent, and (3) provide the means for risk communication and risk strategy across all company levels up to the company board.

Over the last few years, companies have accepted that models need calibration and testing against a company's data and research before implementation of the model in the underwriting process. Rating agencies and regulators require model testing, and model calibration has become complex. Companies often view model calibration and testing as a prime differentiator in their risk-management procedures. This is because of the commoditization and wide sharing of the most commonly used *vendor models* (see Chapters 1 and 2) in the market.

The importance of model calibration means that responsibilities typically reside at the highest levels of a company and have become a significant part of the role of the chief risk officer.

The following questions typically guide model validation and calibration:

1. What are the strengths and weaknesses of the model?
2. What reference hazard and risk models can be used to validate and calibrate the model?
3. What historical loss data can be used to test the model results?
4. Where are frequency, severity, and correlation results biased?
5. What is the most accurate resolution of a model (*and it may not be the highest*)?
6. Where is the model likely to be "overfitted" and hence creating false expectations (and bias)?
7. Where is it likely that our own data and assumptions might not fit the model requirements?
8. For what ranges are the results reasonably accurate?

The list is long and model calibration and validation is "core" for (re)insurance companies, who employ teams of experts (often highly experienced scientists and professors) to create their "secret sauce" in how they create their own models and calibrate the commonly used vendor models. Discussion of model testing and calibration is not part of this chapter, but the process of modeling risk in a company described below requires that model testing and calibration is complete, and the company believes that the used and calibrated catastrophe models are the best possible, given their current state of knowledge and research.

CATASTROPHE MODELS, THE PINNACLE IN THE UNDERWRITING PROCESS

Most insurers, all larger brokers, and all reinsurers use at least one catastrophe model. Their use is mandated by rating agencies and regulators. These models may be developed internally by the (re)insurance company or licensed from third-party vendors. The model provides an underwriting understanding of the risk. From a reinsurance perspective, the model helps the underwriter to understand the likelihood of a particular portfolio of business suffering losses from particular peril types and territories. Comparison of model results across risks enables the selection of more attractively priced deals or risks that fit a portfolio better. In addition, combining model results for selected risks or portfolios creates an overall company risk curve. This enables the (re)insurer to understand his or her own potential for loss. It can also help them to plan, think about growth opportunities, and the best way to increase returns on capital.

EXPOSURE DATA, THE FOUNDATION OF THE MODELING PROCESS

Exposure data, i.e., the information about values, risk types, and risk locations, are the foundation of the modeling process for most companies today. In the early days of modeling, obtaining the appropriate data was one of the largest obstacles to being able to use the models. In the absence of exposure models, premium was used as a proxy for risk. Exposure data was less important and often not even collected throughout insurance companies. Then, if collected, it was often in a format that was not suitable for running models.

With the advent of exposure modeling, clients changed their processes and began collecting exposure data. Exposure became an advantage for clients (the more complete or detailed the exposure information, the greater was the respect or interest in the client), such that more information was expected, decreasing the price for hedging. Reinsurers and the capital markets offer the hedges (reinsurance programs) required by insurance companies. As exposure models become more detailed, reinsurers will demand, in turn, higher exposure quality and resolution. Deemed incomplete exposure was heavily loaded, and brokers and reinsurers put pressure on insurance companies to increase data resolution and completeness further. Today, data quality has improved greatly, reaching the reinsurer after cleansing and in standard format for model analysis.

The insurance company, or cedant (the company that cedes risk into a contract), collects the data, typically through the network of agents and local offices involved at the time of writing the policy. Most companies have either proprietary or third-party software for this. This software allows geocoding of risks and the adding of estimates for property values (with indicators such as square footage), time-element coverages (e.g., business interruption), content, additional living expenses, and other factors. Geocoding is a process that assigns the attributes of the location—city, postal code, and latitude/longitude—to a "physical" location that the model can handle. Some software processes hard and soft information about property risk, including numbers of stories, age, mitigation features, distance to coast, and so on.

The broker, working for the cedant (and in between the cedant and the reinsurer), helps cleanse and analyze the data. Cleansing includes checking for duplicate entries, obvious mistakes, risk undervaluations, location errors, and the like. Brokers also check risk concentration using third party and internal models. How much of the modeled risk is due to single large risk locations? What are the 10% largest risks and are the data for those risks likely to be correct? There have been incidences of clients or brokers "adjusting" exposures to keep model results low and prices down. This is illegal (not only since Solvency II, http://www.lloyds.com/The-Market/Operating-at-Lloyds/Solvency-II/About/What-is-Solvency-II) and can breach a contract ("uberimma fides, utmost good faith" https://en.wikipedia.org/wiki/Uberrima_fides).

(Re)insurers calculate Probable Maximum Loss (PML) values and capital requirements for the (re)insured business from the results. The data are applied to structure the reinsurance program. The reinsurance program helps to optimize capital usage and maximizes return, subject to the underwriter's risk tolerance, and regulatory and rating agency requirements. Capital is allocated for the "bad days," i.e., for years in which premium is not sufficient to pay the losses arising from large natural or man-made catastrophes. Regulations permit (re)insurance companies to diversify risk and hence include "leverage" (https://en.wikipedia.org/wiki/Leverage_(finance)) in their capital allocation. The more diversified a company is, the less capital is (relative to the risk limits underwritten) needed. Reinsurance companies are often global and hence have lower (relative to their risk exposure) capital requirements than local insurance companies. Local insurance company risk is typically concentrated, and each new policy may add further risk concentration and hence might require more capital. Buying reinsurance is typically at lower cost than equity and hence reduces capital cost. The insurance company uses catastrophe models to determine levels of reinsurance purchase in comparison to capital requirements and return.

The data package transferred for the to-be-reinsured program to the reinsurer will typically include a summary of the exceedance probability loss (EP) curve modeled by the broker or cedant. The package will also include the exposure data, which are a listing of policy and location information. This information also includes construction and occupancy attributes as well as mitigation feature information if available. Large regional or overregional insurance companies possess portfolios with millions of locations and transfer exposure data using SQL databases.

The reinsurer uploads the exposures into the models and geocodes the data. A validation process then assesses the accuracy of the exposure data, highlighting outliers. For a mature book of business, where many of the accounts are renewals, the reinsurer will conduct year-on-year exposure comparisons. This will include monitoring changes in risk concentration, average values, single large policy values, and others. Such portfolios are stable, with organic growth potential limited in most areas and any significant changes (such as mergers and acquisitions) known to the market and/or communicated in the data package. The package will also list inflation and overall premium growth and exposure assumptions. If no information from previous year(s) is available, then the reinsurer will compare exposure distributions and values to standard industry metrics.

Standard software is available to assist with the exposure validation for developed countries. This allows comparison of client exposure at the highest possible resolution (up to "4-wall for the United States") to the industry exposure data sets. Validation includes checking the average value per property and providing a probability on whether it is likely that the client has certain exposure with certain construction types in a specific area. When reviewing the exposure data, the reinsurance analyst is looking for particular concentrations of exposure. For example, for wind exposure, the analyst will consider the average distance from the coast of the modeled locations, for earthquake locations, distance from particular risk zones or faults, and for flood, the relative elevation and distances to the coast or any nearby water body such as a river or lake. This information helps the underwriter to understand whether a particular book is more or less concentrated in high-risk areas. In addition, the analyst will review the exposure data for what is not included. For example, does the insurer write auto business that has been excluded from the exposure information? What is the insurer's exposure to infrastructure, cyber, terror, riot, business interruption, and other risks? Are those risks insured and how much exposure is allocated to those risks? Is the insurer subject to assessments by insurance pools, which are common in the United States, or other state mandated frameworks, which might contribute or reduce loss to a reinsurance program such as the coverage mandated by the Florida Hurricane Cat Fund (https://www.sbafla.com/fhcf/)? The analysis will also include any participation in other reinsurance programs, which could reduce expected losses.

The analyst checks peril coverage and processes the data through the models, selecting exposed and covered territory perils such as US hurricane. If the model excludes certain perils, then actuaries or other specialists will model any related coverage separately and then combine their results with those from the model. The analyst may also apply specific analysis options in the model including storm surge, demand surge or "postloss amplification" (including heightened construction prices after an event), time dependence for earthquakes if available, and consideration of average, or warmer than average, sea surface temperatures, or any other change in climate variability forecasted for the next season (if applicable in the model). The model generates loss results, typically by event. These events are associated with rates and assigned to (stochastic) years.

The analyst models the risk to help the underwriter determine the risk-based price to charge for the deal. Historical loss information usually provides the foundation for pricing a transaction; however, such information cannot offer a full picture of risk where the peril is infrequent and potentially catastrophic. Although the data package provided by the broker includes historical loss information, data points will be limited and, in many instances will not include losses going back more than 10 years. Reliance on such a short event history when trying to price against a peril that is infrequent might lead to significant under- or, in the case of recent significant losses, overpricing. To assist with the risk assessment, commonly used models include specific historical event footprints. Losses are calculated for these footprints using fragility/vulnerability curves. Calculated historical losses are then compared with inflated actual losses. However, given the nonlinear nature of fragility (see Chapter 4), judging model accuracy through a few historical events can be misleading, and the deemed best footprints available can still be out in terms of losses by a factor of two or more if not carefully calibrated. This is especially true, given that a few ticks up or down in local intensities can change losses by multiples.

Although referred to as "complete event sets," catastrophe models typically focus on higher severity events and enable analysts to make risk estimates beyond history. Typically, catastrophe models have predefined intensity (event) footprints with exposures and losses scaled to specific risks or insurance programs. Exposure attributes, such as construction and occupancy, refine loss results further. Integrating overall event losses and event frequencies will result in *mean or expected* losses for a portfolio of events. Running each event against each location for a portfolio of risks will create a spectrum of probable losses for a given risk or portfolio of risks.

Expected loss or the *mean* loss from the distribution is a measure of the risk the exposure is likely to undergo during the next year(s). The standard deviation around the *mean* approximates the steepness of the distribution and hence is a measure of the volatility around the *mean*. Within property catastrophe reinsurance, underwriters usually price *layers* or "tranches" of risk. The layers can be area, peril, or line of business specific. This makes managing and sharing/hedging of risk more efficient. For these layers, the underwriter considers loss averages across segments of the overall distribution rather than the distribution as a whole. For example, a layer may cover all losses for a large commercial building for fire and windstorm only above a deductible point (e.g., USD 50 mn) to a specific limit (e.g., of USD 60 mn), thus protecting the building to a maximum loss (USD 110 mn), with the owner (or another layer) paying for losses below USD 50 mn and above USD 110 mn.

Expected losses for each of the different perils covered in a policy (such as fire, windstorm, earthquake, and flood) are, in general, treated as additive. As mentioned earlier, where loss models do not include perils covered, companies have to deploy alternative tools to calculate expected losses and/or define loadings. Modeling and combining internal and external event sets will enable the company to calculate expected annual losses. The calculated expected loss is further loaded for the expenses associated with writing the transaction such as brokerage or commissions. In addition, general and administrative expenses, i.e., the incremental costs of running the operation need to be included in the calculated losses as well. The loading should include the cost of equity, i.e., the return shareholders require for providing equity, as well as the cost of debt (see below). Some companies also include an "uncertainty" loading. Historically a "Kreps factor" (or percentage of standard deviation) has been used as a proxy for the capital cost needed for each risk or portfolio of risks. This works because (re)insurance businesses must allocate capital to the tail of the business (see below). A company can approximate capital requirements using a PML [VaR, tail value at risk (TVaR), or similar, see below]. The PML defines the tail or high-severity losses. If adding a risk or new program to the portfolio changes the tail losses, then the standard deviation changes in proportion to the additional tail risk. More diversified portfolios need relatively less capital and their risk curves are less steep. Less steep curves have a lower standard deviation. Today, the majority of companies measure their capital cost for each deal based on explicit incremental tail losses (see Fig. 10.1).

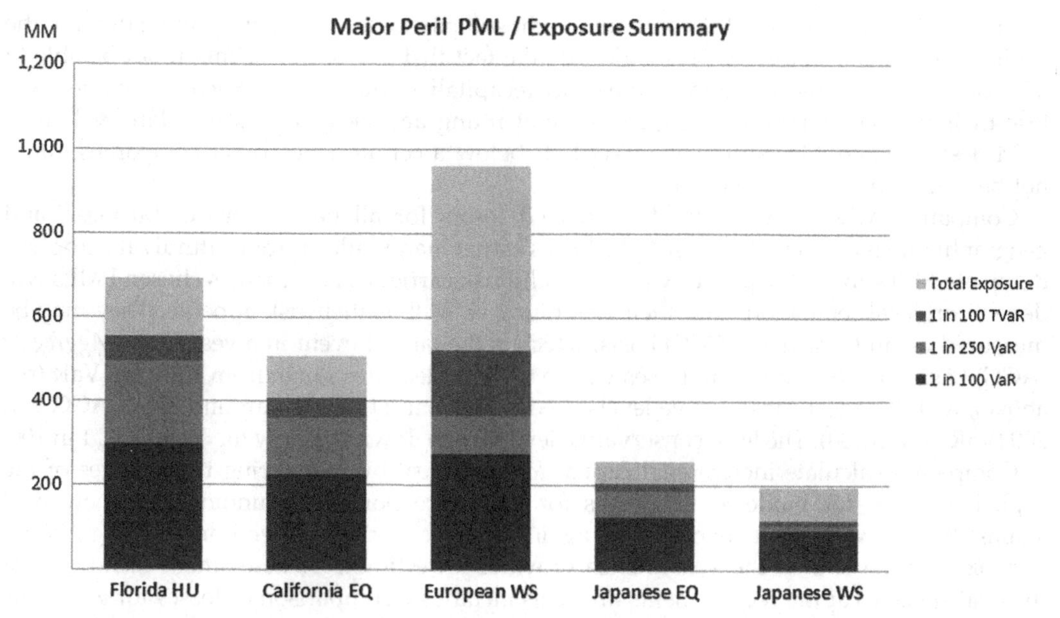

FIGURE 10.1 Value at risk (VaR), tail value at risk losses (TVaR), and exposure numbers as defined in the text. VaR and TVaR numbers are modeled for an assumed midsize reinsurer. It reads as follows: 28% of the overall exposure (limits) assumed for European Windstorm are destroyed by a 1 in 100 event. The TVaR100 number is 1.6 times larger. Depending on the risk appetite and regulation, companies will chose one of these measures. Exposure represents the sum of the layer limits.

PROBABLE MAXIMUM LOSSES, CAPITAL COST AND PORTFOLIO CONSTRAINTS

PMLs are defined using severity losses from EP curves. Cumulative probabilities translate event risk curves into EP curves (see Chapters 1 and 3). PMLs in the insurance industry describe certain return period losses, with return periods being defined as one through the cumulative likelihood of a loss happening in the coming year. This means that a 1 in a 100 event has a probability of 1% and a 1 in a 250 event has a probability of 0.4%. PMLs often follow Value at Risk (VaR), a measure for a specific return period loss as defined by the EP curve. TVaR integrates over the tail section of the EP curve beyond a defined return period (e.g., the 1 in 100). As TVaR losses include losses above a certain return period, these are hence higher than VaR losses for the same return period. PMLs (VaR or TVaR) are measures that define a company's expected loss from a high-severity event (such as a large California earthquake or a Miami Hurricane). Annual premiums or risk fees may not be sufficient to pay for these tail losses, and regulators require companies to hold capital to pay for these unlikely but high-severity events. The associated capital may take different forms—from equity and debt through to risk transfer methods such as reinsurance, derivatives, or other forms of hedging. Depending on the likelihood of the risk transfer–related capital actually being available in the event of the loss—defined as *credit risk*—the company may need to apply loads to capital for this deficiency. Certain PMLs define the level of risk that a company has to be able to sustain. Above that level, the company might still pay losses in full or prorated down if the capital suffers major erosion. This is despite the fact that the company might not be able to take on any new business and/or need to recapitalize after a loss. Such companies will lose their financial strength rating from credit rating agencies such as Standard & Poor's, A M Best, or Fitch. Once a credit rating falls below a certain level, other companies might not be able to trade with that company.

Company PMLs derive from the modeled losses for all perils, lines of business, and geographic areas combined. Catastrophe losses (either man-made or from natural catastrophes) dominate PMLs for many property and casualty risk carriers. A company's chosen PMLs will depend on local regulations and their risk rating as well as their risk appetite. They may be measured on an *Occurrence* (*OCC*) basis, which is the largest event in a year, or an *Aggregate* (AGG) basis, i.e., the sum of all losses within a year. Measures range from VaR to TVaR (see above), with the most conservative levels used in the market ranging around TVaR AGG 1 in 200 (VaR ~ 1 in 500). The least conservative levels reach down to below the VaR OCC 1 in 100.

Companies calculate incremental capital for each deal by monitoring the changes of the capital required for modeled tail losses for the entire portfolio. Running each new deal against the entire risk portfolio and/or against a forecast prospective portfolio (e.g., target portfolio for the end of the year) hence provides an estimate or assessment of how large the deal-specific capital requirements are. This includes a comparison of losses for all events for the portfolio with and without the new risk. For example, where a company has its major risk in California, then all losses outside a California earthquake might be paid from its premium because modeled losses outside of a California earthquake might not reach the tail of the distribution. The opposite holds for a new deal covering California earthquake. Modeled losses for this deal increase the tail risk distribution and hence the amount of capital needed (as a function of VaR or TVaR as defined earlier). The company may divide the derived additional amount of capital for this deal through their overall capital needed and then multiply this ratio with the capital cost to come up with an incremental cost of capital for the transaction. Although often defined at a point (e.g., VaR), calculating capital cost needs to consider a range of events. Return period, at which incremental loss for capital calculations technically start, require careful selection. A too high point leads to a too small number of events triggering capital costs and generate relatively unstable capital measures. If the selected range of incremental capital calculation is too low down, then the company is not considering diversification appropriately and marginal deals start to cost too much capital. No matter which range is chosen, numbers need to be scaled to the final PML VaR or TVaR target levels as discussed earlier. Companies often float their technical capital down to the technical requirements rather than fixing amounts to the actually available capital. This has the advantage of erasing the issue of changing the apparent capital cost for the succession of deals (portfolio deploying capital over time). The downside with this is that excess capital might not be considered accurately.

PORTFOLIO OPTIMIZATION AND REPORTING

Portfolio optimization requires a process of making select changes in risk shares for different perils and territories to simulate incremental changes in portfolio returns. These simulations (see Chapter 11) are typically the outputs of Monte Carlo–style testing, where the modelers change shares iteratively, with the model calculating the associated probable returns for all or for an optimized set of realizations. Portfolio optimization in general considers the underwriter's expectation of ranges in shares that may be achievable for each deal. Many reinsurance companies have integrated portfolio optimization into their decision-making processes.

Company senior management typically requires at least quarterly reporting of PMLs (some monthly or even more frequently). This is because PMLs drive capital and hence shareholder needs. Internal reports usually include an extended range of numbers, which comprise return period losses (both single occurrence losses as well aggregate or cumulative losses in one or several years) from the 1 in 2-year losses, all the way up to the highest tail PMLs—around the 1 in 1000 years or beyond (Table 10.1).

Combining PMLs with expected and actual earnings patterns in enterprise risk management tools enables companies to provide the most likely net returns for this or the next year(s) and/or expected losses for high-loss years, e.g., with large catastrophic events. Companies also model catastrophe hedging in this way. If a company buys cover, then it will subtract the cost of that cover from all relevant event losses modeled. If hedging includes parametric triggers, then PMLs must take account of *basis risk*, which, as defined here, means the expected difference between what the parametric trigger might pay out versus what the actual (*real-world*) portfolio loss might be at the trigger level. Model error (parameter risk) is part of all modeled portfolios or transactions. Parameter risk is often divided into an *aleatoric* or random part as well as an *epistemic* or knowledge part. Although the random part might seem easy to assess, the complex interplay of errors and skews loosely defined in error patterns may be difficult to unravel. Taking account of epistemic errors in models is highly subjective. European regulation now requires companies to understand as much as possible about both the random as well as the knowledge error.

Company senior management will regularly review PMLs and expected earnings and losses provided by the modeling team. Senior management will also create risk constraints and underwriting guidelines according to their risk appetites and tolerances for the company. In doing this, three questions will be considered: (1) how much return versus risk do we seek, given the prevailing market conditions?; (2) at what return period do we allow our premium to be lost? (1 in 5 for high risk takers, 1 in 20 or beyond for highly diversified, and more risk averse companies); (3) at what level are we prepared to completely lose our capital? The higher a portfolio's volatility, the higher the requested return in good years because shareholders will start to lose their capital in bad years. Even the most conservative companies might, however, lose their capital somewhere along the risk curve as capital becomes otherwise too large and expensive or risk-management costs begin to exceed profit. Balancing capital and return is the task of the CEO and the CRO. A company may fail because the CEO or CRO is too risk averse and returns become unsustainably low as much as it may fail because the CEO or CRO is too risk tolerant. In addition, companies are becoming more likely to fail because of: (1) increasingly competitive pricing and high-frequency losses that erase profit over several consecutive years, (2) unsuccessful M&A activities, or (3) too little risk appetite, which leaves too much fallow capital on the balance sheet earning too little return. Given the current levels of regulation and the increasingly sophisticated external and internal models available, company failure due to lack of capital has become increasingly unlikely in recent years. Catastrophe models guide all of this.

TABLE 10.1 TVaR Contribution Report for Florida

#	Cedant	Program	Limit	Attachment	ROL (%)	EL	Share (%)	TVaR PML Contrib.	US Hurricane Ev1	US Hurricane Ev2	US Hurricane Ev3
1	ABC Insurance Company	Cat XL	50,000,000	50,000,000	12.25	4,675,000	10	4,800,000	5,000,000	4,400,000	5,000,000
2	Albatross Ins Co	Cat XL	30,000,000	35,000,000	5.25	1,080,000	25	2,500,000	7,500,000	–	–
3	Albatross Ins Co	Cat XL	15,000,000	20,000,000	7.31	985,300	25	1,250,000	3,750,000	–	–
4	Albatross Ins Co	Cat XL	10,000,000	10,000,000	10.25	956,250	25	2,500,000	2,500,000	2,500,000	2,500,000
5	CLM Insurance Company	Cat XL	30,000,000	45,000,000	17	124,950	7	1,400,000	2,100,000	2,100,000	0
6	CLM Insurance Company	Cat XL	20,000,000	25,000,000	22	1,728,000	12	1,440,000	1,920,000	2,400,000	–
7	Everpaid Insurance Co	Cat XL	25,000,000	25,000,000	4	57,000.00	15	3,750,000	3,750,000	3,750,000	3,750,000
8	Everpaid Insurance Co	Cat XL	10,000,000	15,000,000	12	145,200.0	22	1,173,333	0	1,540,000	1,980,000
9	Everpaid Insurance Co	Cat XL	7,500,000	7,500,000	18	149,175	17	964,750	535,500	1,083,750	1,275,000
10	Everpaid Insurance Co	Cat XL	4,000,000	3,500,000	30	120,960	14	392,000	56,000	560,000	560,000

Numbers are in USD. Programs are catastrophe excess of loss programs with layers (tranches discussed earlier) with limits up to 50 mm and retentions down to as low as 3.5 mm. The rate on line (ROL) is defined as the price through the limit. A 12.25% ROL (first row) means that the premium for this deal is 6.125 mm. The expected loss (EL) is the technical mean or expected loss for this deal. The TVaR contribution stands for the Tail Value of Risk USD contribution this deal has to the tail risk of the overall portfolio run by the reinsurer. US hurricane event losses 1–3 stand for losses these deals would have to three specifically defined hurricane events (such as a Cat 3 making landfall in a specific territory). EL, TVaR, and event results are all coming out of a model that uses exposure data to run these programs and financial perspectives.

The table is from a reinsurer analyzing 10 Florida companies (not actual names) in her portfolio.

USING A MODEL: INSIGHTS OF A PRACTITIONER

When analyzing model output, it is important to remember that the model provides an estimate and perhaps a technocratic view of risk, which will still contain inherent bias no matter how sophisticated the modeler or expert who created the model might have been. The need to combine real-life underwriting experience with model results is important (see Chapter 15). Once the model has provided a number, there is always the inclination to believe it, no matter how complex or uncertain the underlying risks are and hence how unlikely it is that the number is accurate. In addition, most models assume that the past and the longer-term average approximate the future for the next year, an assumption that is likely to be incorrect.

It is important to note that all models are not suitable for all types of analysis. Each model has areas of strength and weakness. Knowledge of all of the exposure details and mitigation factors for a single risk does not mean that the model will provide the right answer because most often the focus will be on the strength of the model and not on its limitations. If an underwriter considers a book of Asian risks and only analyses the commercially available modeled output, then the price would exclude allocation for those perils that the model excludes. This approach would not have included a Thai flood peril because it is not included within the model suite (see Thai flood losses in 2011, https://en.wikipedia.org/wiki/2011_Thailand_floods). To take this example a step further, even where a peril such as US hurricane—probably the best-modeled peril around the globe—is included in the model suite, the model can still misestimate loss. Take for example, Hurricane Katrina in 2005, where modeled losses were significantly less than the actual losses. In some instances, models misestimated the vulnerability of particular types of structures to wind speeds, whereas other models underestimated the storm surge risk or the susceptibility of commercial property vulnerability to high wind speeds. Moreover, models struggled (and continue to struggle) with the assessment of losses because of contingent business interruption (indirect losses, see Chapters 5 and 6). Current models also continue to struggle with modeling tornado or flood losses.

Typically, (1) the more complex the financial structure of a transaction, (2) the larger the average claim size, or (3) the more postevent human interaction might change the loss, the greater the likelihood of a model producing inaccurate results. Models work well for relatively homogeneous books of business such as homeowner risks, where each risk is similarly insured, and will typically behave in a similar manner in a catastrophe. The lower model error for homeowner risks has reduced margins in this business and has driven companies to concentrate on vulnerable homeowner risks around the globe. As risks become more complex, and/or if the average claim size increases, modelers must make more assumptions about the performance of the risk and its potential characteristics. The more detailed a portfolio or model is, the more it may remove the potential bias created by exposure peculiarities. However, models that are more detailed are also more complex, and the data set for calibrating the model might be small in comparison to the assumptions required to achieve the fit. Hence, the highest model resolution can also be the least accurate resolution, and modelers have to carefully balance model resolution with the *necessary* detail required to run the model.

Models have clearly "learned" over time and catastrophe models are becoming better with new loss experience and new science introduced. These features are, however, only a small part of a model. The assumptions and uncertainties that drive a model often provide more insight than that learned from a few new events over a small number of years. There are areas where models have become better at erasing past model error. There is, however, no clear evidence that complex risk models provide more accurate risk results in general now than they did 5 years ago. The fact that some recent releases of the same vendor risk models have industry losses going up and down without considering any short-term changes in hazard may be taken as evidence for this. Using models as trading platforms might make it necessary to run at loss resolutions that are much higher than what the uncertainty of the model would actually allow. Capital models, however, are required to be stable and a new model version changing a company's capital needs with unchanged exposures and at the same overall hazard conditions could mean that the company running the model might not have considered model uncertainty appropriately.

CHAPTER

11

Portfolio Optimization Using Catastrophe Model Results, a Use Case From the Insurance Industry

Augustin Yiptong[1], *Gero W. Michel*[2]

[1]Sompo International, Carlisle, MA, USA; [2]Chaucer Copenhagen, Copenhagen, Denmark

INTRODUCTION

Portfolio optimization attempts to maximize portfolio returns by selecting or increasing those risks in a portfolio that have higher premium to return ratios and/or lower tail risk for a given set of deals. Portfolio optimization has been widely used throughout the financial market, and several of its creators won Nobel Prizes for the theory (Markowitz, 1991).

Real-world portfolio optimization scenarios, however, do not always fit the Markowitz mean-variance model (Markowitz, 1952) because of complex constraints and formulations of desired return. This had led to the application of heuristic approaches (Gilli and Kellezi, 2000; Estrada, 2007), simulated annealing (Crama and Schyns, 2003), and genetic algorithms (Lin and Liu, 2008).

Insurance companies have been using portfolio optimization methods since over two decades trying to optimize both their underwriting and asset portfolios. Optimizing large underwriting portfolios, however, often lacked consistent modeling results, and/or insurance companies' portfolio optimization has often been restricted to traditional standard deviation measures. These methods take advantage of the fact that the standard deviation around the mean decreases if the tail risk decreases. Checking the standard deviation after adding or removing risks in a portfolio is hence a proxy for how the overall portfolio changes with each new risk. Procedures were often constrained within the optimization based on single risk frontiers herewith, restricting a company's ability to optimize across differences in risk appetite and/or across competing portfolios.

With the advent of catastrophe risk modeling in the 1990s, insurance companies started to use "complete" sets of loss events for a variety of territories and perils. An event is herein

Risk Modeling for Hazards and Disasters
http://dx.doi.org/10.1016/B978-0-12-804071-3.00011-2

considered "physical representation" of the loss footprint of a single earthquake, flood, windstorm, terror attack, or alike. These models have been producing risk tensors (used and defined as vectors below) for each modeled risk in a portfolio. These constructs are made out of a set of incremental losses for each risk and event such that the sum of all incremental risk losses in a portfolio defines any one loss event for the final modeled portfolio. Because models are deemed complete, integrating over a risk curve (i.e., the sumproduct of losses and likelihood for all events modeled) therefore results in the mean or average annual loss for a portfolio. Tail risk is derived by sorting event losses by size and considering the cumulative likelihood of losses. As long as correlation of underlying risks is taken into account for each event (and deemed complete), optimizing a portfolio simply means varying the shares for each possible risk and comparing the overall portfolio returns with a variation of shares for each risk. The shares for the best returning portfolio are then selected for the portfolio strategy and used as guides for underwriting.

This contribution describes details of a method used in the reinsurance industry. We have decided to leave out any knowledge that is deemed proprietary. This includes specific portfolio and/or client data. We have also opted to rename specific parts of the models used not to confuse the reader with company-specific nomenclature.

The optimization procedure presented can be applied to all those portfolios that can be modeled with complete event sets and incremental losses for each risk and event. We have, however, chosen to describe the procedure using a portfolio of insurance policies herewith explaining the methods using a hands-on example of an insurer or reinsurer who would apply this method. The paper is divided into three sections including (1) an introduction to the method, (2) portfolio results and case studies, and (3) a discussion around the value of the results for decision-making as well as how decision-making is restricted by assumptions made in the underlying models. The final section then discusses ideas that might help for the future of portfolio optimization methods in the insurance industry.

THE PROCEDURE

The purpose of this paper is to describe how portfolio optimization is used in the insurance industry, and how a specific algorithm can be used to optimize a portfolio of complex risks in a reasonable amount of time. This paper hence describes a method that has been used in a company to select risks and hence serves a practical example of how insurance or reinsurance companies make portfolio decision using catastrophe models. This paper is written as guide for anyone interested to reprogram and use the algorithm.

We have chosen a portfolio of risk layers for the portfolio optimization procedure described below. Risk layers are reinsurance treaties that cover loss slices or sections of individual portfolios (covering several risks) or single risks. A layer can be understood to, e.g., cover only losses above a certain retention or deductible up to an agreed maximum limit as defined in an insurance contract. These layers are common in commercial and facultative insurance as well as reinsurance. Insurance and reinsurance companies define risk for their portfolios based on simulated years, each of which could be the next year. We have 100,000-year simulations for the procedure below. 100,000 years is commonly used for insurance and reinsurance portfolio risk management although there is no agreed rule to any number of years and companies chose to simulate between several thousand and several million years depending on risk appetite/tolerance and on the portfolio mix. Steep risk curves for, e.g., intraplate earthquake risks require higher numbers of simulated years, whereas, e.g., portfolios that are highly diversified and/or cover mostly high-frequency convective storm risks might be fine with simulating only 1000 loss years. For the selected portfolio and in each year, a number of events of different perils occur in different locations as modeled using a suite of catastrophe models. For this example, the total number of events is ~8 million.

The following abbreviations are used in the procedure below:

GUL
Cat models are used to calculate ground-up losses (GULs) for each event for each layer. GULs do not consider policy conditions and are void of deductibles and so are losses from the ground-up.

ELT
Given the layer terms, our financial module calculates the layer loss for each event in each year. This is the layer event loss table (ELT). The layer ELTs can be summed to get the portfolio ELT. This means that ELTs as used herein—and other than GUL—consider policy conditions.

YLT
We can also sum the event losses in each year to get the year loss table (YLT). This can be done for each layer and also for the entire portfolio.

With the layer and portfolio ELTs and YLTs, risk metrics can be calculated, such as occurrence exceedance probabilities (the largest single loss in a year), aggregate exceedance probabilities (i.e., the sum of losses in a year), expected loss or the mean or integral losses for a portfolio, the expected profit (premium minus losses and expenses), the VaR (value at risk) or PMLs (probable maximum losses), the TVaRs (tail values at risk), or the market share in case that an industry ELT is run in parallel to the target portfolio. Given the assumption of event definition by a certain geographical resolution and considering event completeness, these metrics can also be calculated for different geographic zones or perils.

Probable Maximum Loss or Tail Value at Risk Constraints

Insurance companies base their capital constraints commonly as functions of the PMLs (VaRs) or TVaR results for return periods between a 1 in 100 or 1 in 200 depending on regional regulation and specific company risk tolerance (see Chapter 10). A given amount of capital hence restricts the amount of risk that can be taken in a portfolio.

Optimization

Optimization is generally defined as maximizing a utility function in a parameter space while satisfying one or more constraints. We can alternatively minimize a cost function, which can be thought of as just the negative of the utility function.

The following scenarios are part of our first example: Starting from our existing portfolio, we want to maximize the expected profit (this is our utility function). Our parameter space is defined by the range of values for each layer-signed share. In our example, layer-signed share can be doubled, left unchanged, or dropped. Our constraint is to have an optimized portfolio 100-year TVaR less than the current portfolio 100-year TVaR.

Algorithm

Our portfolio has 833 layers, and so our parameter space has 833 dimensions and a total number of 3^{833} discrete points, which is $>10^{397}$. This space is much too large for an exhaustive search or direct Monte Carlo sampling. Only 4.32×10^{17} or so seconds have elapsed since the big bang! In addition our utility function is not guaranteed to be convex, i.e., it may contain local maxima, see Fig. 11.1 below. This excludes using convex optimization algorithms or hill-climbing algorithms. We hence propose using the simulated annealing metaheuristic (Kirkpatrick et al., 1983).

Trial Portfolio

A trial portfolio is defined by the signed share of each layer in the portfolio. Thus the 833 element vector of layer-signed shares defines the location of the portfolio in parameter space. We can compute the trial portfolio ELT (~8 million element vector) by multiplying the layer-signed share vector by the matrix of layer ELTs where each row of the matrix is a layer ELT.

$$\text{Portfolio ELT Vector} = \text{Signed Share Vector} \times \text{Layer ELT Matrix}$$

Utility Function

For each layer in our trial portfolio we can calculate the expected layer loss from the layer YLT. Given the layer premium we can calculate the layer profit. Multiplying each layer profit by the layer-signed share in the trial portfolio and taking the sum over all layers results in the expected profit of the trial portfolio. This is our utility function.

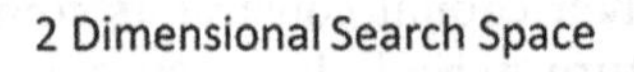

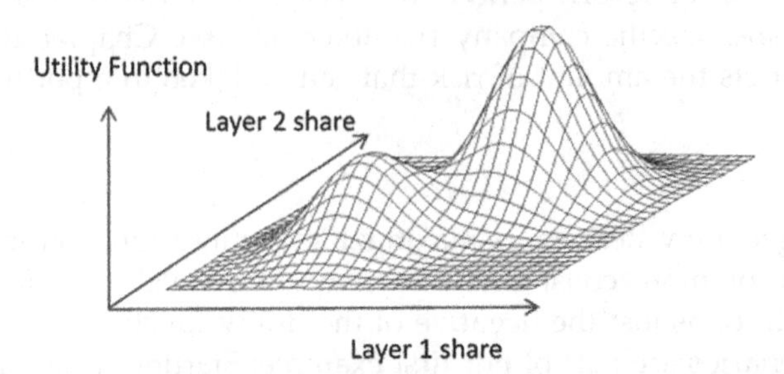

FIGURE 11.1 Search space landscape may contain local maxima.

Constraint

From the trial portfolio ELT we can calculate the YLT and TVaR 100. This tells us whether our trial portfolio is within our TVaR 100 constraint or not (see previous discussion).

We start the simulated annealing metaheuristic with the current portfolio. The next step includes generating a trial portfolio by selecting a number of layers at random (5 is a number that works well in our example) and giving each selected layer a signed share picked randomly from 0x, 1x, or 2x the starting share.

Temperature

The "temperature" is calculated at each iteration i using this formula:

$$T[i] = e^{-\text{Cooling Rate} \times \frac{i}{n \text{ Iterations}}} - e^{-\text{Cooling Rate}}$$

where a value of 6.0 is used for the cooling rate and n iterations is the total number of iterations in the simulated annealing run, 100,000 works for our example. This results in a "temperature," which cools exponentially from an initial value close to 1.0 to a final value of 0.0.

Acceptance/Rejection of Trial

If the constraint set earlier is not satisfied, the trial portfolio is rejected.

If the constraint is satisfied, we evaluate the utility function. If the utility function is greater than the previous trial portfolio, the trial portfolio is accepted. We can also accept the trial if the utility function is smaller than the previous trial depending on the difference in value of the utility function and the temperature at that step. The formula used for acceptance is:

$$\frac{\text{Utility Function}[i-1] - \text{Utility Function}[i]}{\text{Utility Function}[i]} \times \frac{1}{1000} < \text{Annealing Temp}$$

If the trial portfolio is accepted, it is used to generate the next trial portfolio. Otherwise, it is discarded and the previous trial portfolio is used.

Iterations of generating trial portfolios are continued until we reach a total of 100,000 iterations or our utility function meets our convergence criterion. The utility function is checked every 10,000 iterations, and we stop the simulation if it remains unchanged from the last check (Fig. 11.2).

Global Versus Local Optimum

Several independent simulated annealing runs are performed, starting from randomly chosen trial portfolios in parameter space. All the runs converge to similar values of utility function and optimum portfolios. This gives us confidence that the simulated annealing metaheuristic performs as expected in finding the global optimum and does not get stuck in local optima.

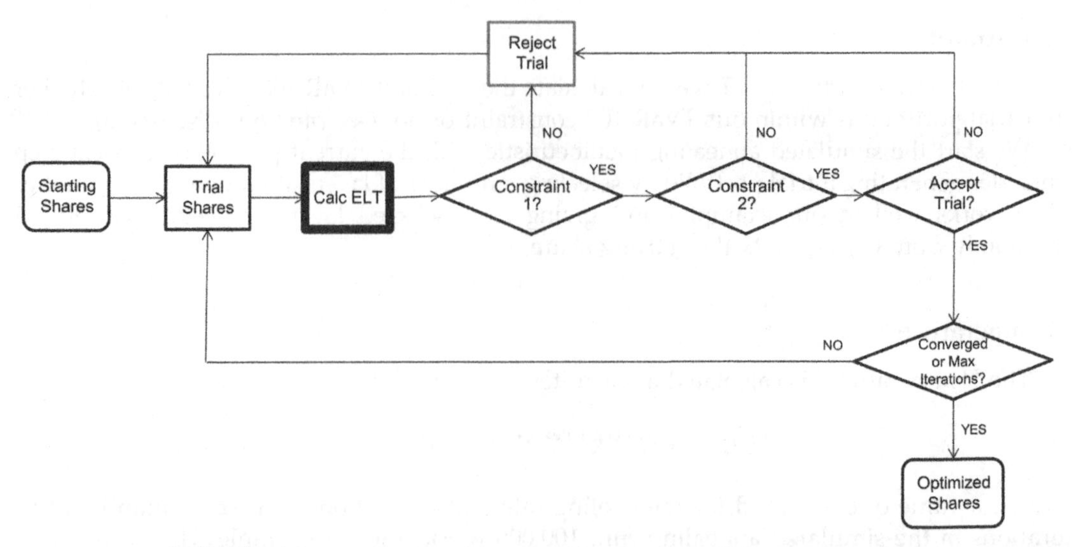

FIGURE 11.2 Simulated annealing algorithm flowchart. *ELT*, event loss table.

Stochastic Hill Climbing

We have used stochastic hill climbing as an alternative to simulated annealing. Stochastic hill climbing is achieved by systematically rejecting trial portfolios with worse utility functions. Another way to look at the stochastic hill climb is simulated annealing with the annealing temperature fixed at zero. The stochastic hill climb is more computationally efficient than simulated annealing, i.e., it converges to a solution within fewer iterations. Typically the solution is almost as good as the one found by simulated annealing, which indicates that our utility function does not have many deep local minima. Stochastic steepest descent is a similar algorithm with the difference of minimizing a cost function instead of maximizing a utility function (Fig. 11.3).

In the example, we compared four simulated annealing runs to four stochastic hill climb runs. All the results are between 11.924% and 11.962% for the portfolio return on equity (ROE) (defined as portfolio profit divided by TVaR 100, i.e., assuming that TVaR 100 is used as a measure for the capital needed), with the simulated annealing on average only negligibly better than stochastic hill climbing (by 0.018%) (Fig. 11.4).

PLOT EXPECTED PROFIT VERSUS TVaR 100

We can perform multiple optimization runs with different TVaR 100 constraints, i.e., allowing TVaR and hence capital to increase or decrease.

Fig. 11.5 shows that the selected base portfolio is suboptimal, and we could increase the return without changing TVaR 100 and or required capital if we could select layers differently. The efficient frontier in Fig. 11.5 allows management to choose profit against TVaR 100. The depicted relative returns on TVaR are higher for lower TVaR values because the

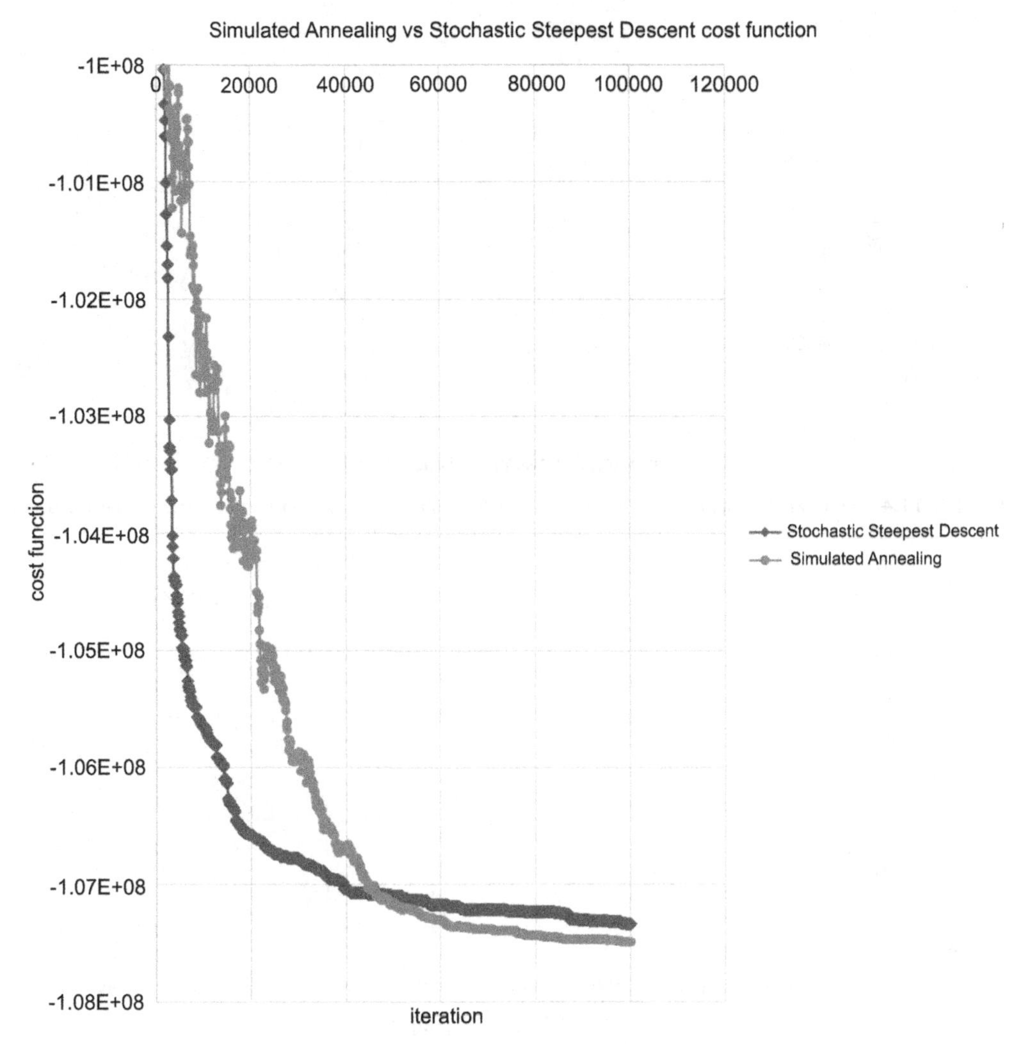

FIGURE 11.3 Simulated annealing typically takes longer to converge compared to stochastic steepest descent, but converges to a better solution.

relative abundance of well performing layer shares decrease with portfolio size, i.e., the larger the portfolio chosen from the available layers, the smaller is the relative return.

Robustness of Optimum Portfolio

Portfolio optimization is based (here) on modeled results that are subject to model error and uncertainty (see discussion). In addition, execution of the optimum portfolio is likely to be imperfect because of market or underwriting constraints (see discussion). To make choices valuable it is therefore important to find an optimum portfolio where the utility

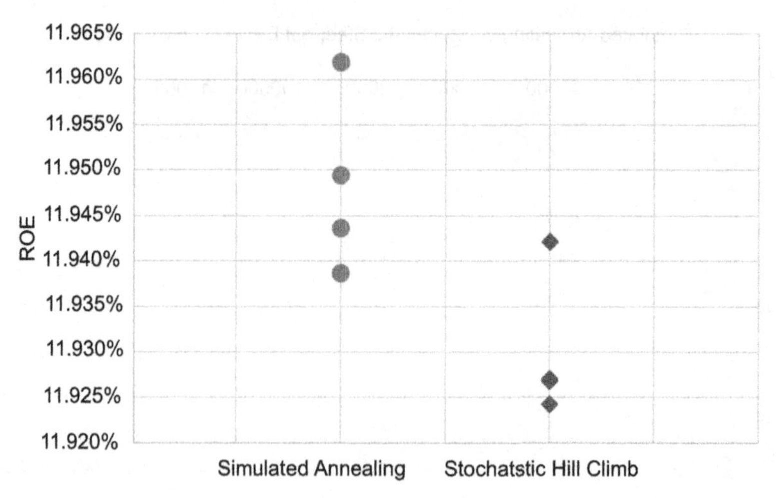

FIGURE 11.4 Simulated annealing typically finds a better solution compared to stochastic steepest descent.

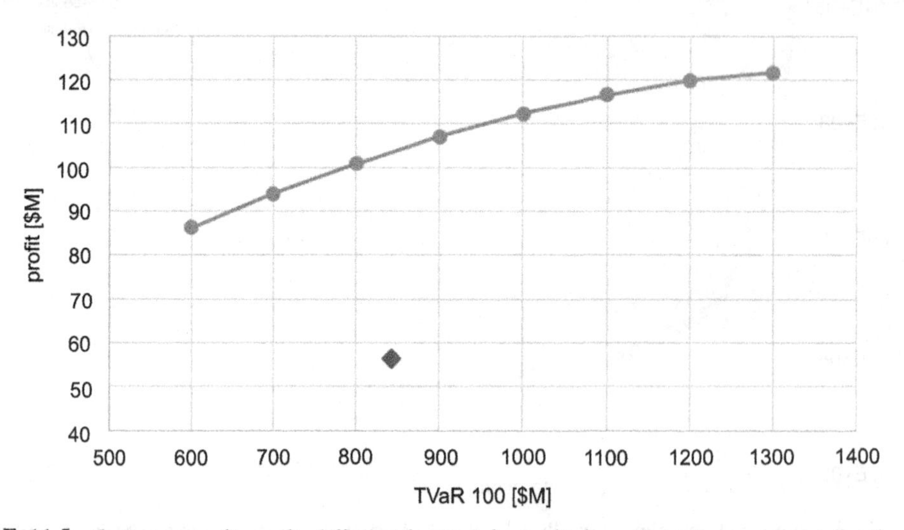

FIGURE 11.5 Optimum profit results (efficient frontier) for multiple optimizations against tail value at risk (TVaR) (line) versus the selected base portfolio (point).

function does not decrease steeply if we deviate slightly from the optimum in parameter space. To do this we performed stress tests for which each layer risk (signed share) was increased or decreased by 5%. The ROE dropped from 11.94% to 11.92% in the worst case. We repeated tests at 10% and 15%. Results are plotted below (Fig. 11.6).

Other Optimization Work

Convex optimization can be used in certain special cases if the utility function and constraints can be proven to be convex or transformed to convex functions (Tuncel and Dalis).

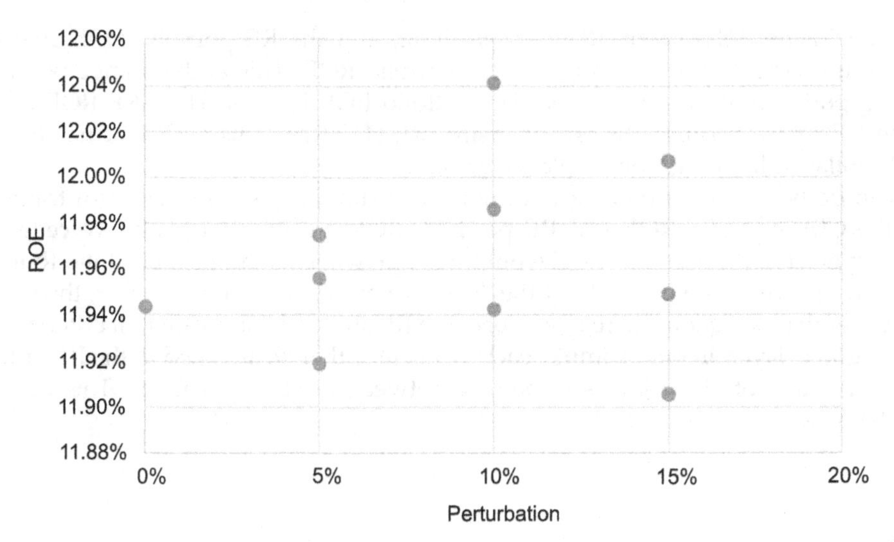

FIGURE 11.6 Randomly perturbing the layer signed shares of the optimum portfolio does not significantly reduce the portfolio return on equity (ROE).

Convex optimization algorithms are much more computationally efficient. The advantage of simulated annealing is the greater flexibility in defining utility functions and constraints.

A genetic algorithm (Holland, 1975; Melanie, 1999) can be used as an alternative to simulated annealing, which can be considered a special case of a genetic algorithm with a "population" of one.

Win–Win, Optimizing Two Portfolios Based on Different Performance Constraints

In our second example we consider the following scenario: A reinsurance company wants to cede out a subportfolio from their main reinsurance entity (RE) portfolio to another reinsurer. This could be an insurance partner who reinsures our portfolio or it could be, e.g., a fund in the capital market with, e.g., a collateralized reinsurance entity (Ceded Re, CRE). For this example we have chosen a collateralized entity, i.e., a partner that cannot diversify risk in its insurance portfolio. Collateralized entities come from the banking sector with constraints and return requirements different to the constraints in the insurance industry. Different constraints give an option to optimize portfolios simultaneously herewith allowing for win–win solutions, i.e., for a solution that both mutually profit from. We assume that CRE has a capital of $500M. We would like the ROE for both portfolios to be as high as possible. ROE for the RE portfolio is defined as the profit through TVaR 100, whereas in the CRE portfolio case it is defined as profit through a fixed amount of $500M. The CRE portfolio sum of layer limits cannot exceed $500M.

A commonly used way to cede business from RE to CRE is to simply give CRE a (quota) share of the RE book, i.e., let CRE participate proportionally on all layers in the RE book such that the CRE portfolio limit equals $500M. In this case the CRE portfolio ROE would be simply the RE portfolio profit/RE portfolio limit, and the RE portfolio ROE after ceding would be unchanged from the original RE portfolio ROE. This is the orange point on the plot: CRE portfolio ROE is 1.8%, and RE portfolio ROE is 6.7%. The CRE ROE is smaller than the RE return because RE can diversify risk (TVaR) versus CRE that has to account for each dollar of limit with one dollar of capital.

We can do better by optimizing both portfolios simultaneously. There is a trade-off between the CRE portfolio ROE and RE portfolio ROE. In other words, if we cede out the "worst" layers, the RE portfolio ROE would be high, whereas the CRE portfolio ROE would be low and vice versa if we ceded out the "best" layers. As mentioned earlier, the reason we can do better than the quota share case is because RE and CRE have different criteria for what makes a "good" layer. In our example, each layer can either be assigned to the RE or the CRE portfolio. We do not allow layers to be split between the two portfolios. This defines our parameter space.

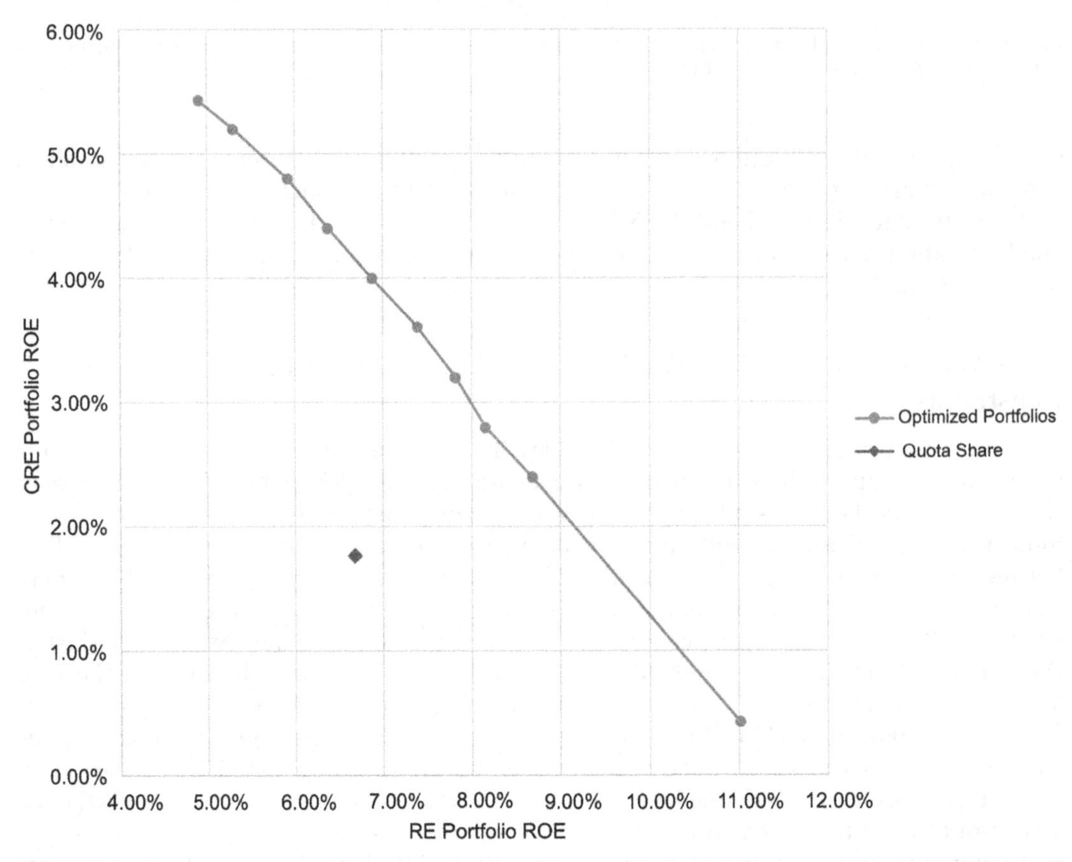

FIGURE 11.7 Optimization of 2 portfolios is a trade-off between the 2 portfolio returns on equity (ROEs) but still significantly better than the unoptimized quota share solution. *CRE*, collateralized reinsurance entity; *RE*, reinsurance entity.

The first run optimizes the CRE portfolio only, without considering the RE portfolio. The utility function is CRE portfolio profit and constraint CRE portfolio limit not exceeding $500M. This gives the leftmost point on the plot. CRE portfolio ROE is maximized at 5.4% at the expense of the RE portfolio ROE, which drops to 4.9%.

The second run optimizes the RE portfolio only. The utility function is RE portfolio ROE and the constraint is CRE portfolio limit not exceeding $500M. This gives the rightmost point on the plot. RE portfolio ROE is maximized at 11.0% at the expense of the CRE portfolio ROE, which drops to 0.4%.

To get the points in between those two extreme cases, we run several optimizations where the utility function is the RE portfolio ROE. Now we have two constraints: the CRE portfolio limit cannot exceed $500M, and the CRE portfolio profit must be at least X, where the value X is $12–$26 M to generate the point on the plot.

Fig. 11.7 shows that it is possible to beat the quota share case and have a win–win scenario. It is up to the management to decide where on the curve they want to be. For example, they can cede a CRE portfolio with an expected ROE of 4.0% while simultaneously improving the RE portfolio ROE from 6.7% to 6.9%.

COMPUTATION DETAILS

We have performed computations on servers equipped with ivy bridge CPUs. One optimization run would take on the order of a few hours using 8 CPU cores. Our high-performance computing environment enables us to queue jobs up and run jobs in parallel subject to availability of compute cores. The algorithms are implemented in Python/NumPy code. R programming language (Xiang et al., 2013) or MATLAB (Mathworks) could be used also.

DISCUSSION

Portfolio optimization using detailed output from catastrophe exposure risk models has been evolving in our industry over the last 5 years. The procedure described here has provided practical help for both primary and reinsurance underwriting as well as portfolio strategy and management throughout a company. Input from catastrophe models is sufficiently detailed to do the study at almost any targeted resolution, and results are scalable covering small local insurance portfolios to global reinsurance books. The two most important downsides for the use of the procedure above and algorithm are (1) the ability of underwriters to perform and follow the targeted/optimized share strategy and (2) that model uncertainty and model risk have not been addressed in the described procedure. The practical approach for the latter was to (1) calibrate the used model to the specific needs and data available for the targeted book before optimization and (2) concentrate the optimization on the use of mostly "simple risks" (homeowners) portfolios for which model risk is deemed manageable. We then assumed that model risk is negligible during the procedure. This is an oversimplification of the issue because model risk (even for homeowners' books) might be larger than the event losses used to optimize. Model risk includes (1) an aleatoric or random part that is (to a certain degree) measurable and included in the model output, i.e., could be used during the optimization. The (2) epistemic or knowledge part of the uncertainty is, however, more challenging to implement and likely to drive the model risk. This includes mismatch in the processes assumed in extrapolating smaller historical events to potential large future events, fallacy in the idea that the past suffices to forecast the future, or the fact that most models used are created for the long term rather than trying to include the specific conditions for the next year relevant for the single-year policies optimized for most insurance and reinsurance portfolios. Model results are, in addition, scale-dependent, and accuracy changes with resolution. Higher resolution results are only better if the higher complexity that comes with the higher resolution is backed by the data used to calibrate the model, which is often not the case. The number of events often allows only a crude view of the risk, and optimizing single large risks portfolios is unlikely to be promising using the commonly used catastrophe models. Involving uncertainty in the optimization in the future is probably the most important and most promising aspect currently missing in our procedure.

Underwriting strategy and performance: Let us consider that the optimal outcome for a certain deal was 10%, but the underwriter could only get 3%, or to take the planned 10%, the underwriter had to take another share for another risk of the same company that is rather suboptimal for the book. How can this be dealt with using the above procedure? In addition, prices are not fixed, and the targeted price for a certain risk changes because of market conditions. Even if we run the portfolio optimization quasi real time, i.e., rerun it after assuming each new deal, the procedure needs to be forward-looking and "holistic." This means that changes in one part of the portfolio might require subsequent changes in other parts to achieve an optimum result. Forecasting price changes or having a good idea of what is available is hence paramount in addition to trying to stick to an overall underwriting strategy and plan. To get around the underwriting and market issues, underwriters were predicting certain ranges of possible shares for a targeted book before the renewal season. Rather than allowing any share to be taken for any of the targeted risks, the procedure was hence running with forecasted share ranges such as 5%–15% for one risk or 20%–35% for another according to underwriters' judgment and experience. Price forecasts were, in addition, varied and portfolios were optimized based on ranges of anticipated price changes. Results were compared, and the underwriting strategy was based on a combined view herewith checking the robustness of the measures with respect to underwriting uncertainty. In addition large year-on-year swings in shares are unlikely at least for mature portfolios, and upfront portfolio changes are often incremental (changes of 10%–20%). Insuring or reinsuring clients is a long-term service, and changing your portfolio radically over a year will not help your long-term relationship with clients.

This is, however, different when it comes to the back end of a portfolio, and portfolio optimization has its deemed largest gains in hedging risk using reinsurance, retrocession including indemnity deals, triggers, swaps, and alike. This means that risk in a portfolio can be selectively changed by moving specific risk to another entity in the insurance or capital market. The other entity might have a very different risk appetite (see ROC measure example early in the chapter). Deals that allow detailed changes in risk returns of a portfolio include surplus deals (reinsurance of selective shares for a portfolio of risks), facultative business (hedging parts of individual large risks), "excess of loss" reinsurance (taking out certain layers for certain perils), or alike.

The herein-described portfolio optimization procedure using catastrophe models was used in a company that specialized in both insurance and reinsurance throughout the globe. We have kept some proprietary details out and herewith made the procedure more commonly applicable over a wider market. To implement the methodology we recommend companies to think specifically about their own process and how this might require certain details of the code and procedure to be adjusted. Some specific companies such as Analyze Re, AIR, and others have been involved or specializing in portfolio optimization and can help companies to do the work. Codes and optimization can, in addition, follow different procedures, and it was not the intention of this paper to provide a comprehensive overview of what exists but rather help with a practical use case of a procedure applied in the market.

CONCLUSION

We have demonstrated how reinsurance portfolios can be optimized using simulated annealing and stochastic steepest descent algorithms. The simulated annealing algorithm generally converges to a better solution at the expense of a longer convergence time. These algorithms, although computationally expensive, are quite generic, allow great flexibility in the choice of utility function, and permit multiple constraints to be defined. It is important to note that we use the full matrix of all layer event losses in our 100,000-year simulation. This enables us to calculate "exact" metrics for our utility function and constraints of choice. We have found that the return of an insurance portfolio can be increased substantially, over 50% in a theoretical scenario, while maintaining a similar level of capital requirement. The optimized portfolio has been stress tested and found to be resilient even if the portfolio execution is suboptimal. In the exercise of ceding a subportfolio from a reinsurer to a collateralized entity, we have shown that we can achieve a win–win scenario, doubling the ceded subportfolio returns while maintaining the reinsurance portfolio returns, compared to the straight quota share result.

References

Crama, Y., Schyns, M., 2003. Simulated annealing for complex portfolio selection problems. European Journal of operational Research 150 (3), 546–571.

Estrada, J., 2007. Mean-semivariance optimization: a heuristic approach. SSRN Electronic Journal. http://dx.doi.org/10.2139/ssrn.1028206.

Gilli, M., Kellezi, E., 2000. Heuristic approaches for portfolio optimization. In: Sixth International Conference on Computing in Economics and Finance of the Society for Computational Economics. Barcelona, 2000.

Holland, J., 1975. Adaptation in Artificial and Natural Systems. The University of Michigan Press, Ann Arbor.

Markowitz, H., 1952. Portfolio selection. The Journal of Finance 7 (1), 77–91.

Markowitz, H.M., 1991. In: Frängsmyr, T. (Ed.), Autobiography, The Nobel Prizes, 1990. Nobel Foundation, Stockholm.

http://www.mathworks.com/discovery/portfolio-optimization.html.

Melanie, M., 1999. An Introduction to Genetic Algorithms, vol. 3. Fifth printing. Cambridge, Massachusetts, London, England.

Kirkpatrick, S., Gelatt, C.D., Vecchi, M.P., 1983. Optimization by simulated annealing. Science 220 (4598), 671–680.

Lin, C.-C., Liu, Y.-T., 2008. Genetic algorithms for portfolio selection problems with minimum transaction lots. European Journal of Operational Research 185 (1), 393–404.

Tuncel and Dalis. http://www.air-worldwide.com/Publications/AIR-Currents/2015/Optimizing-Your-Cat-Bond-Portfolio-with-a-New-LP-Framework.

Xiang, Y., et al., 2013. Generalized Simulated Annealing for Global Optimization: The GenSA Package.

SECTION IV

MODEL RISK, RESILIENCE AND NEW CONCEPTS

CHAPTER

12

Parsimonious Risk Assessment and the Role of Transparent Diverse Models

Patrick McSharry[1,2,3]

[1]ICT Center of Excellence, Carnegie Mellon University Africa, Kigali, Rwanda; [2]University of Rwanda, Kigali, Rwanda; [3]Smith School of Enterprise and the Environment, University of Oxford, Oxford, UK

INTRODUCTION

Extreme weather events such as floods, heat waves, and drought are some of the greatest global risks to humanity according to the World Economic Forum (WEF, 2017). As with many natural disasters, it is not simply the hazard itself but the geographical location, vulnerability of communities, and collective actions of human beings that determine the magnitude of the disaster in terms of the number of deaths, injuries, and economic losses. Indeed, many have argued that it is the negligence of humans, through failure to take appropriate preventive measures and consequences of urbanization, which transform natural disasters into unnatural disasters (World Bank, 2010). Big data from insurance claims is helping to provide insights into the generation of insurance losses and to suggest how vulnerabilities might be reduced. In the case of earthquakes, it was found that 83% of all deaths from building collapse over the past 30 years occurred in countries that are anomalously corrupt (Ambraseys and Bilham, 2011).

A number of recent trends in quantitative modeling for forecasting and risk assessment, recently enshrined in the field of machine learning, serve to offer a perspective on how insurers can better assess the risks of losses due to natural disasters and adverse weather events. In the case of catastrophe (CAT) models used for pricing risk, there is growing evidence for the need to promote openness to sharing data and greater transparency and diversification of modeling techniques. The following sections describe the history of ensemble modeling, advantages of parsimonious models, and potential opportunities from

Risk Modeling for Hazards and Disasters
http://dx.doi.org/10.1016/B978-0-12-804071-3.00012-4

big data and offer some recommendations for how these developments might improve the performance of both index insurance products and catastrophe models.

PARSIMONIOUS MODELS

The concept of seeking parsimonious models is not a new one and is indeed a recommended approach from multiple perspectives such as transparency, ease of understanding, and the ability to generalize to new situations. Occam's razor is a philosophical principle that suggests that one should select the simplest theory with the least assumptions that explains the available observations. This principle is firmly embedded in statistical learning where one attempts to select a model that does not overfit the historical data and is expected to perform equally well on future data. A model can be said to be parsimonious when its level of complexity (number of parameters) is statistically justified, especially relative to the amount of data available for estimating the parameters. Parsimonious models have been shown to outperform other candidates in applications such as forecasting economic output (Arora et al., 2013) and electricity demand (Taylor and McSharry, 2007). In most situations where predictions of the future are required, it is recommended to seek a parsimonious model.

Machine learning practitioners use a technique known as cross-validation to obtain a realistic measure of likely future performance. The essential idea is to ensure that distinct data sets are always used for estimation and evaluation, thereby giving out-of-sample statistics results. As a result of avoiding the problem of overfitting, cross-validation aims to identify a parsimonious model that is consistent with the data. In this way, the performance on past data should provide a good indication of likely future performance.

ENSEMBLE MODELING

In 1907, Francis Galton discovered, to his surprise, that the average of all the entries in a "guess the weight of the ox" competition at a country fair was amazingly accurate and better than so-called cattle experts. There is now a growing body of scientific evidence that suggests seeking a diverse set of opinions is not only advisable for avoiding surprises but may offer the best means of making predictions. Analysis of entries to forecasting competitions has demonstrated that combined forecasts are usually better than those from any individual model (Clemen, 1989; Makridakis and Hibon, 2000). It is important to collect as diverse a set of models as possible when creating a combined forecast. The challenge is therefore to find an efficient means of obtaining an ensemble of models (Fig. 12.1). This can be achieved in a number of distinct ways, each motivated by the need to improve performance for a specific application. Interestingly all these developments in disparate fields provide growing empirical support for the benefits of ensemble modeling.

Meteorologists were first to explore the value of creating ensemble weather forecasts by running multiple versions of a low-resolution model instead of a single high-resolution deterministic model. Inspiration came from the realization that nonlinearity in the atmosphere and potential chaotic dynamics causes sensitivity to initial conditions whereby the forecast

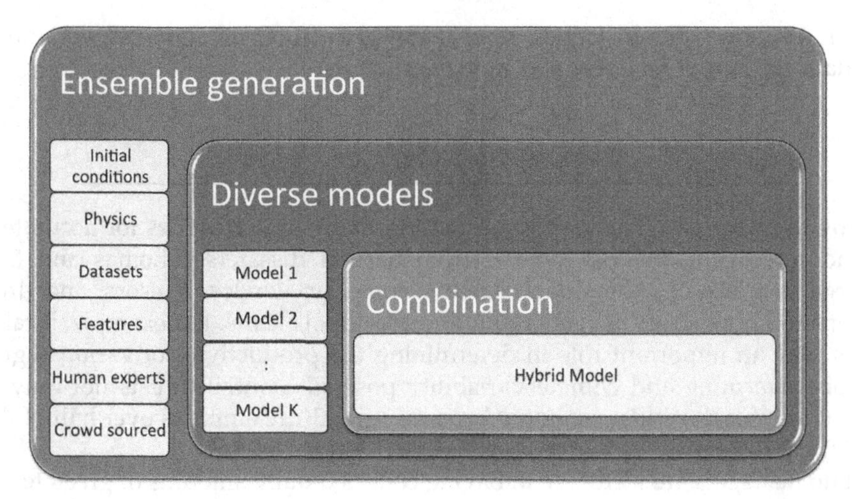

FIGURE 12.1 Generation of an ensemble of diverse models.

uncertainty grows with time (Lorenz, 1963). To account for this uncertainty, each member of the weather ensemble forecast is generated by starting at a separate initial condition. The average of the weather ensemble has been shown to be superior to the high-resolution point forecast when using wind speed forecasts for wind power generation (Taylor et al., 2009). The accuracy of the forecasts was improved further when using a statistically calibrated ensemble. Climate scenarios are now typically generated by considering not only different initial conditions but also perturbations to the physics and a range of forcings based on the amount of carbon emissions assumed.

The private sector has embraced the idea of inspiring many researchers to help find alternative models through crowdsourcing. In October 2006, the online DVD rental firm Netflix offered a million dollar prize to anyone who could improve on their proprietary *Cinematch* model for predicting user ratings by more than 10%. By September 2009, two teams had achieved sufficient improvements, and a winner was declared. The Netflix experiment proved that public prediction competitions have merit in making scientific progress. Perhaps, more importantly, the competition demonstrated that the substantial advances made by both the winning and second-place teams were achieved by combining models from different competitors who agreed to collaborate. Motivated by the success of Netflix, the prediction competition platform Kaggle was established to facilitate organizations that wish to take a crowdsourced approach to sharing data and identifying models.

The field of machine learning has made considerable progress in the use of ensemble approaches for statistical learning. The concept of using an ensemble of "weak learners" to improve predictions is now well established and has been successful in a wide range of applications. Indeed, there are parallels between the creation of weak learners and the search for parsimonious models. For example, the technique known as random forests relies on ensembles of decision trees (Breiman, 2001). The key finding is that ensembles of similar models are unlikely to offer an improvement and that diversification is important. Ensemble approaches therefore attempt to find a means of generating models with variations in the learning data sets, input features, and model structures.

BIG DATA

The growth of big data is certainly presenting many opportunities for accurately quantifying and forecasting the risk arising from natural disasters (Thomas and McSharry, 2015). Large volumes of claims data, satellite imagery, wireless sensors, and drones can help understand how losses are generated for individual clients. For example, local weather conditions play an important role in determining the productivity of various agricultural crops. Global warming and climate variability pose substantial threats not only for food security but also for the wider economy because agriculture employs over half of the workforce in Africa.

Tea estate managers are fortunate in having collected daily amounts of green tea leaves at each factory over several years. Machine learning was used to identify relevant features and a model for forecasting tea yield at several factories in Rwanda (McSharry et al., 2016). Weather observations from ground-based stations are often limited because of large distances from tea estates, short historical records, missing data, and sensor failure. Fortunately, weather estimates from satellite imagery offer substantial potential for obtaining information about local weather conditions at exactly the locations where tea is grown. The average of different satellite-based estimates for rainfall was found to improve predictability, providing evidence for ensemble approaches. Furthermore, additional satellite products such as surface, soil, and root moisture content are strongly correlated with tea productivity and helped to improve the model.

INDEX INSURANCE

An important application of risk modeling can be found in the development of index insurance products for farmers in developing countries. The theoretical concept of index insurance is extremely attractive, with the promise of reduced costs and timely payouts, but the practical implementation is not straightforward. The quality and duration of historical weather observations and crop yield or livestock data can make it difficult to identify adequate models for constructing the index. A simple example of a weather index would be to trigger a payout when the rainfall falls below a specific threshold. There is also potential for using multiple variables, and the objective should be to forecast the actual losses suffered by farmers. Failure to accurately forecast losses leads to basis risk, representing a low correlation between insurance claims and payouts.

The lack of transparency about the construction of indices and absence of a formal and accepted approach to define and measure basis risk is proving a challenge. Although basis risk is widely discussed, independent evaluations of basis risk for actual products are scarce. Clarke et al. (2012) found a correlation of −0.13 between claims and payments for weather index insurance in India. Jensen et al. (2016) found that an index-based livestock insurance product in northern Kenya left the policyholders with an average of 69% of their original risk because of high-loss events. The few studies of index insurance products that exist suggest that basis risk is high, and this represents a reputational issue for the industry.

On the positive side, there is no doubt that big data and specifically improvements in the quality and spatial resolution of data obtained through remote sensing, drones, and wireless sensors will eventually contribute to the reduction of basis risk. In the meantime, it is recommended to undertake an independent evaluation of basis risk and wait until the product is appropriate for those farmers who need access to affordable insurance.

CATASTROPHE MODELS

Catastrophes represent a substantial challenge to scientific modeling given that a deep understanding of both the physical and socioeconomic systems are required. Although little data exist for catastrophes, because by definition these are rare extreme events, substantial information does exists about the humans who are affected. For insurers, geocoded claims data offer enormous insights not only into the catastrophes themselves but also into the impact they have on society.

In 1986, the pioneer of catastrophe modeling proposed a computer-based simulation approach to estimate the probability distribution of losses (Clarke, 1986). These so-called catastrophe models have become the norm in reinsurance when attempting to price the risk of a whole host of perils and are invaluable for decision-making and planning. The essential idea is to create modules for the hazard, exposure, and vulnerability and combine these to estimate the distribution of losses. From floods to windstorms and terrorism attacks to hailstorms, insurers are reluctant to offer cover without access to a decent CAT model. Over the years, the number of CAT model providers has grown in number, and the models have become increasingly sophisticated, yet not necessarily diversified. Not content with the three established vendors of CAT models, AIR-Worldwide, RMS, and EQECAT, the marketplace has been calling for greater transparency, a larger variety of models, and extended geographical coverage.

The evidence in favor of parsimonious models and the advantages of using ensembles of diverse models pose a challenge for the purveyors of complicated black-box proprietary CAT models. By design these CAT models are not transparent because of the need for vendors to protect their intellectual property. Although the general recipe for building a CAT model is common knowledge, it is usually difficult to modify any of the ingredients of a commercial model.

Frustrated with the limitations and constraints associated with this commercial approach to modeling risk, many insurers have decided to financially support an open-source catastrophe modeling platform known as Oasis. Described as the democratization of catastrophe risk information by Oasis CEO, Dickie Whitaker, Oasis encourages data sharing and promotes a plug-and-play approach to CAT modeling. To address the lack of available CAT models for developing countries, the Organisation for Economic Cooperation and Development initiated a public–private partnership called the global earthquake model (GEM). Through its open-source software, GEM promotes collaboration around data sharing, best practice, and applications to improve the understanding, assessment, and management of risk. These are all positive indications that the insurers are demanding the necessary changes to obtain the best risk estimates available.

CONCLUSION

The role of scientific modeling is gaining importance in helping support decisions in a wide range of business sectors. With the advent of big data comes the expectation that predictive analytics will offer enormous potential for managing the risks associated with natural disasters. Although access to increasing amounts of high-quality data is certainly a positive factor, guidelines are required to help managers and policymakers determine when and when not to rely on models. Catastrophes are, by definition, rare events and as such the historical data about these events are generally scarce. Faced with the high likelihood of overfitting to these few events, it is crucial to use cross-validation techniques to establish efficacy and to be aware of the dangers of overfitting as recommended by Occam's razor. The need to independently evaluate models and compare with appropriate benchmarks is greater now than ever as the opportunities of using models to assess risks are increasing. Whether for evaluating the economic impacts of climate change or the creation of an index insurance product, the quality of the relationship between the hazard and losses must be formally evaluated using a statistical approach. Without guidelines and independent evaluation protocols, the consumers of risk models, who rely on accurate risk assessments to make informed decisions about the threats of future natural catastrophes, are likely to have their expectations dashed.

References

Ambraseys, N., Bilham, R., 2011. Corruption kills. Nature 469, 153–155.

Arora, S., Little, M.A., McSharry, P.E., 2013. Nonlinear and nonparametric modeling approaches for probabilistic forecasting of the US gross national product. Studies in Nonlinear Dynamics & Econometrics 17 (4), 395–420.

Breiman, L., 2001. Random forests. Machine Learning 45 (1), 5–32.

Clarke, D.J., Mahul, K.N., Rao, O., Verma, N., 2012. Weather Based Crop Insurance in India. World Bank Policy Research Working Paper No. 5985.

Clarke, K., 1986. A formal approach to catastrophe risk assessment and management. Casualty Actuarial Society Proceedings LXXIII Part 2 (140), 69–92.

Clemen, R.T., 1989. Combining forecasts: a review and annotated bibliography. International Journal of Forecasting 5, 559–583.

GEM, Global Earthquake Model. www.globalquakemodel.org.

Jensen, N.D., Barrett, C.B., Mude, A.G., 2016. Index insurance quality and basis risk: evidence from Northern Kenya. American Journal of Agricultural Economics 98 (5), 1450–1469.

Lorenz, E.N., 1963. Deterministic nonperiodic flow. Journal of the Atmospheric Sciences 20 (2), 130–141.

Makridakis, S., Hibon, M., 2000. The M3-competition: results, conclusions, and implications. International Journal of Forecasting 16, 451–476.

McSharry, P.E., Swartz, T., Spray, J., 2016. Index Based Insurance for Rwandan Tea. International Growth Centre, London School of Economics and Political Science & University of Oxford, London, UK.

Oasis, Oasis Loss Modelling Framework. www.oasislmf.org.

Taylor, J.W., McSharry, P.E., Buizza, R., 2009. Wind power density forecasting using ensemble predictions and time series models. IEEE Trans Power Conversion 24 (3), 775–782.

Taylor, J.W., McSharry, P.E., 2007. Short-term load forecasting methods: an evaluation based on European data. IEEE Transactions on Power Systems 22 (4), 2213–2219.

Thomas, R., McSharry, P.E., 2015. Big Data Revolution: What Farmers, Doctors and Insurance Agents Teach Us. John Wiley & Sons, London, UK.

World Bank, 2010. Natural Hazards, Unnatural Disasters: The Economics of Effective Prevention.

WEF, 2017. World Economic Forum, Davos, Switerland. https://www.weforum.org/events/world-economic-forum-annual-meeting-2017.

CHAPTER

13

Creating a Probabilistic Catastrophe Model With the Characteristic Event Methodology

Karen Clark

Karen Clark & Company, Boston, MA, USA

INTRODUCTION

Catastrophe models have become very important tools for insurers and reinsurers—the models are used for pricing, risk transfer, and portfolio management decision-making. The fundamental structure of the models has not changed since the first catastrophe model was developed and brought to market in the late 1980s. For every peril region, the models have the same four components—event generation, intensity calculation, damage estimation, and insured loss calculation—and produce the same output—the Exceedance–Probability (EP) curve.

The EP curve provides estimates of the probabilities of exceeding losses of different amounts on a single property or portfolio of properties. It has traditionally been derived using a large sample of hypothetical events created in the event generation component. In the traditional models, the event generation component uses random simulation techniques, such as Monte Carlo simulation, to produce the sample of events.

The EP curves based on large random samples provide valuable information, but they do not answer all the questions decision-makers have with respect to catastrophe loss potential. For example, the EP curve does not highlight where a company may be over exposed to large and potentially outsized losses relative to peers. It does not answer the basic underwriting questions around where a company should be writing more versus less property business. It does not provide concrete information desired by senior management, such as loss estimates for the 100 year–type events.

The other drawbacks of randomly generated samples include a subset of events that may be unrealistic and not likely to occur in nature, lack of transparency on the key model assumptions driving the loss estimates, and the inability to conduct efficient sensitivity analyses on the model assumptions.

Risk Modeling for Hazards and Disasters
http://dx.doi.org/10.1016/B978-0-12-804071-3.00013-6

This paper illustrates a new approach to event generation that achieves the benefits of a large random sample and provides additional benefits to insurers, such as:

- Identification of "hot spots" and areas where insurers may be susceptible to surprise solvency-impairing losses from infrequent but not unlikely events
- Intuitive and operational information for underwriting, portfolio management, and capital allocation

Called the Characteristic Event (CE) methodology, this approach starts by defining return period events for specific geographical areas, e.g., segments of coastline for hurricanes and seismic source zones for earthquakes. These scientifically defined return period events are then systematically "floated" within the geography of interest to estimate the losses on a portfolio of properties and for individual properties.

This paper illustrates the approach using a hurricane example and demonstrates how the CE methodology leads to a robust and operational sample of events for decision-making. With the CE approach, insurers obtain all of the traditional EP curve metrics along with new insightful information on losses from different return period events.

STRUCTURE OF A CATASTROPHE MODEL

Catastrophe models have four primary components as shown in the figure below.

Model Component	Description
Event generation	Defines event parameters including frequency and physical severity by geographical region
Intensity calculation	Estimates the intensity experienced at each location in the area affected by each event
Damage estimation	Estimates the damages to building, contents, and time element exposures (may also estimate casualties)
Loss calculation	Applies insurance policy conditions and reinsurance terms to estimate insured losses

The primary model output is the Exceedance–Probability (EP) curve as shown below.

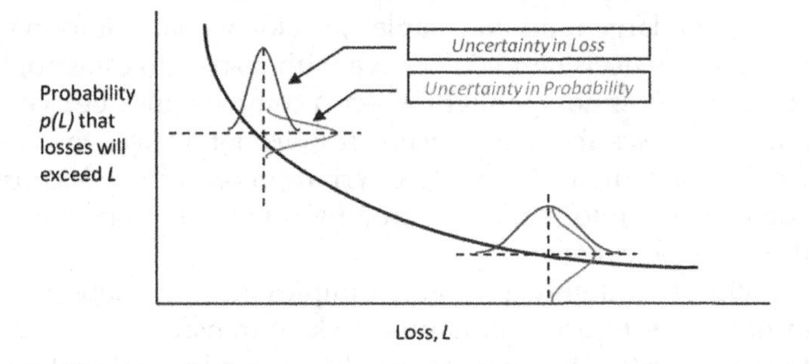

Event generation is the most important component for defining the probabilities of exceeding losses of different amounts. The event generation component answers the questions of where and how frequently events of different magnitudes are likely to occur.

In the traditional models, the events themselves do not matter too much because they are used primarily to generate the EP curve. Because the events are randomly generated, the models require a large sample to converge to a stable EP curve, eliminate areas of over sampling and under sampling, and generate enough large losses to estimate the "tail" of the distribution.

The frequency, severity, and other assumptions underlying the event generation component are highly uncertain and, for many geographical areas, are based on very little reliable data. This means that a lot of subjective judgment goes into this part of the model construction. Model differences and the volatility of the model output frequently result from different scientific opinions on the event frequency–magnitude relationships.

Increasing the sample size does not eliminate the inherent uncertainty in the event generation component. Contrary to popular belief, larger samples do not lead to the model capturing all possible events. There are an infinite number of combinations of event characteristics, and no sample can capture all possibilities. No matter how large the sample is, it is likely that the next actual event will not be in that sample.

The larger the event catalogs become, the smaller the probability of each event in the catalog. The rates become so low that the individual events do not have meaning to decision-makers. This paper focuses on the event generation component and how the CE method is applied to provide new model output for decision-makers.

US HURRICANE EXAMPLE

The hurricane event generation component defines the frequency-severity distributions for hurricanes all along the Gulf and East Coasts. In the traditional models, the historical hurricane data are fit to statistical distributions and the parameters estimated for coastal segments or "gates" usually 50 or 100 miles in length. These distributions are then used to generate the random events.

The first problem to tackle in event generation is the selection of the distributional form for each storm parameter. Because there have been less than 200 hurricanes to impact the US coastline in the historical record, the data are quite sparse for most coastal segments and are not enough to have confidence in the true shape of the distribution. Along with the parameter distributions, the model developers must make assumptions about the tails of the distributions. Most of the distributions must be truncated to minimize the generation of unrealistic and implausible events.

All of this means that many assumptions go into the event generation component, and most of these assumptions rely on expert judgment rather than actual data (see also Guin et al., this volume). If any of these assumptions need to be changed, the entire event catalog needs to be regenerated, which contributes to the model development time and model volatility.

Event Generation Using the Characteristic Event Methodology

In the CE methodology, model assumptions are made using a more direct and transparent process. Rather than letting the random simulation process determine the frequency-magnitude distributions by coastal segment, the model developers carefully select the "typical" event characteristics by return period for specific coastal regions.

For example, the chart below shows the peak wind speed at landfall for the historical landfalling hurricanes in southern Florida—the highest frequency coastal region. This variable is the most important storm parameter and is used to define the intensity of the storm (and Saffir-Simpson category). The blue dots show the historical wind speeds plotted on a cumulative graph, and the red dots show the peak winds that could be selected for different return period events in the CE approach.

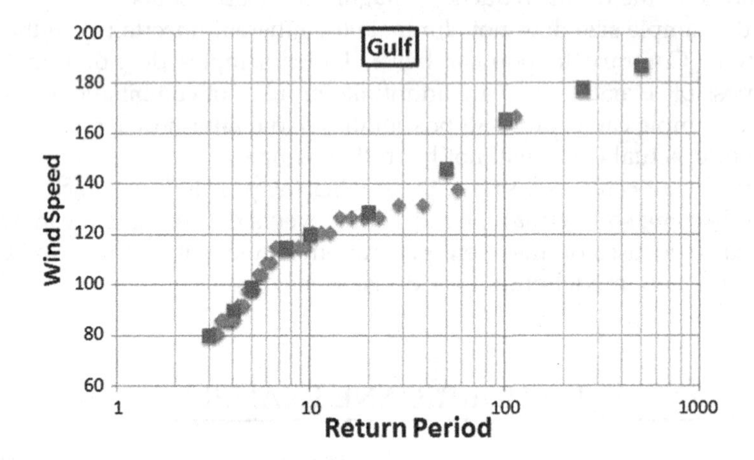

The wind speed for the 100 year CE in the chart is 165 mph—in line with the peak winds for Hurricane Camille, the most intense storm at landfall in the historical record for this region. The 250-year CE has 180 mph peak winds corresponding to events that have occurred in the Atlantic, but either did not make landfall along the US coast or did so at a lower intensity. The chart shows that for lower return periods the CE intensities line up with the historical data as one would expect.

Once the peak intensities are defined for the return period events, other storm characteristics can be determined. For example, intense storms in this region tend to be tightly wound and would have a small radius of maximum winds (Rmax)—another very important storm parameter. Other characteristics such as forward speed and filling rate are assigned based on the region and storm intensity.

Another required storm characteristic is storm track. Every storm track is unique so there is an infinite number of potential storm paths for each event. The hurricane model cannot cover all possibilities, but the model must provide full coverage at every location and not over or under sample any geographic area. To do this (particularly for robust location level loss estimates), the traditional models must generate thousands of tracks for each coastal segment following the random simulation technique.

With the CE method, for each coastal segment, the historical data are used to develop the typical track for that segment. This track is then "floated" along the coast at 10-mile increments to provide full spatial coverage with fewer events. This approach also ensures that individual locations will not be penalized because of too many intense events in that location for the particular simulated sample. On the other hand, it ensures individual locations will not appear to be less risky because of insufficient intense events in that location for the simulated sample.

Once multiple return period CEs are created for each 10-mile increment, there is, in effect, a superstratified sample for each coastal segment. There will be multiple events for each segment reflecting the appropriate frequencies of event severity for that segment.

After the event catalog is defined for the entire coast, wind footprints, damages, and losses can be calculated for each event. Other components of the loss estimation process—intensity calculation, damage estimation, and loss calculation—remain the same as the traditional models.

Developing the Probabilistic Output With the Characteristic Event Method

Developing the EP curve probabilistic output is straightforward in the CE approach. A rate is assigned to each event using the inverse of the return period and the number of floats, and the rates thus calculated along with the losses estimated for the events make up the Expected Loss Table (ELT).

A simulation is then run to assign events to "years" using the event rates and a negative binomial distribution fit to the historical hurricane data. Simulated years can have none, one, or several hurricane occurrences. Model output thus generated is typically referred to as a Year Loss Table (YLT) from which annual occurrence and aggregate EP curves can be derived.

The CE-derived rates can be benchmarked against the historical data. The chart below shows the simulated annual rates of occurrence versus historical by region.

	Categories 1 and 2		Categories 3–5		All Categories	
Region	**Historical**	**Characteristic Event (CE)**	**Historical**	**CE**	**Historical**	**CE**
Northeast	0.05	0.03	0.03	0.03	0.08	0.06
Mid-Atlantic	0.03	0.03	0.00	0.00	0.03	0.03
Southeast	0.24	0.24	0.04	0.05	0.27	0.29
Florida	0.43	0.40	0.20	0.23	0.64	0.63
Gulf	0.20	0.21	0.15	0.16	0.35	0.37
Texas	0.19	0.19	0.12	0.13	0.31	0.32
All regions	1.13	1.09	0.56	0.60	1.68	1.70

Characteristic Event Profiles

Traditional risk metrics from the EP curves such as the 1% and 0.4% loss exceedance probabilities (also known as 1 in 100 and 1 in 250 year PMLs) are commonly used mathematical numbers, but they do not provide full insight into an insurer's large loss potential. They are not operational or intuitive risk metrics and are not easily projected forward for risk-management purposes.

The PMLs can also give a false sense of security because they are frequently misinterpreted as the 100-year events. The PMLs are not associated with specific events, but rather show the loss amount for which there is a 1% chance of exceeding. The CEs show management where, geographically, event losses are most likely to exceed the PML. The CE methodology generates the losses from the 100-year (and other return period) events along with the PMLs and thus provides a complete picture of an insurer's large loss potential.

The chart below illustrates the 100-year event losses for a hypothetical company. The CE profile clearly shows where the insurer can have surprise losses well in excess of their 100 year PML and their reinsurance program—from events that would not be surprising from a meteorological perspective. By adding the market share of loss to the profile, senior management can see where they can have outsized losses relative to peer companies.

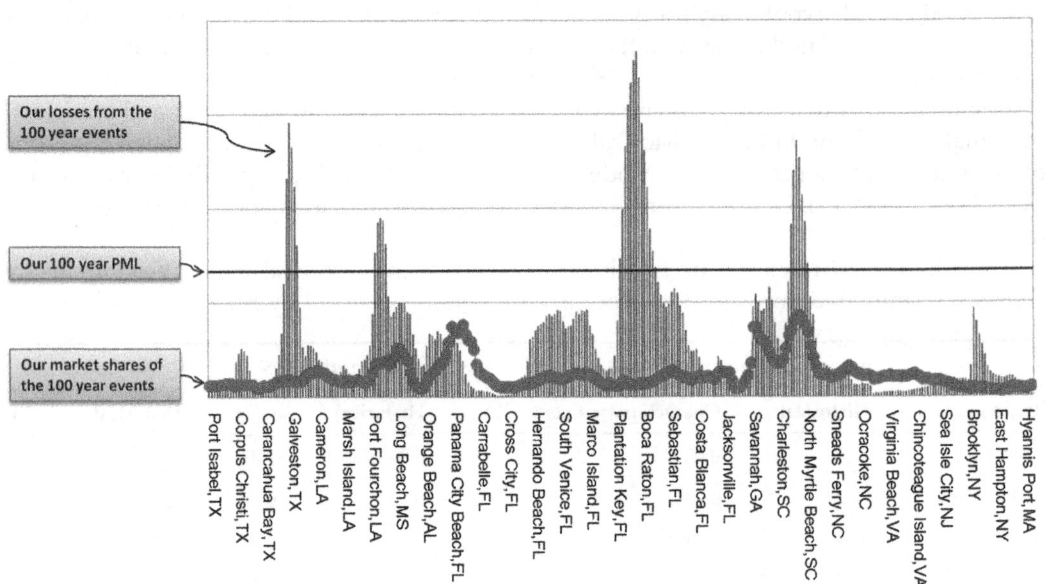

The CE approach generates this information for any return period of interest, and it is applicable to all perils. Because major earthquakes have surprised scientists by occurring on previously unknown faults, it is equally important for insurers to have visibility on losses from the 100-year earthquakes occurring in different locations.

Catastrophes are like real estate—it is about location. A major event can occur in a sparsely populated area and cause few losses whereas a moderate hurricane or earthquake occurring in a densely populated area can cause catastrophic losses. The CE method enables insurers to systematically monitor exposure concentrations and to compare their potential large losses geographically.

DIFFERENT RISK METRICS FOR DIFFERENT PURPOSES

The EP curve and associated risk metrics are necessary for pricing catastrophe risk. But the new CE metrics may be more useful for other risk-management purposes as explained below.

Underwriting

Underwriters require stable risk metrics for risk selection and to allocate capacity geographically and to specific accounts. If risk metrics are volatile, robust underwriting guidelines cannot be developed, and the quality of the portfolio is likely to suffer. Because the CEs represent typical return period events, there is no need for the CEs to be updated or to change from year to year.

The CEs provide information on where a company may be overexposed to large losses that can inform underwriting strategies and decisions. Because the CEs are specific events, insurers can drill down to see every policy's contribution to each CE loss. The expected profitability of each account can be compared with its CE contribution to determine which accounts to eliminate from the portfolio to reduce unwanted large loss potential. (It should be noted here that the "expected profitability" of an individual account exposed to catastrophe losses is highly uncertain—and currently driven by volatile model output—whereas estimating the account's contribution to a specific loss amount involves less uncertainty.).

CEs also illustrate where a company is underexposed and has the capacity to write more property business along with guidance on *how much* new business can be written. The EP curve metrics alone do not provide this insight.

Capital Allocation

Capital allocation on a portfolio basis for a large complex company involves many decisions and assumptions. Again, some of those assumptions are currently driven by model output that is highly uncertain and subject to wide swings. The CEs can provide more stable and operational risk metrics for allocating capital between business units and divisions.

Risk Transfer

Traditional catastrophe reinsurance is not efficient for high layers. There is usually a minimum price for a layer and if an insurer has just one or two narrow CE "spikes" above the PML, it may not be cost-effective to purchase that layer.

The new CE method is sparking innovation in risk transfer transactions. Parametric-triggered catastrophe bonds with the payouts tailored to an insurer's loss potential at each

10-mile CE "gate" have already been constructed and transacted. This type of transaction meets the investor's desire for parametric bonds with the insurer's desire to minimize basis risk.

ADDITIONAL BENEFITS OF THE CHARACTERISTIC EVENT APPROACH

The CEs are visual and intuitive to decision-makers, and this transparency makes the CEs easier to communicate with internal and external stakeholders. Risk-management strategies based on CEs can be communicated from the executive level to the line underwriter all in the same language. Each CE has a unique identifier so they provide a robust "common currency" for decision-making.

CEs make it straightforward to conduct sensitivity tests of new scientific research and to answer questions that arise from the board. For example, senior executives may want to know the portfolio impacts of climate change and a 5% increase in the intensities of hurricanes—a projection in the range provided by the most recent report of the Intergovernmental Panel on Climate Change (IPCC). The hurricane CE peak winds can be adjusted up by 5% and the CE profiles easily recreated with the new assumptions.

The most recent report of the US Geological Survey (USGS) indicates that a 100-year type "background" earthquake in California is a magnitude 7 event. The USGS uses background seismicity to account for unknown faults, and while a magnitude 7 is a moderate-sized earthquake, if it occurs near downtown Los Angeles, the losses may be greater than the larger magnitude events the USGS estimates for known faults.

The CE approach helps insurers deal more effectively with the model uncertainties and the scientific "unknowns." Rather than trying to predict exactly where and how frequently different events will occur, the CE approach allows for events that are typical of different return periods (from the hazard perspective) to happen anywhere in an exposed region. This will help to reduce future surprise losses.

CONCLUSIONS

This paper discusses a new approach to generating event catalogs for the catastrophe models. The traditional approach of simulating tens of thousands of randomly generated events produces large catalogs containing many events that may not be important to decision-makers. On the other hand, the catalogs may not include or highlight the events decision-makers do care about.

Larger event catalogs do not lead to more accurate models because a model can only converge to the assumptions underlying that model. The catastrophe model assumptions are highly uncertain because of the paucity of data and the many scientific unknowns.

The CE methodology produces a superstratified sample of events that can be easily understood by underwriters, CEOs, and boards because the model assumptions are explicit and transparent. A robust EP curve and loss estimates credible at location-level resolution can be developed with fewer events than with random simulation techniques.

While the full EP curves are necessary for pricing purposes, the return period CEs are valuable risk metrics for managing large loss potential. The transparency, stability, and intuitiveness of the CEs mean they can be understood and communicated from the senior management to the line underwriter level. They can also be used to monitor loss potential over time.

Catastrophe loss potential continues to increase because of growing concentrations of property values in areas exposed to extreme events. Over time, more of the premium dollar will go to funding catastrophe losses. Multiple risk metrics and methodologies help insurers to understand more fully and to manage more effectively catastrophe risk.

Further Reading

The 100 Year Hurricane, May 2014. Karen Clark & Company. White Paper.

Characteristic Events for Catastrophe Risk Management, January 2012. Karen Clark & Company. White Paper.

Managing Hurricane Risk with Characteristic Events, May 2014. Karen Clark & Company. White Paper.

CHAPTER

14

From Risk Management to Quantitative Disaster Resilience: A New Paradigm for Catastrophe Modeling

Slobodan P. Simonovic

Department of Civil and Environmental Engineering and The Institute for Catastrophic Loss Reduction, The University of Western Ontario, London, ON, Canada

INTRODUCTION

The most common approach to catastrophic risk management is focused on the assessment of vulnerability, which when combined with hazards, provides for risk evaluation. Based on Cutter et al. (2008), "vulnerability describes the preevent, inherent characteristics or qualities of systems that create the potential for harm. Vulnerability is a function of who, or what, is at risk and sensitivity of system (the degree to which people and places can be harmed). On the other side, resilience is the ability of a complex system to respond and recover from system disturbance and includes those conditions that allow the system to absorb impacts and cope with an event, as well as postevent, adaptive processes that facilitate the ability of the system to reorganize, change, and learn in response to a threat" (paraphrased by the author). It is important to understand the difference between concept of risk management and resilience. They both offer different perspectives for handling impacts of system disturbance. Table 14.1 gives one potential comparison between the risk and resilience (Park et al., 2013).

There are many definitions of resilience, from general: (1) The ability to recover quickly from illness, change, or misfortune; (2) buoyancy; (3) the property of material to assume its original shape after deformation; (4) elasticity; to ecology-based (Gunderson and Holling, 2001): (1) The ability of a system to withstand stresses of "environmental loading"; to hazard-based (UNISDR, 2009): (1) Capacity for collective action in response to extreme events; (2) the capacity of a system, community, or society potentially exposed to hazards to adapt, by resisting or changing, to reach and maintain an acceptable level of functioning

Risk Modeling for Hazards and Disasters
http://dx.doi.org/10.1016/B978-0-12-804071-3.00014-8

TABLE 14.1 Comparison of Concepts of Resilience and Risk

	Risk Management	Resilience
Design principles	Minimization of risk of failure	Adaptation to changing conditions without permanent loss of function
Design objectives	Minimization of probability of failure	Minimization of consequences of failure
Design strategies	Armoring, strengthening, oversizing, resistance, redundancy, isolation	Diversity, adaptability, cohesion, flexibility, renewability, regrowth, innovation, transformation
Relation to sustainability	Security, longevity	Recovery, renewal, innovation
Mechanisms of coordinating response	Centralized, hierarchical decision structures coordinate efforts according to response plans	Decentralized, autonomous agents respond to local conditions
Modes of analysis	Quantitative (probability-based) and semiquantitative (scenario-based) analysis	Scenario-based consequence analysis with possible involvement of scenarios with unidentified causes

After Park, J., Seager, T.P., Rao, P.S.C., Convertino, M., Linkov, I., 2013. Integrating risk and resilience approaches to catastrophe management in engineering systems. Risk Analysis 33, 356–367.

and structure; (3) the capacity to absorb shocks while maintaining function; (4) the capacity to adapt existing resources and skills to new situations and operating conditions. In economic literature, resilience has been defined as the "inherent ability and adaptive response that enables firms and regions to avoid maximum potential losses" (Rose and Liao, 2005).

The common elements of these definitions include (1) minimization of losses, damages, and system disruption; (2) maximization of the ability and capacity to adapt and adjust when there are shocks to systems; (3) returning systems to a functioning state as quickly as possible; (4) recognition that resilient systems are dynamic in time and space; and (5) acknowledgments that postshock functioning levels may not be the same as preshock levels.

Generally, resilience is measured in terms of (1) the amount by which a system is able to avoid maximum impact (i.e., static resilience, robustness, the opposite of vulnerability) and (2) the speed at which the system recovers from a disruption (i.e., dynamic resilience, rapidity, recoverability). In spite of the fact that the resilience is a dynamic process, for measurement purposes it is often viewed as a static phenomenon (Cutter et al., 2008).

There is a significant literature that covers qualitative conceptualizations of disaster resilience, but very little research that focuses on resilience quantification and its implementation in the management of disasters. Even though Bruneau et al. (2003), Chang and Shinozuka (2004), and Cutter et al. (2008) propose conceptual frameworks for resilience assessment in seismic context of response and recovery, challenges remain in developing standard metrics that can be used in quantifying and assessing disaster resilience. Simonovic and Peck (2013) and Simonovic (2014) made a major contribution by moving beyond conceptualizations and into actual resilience quantification.

In this chapter the concept of resilience proposed by Simonovic and Peck (2013) is expanded to capture the process of dynamic resilience simulation in both time and space. The rest of the chapter is organized as follows. Firstly the mathematical framework for the concept of resilience based on Simonovic and Peck (2013) is presented. The following section provides details of dynamic resilience calculation. Case study of railway under flooding is used next to illustrate the application of dynamic resilience. The chapter ends with the conclusions.

QUANTITATIVE MEASURE OF RESILIENCE

The quantification framework recommended by Simonovic and Peck (2013) has two qualities: inherent (functions well during nondisturbance periods) and adaptive (flexibility in response during the system disturbance) and can be applied to physical environment (built and natural), social systems, governance network (institutions and organizations), and economic systems (metabolic flows). An original space–time dynamic resilience measure (STDRes) of Simonovic and Peck is designed to capture the relationships between the main components of resilience; one that is theoretically grounded in systems approach, open to empirical testing, and one that can be applied to address real-world problems in various problem domains.

STDRes is based on two simple concepts: level of system performance and system adaptive capacity (*AC*). They together define resilience. The level of system performance integrates various impacts (i) of disturbance on system performance. Various impacts can be considered—for example, physical, health, economic, social, and organizational. They represent units of resilience (ρ^i). Measure of system performance $P^i(t,s)$ for each impact (i) is expressed in the impact units. Physical impact may include, for example, length (*km*) of road being affected; health impact may be measured using an integral index such as disability adjusted life year; and so on. This approach is based on the notion that an impact, $P^i(t,s)$, which varies with time and location in space, defines a particular resilience component of a system, see Fig. 14.1 adapted from Simonovic and Peck (2013). The area between the initial performance line $P^i_0(t,s)$ and performance line $P^i(t,s)$ represents the loss of system resilience, and the area under the performance line $P^i(t,s)$ represents the system resilience ($\rho^i(t,s)$). In Fig. 14.1, t_0 denotes the beginning of the disturbance, t_1 the end of the disturbance, and t_r the end of the recovery period.

In mathematical form the loss of resilience for impacts (i) represents the area under the performance graph between the beginning of the system disruption event at time (t_0) and the end of the disruption recovery process at time (t_r). Changes in system performance can be represented mathematically as follows:

$$\rho^i(t,s) = \int_{t_0}^{t} \left[P^i_0 - P^i(\tau,s)\right] d\tau \quad \text{where } t \in [t_0, t_r] \tag{14.1}$$

When performance does not deteriorate because of disruption, $P^i_0(t,s) = P^i(t,s)$ the loss of resilience is 0 (i.e., the system is in the same state as at the beginning of disruption). When all of system performance is lost, $P^i(t,s) = 0$, the loss of resilience is at the maximum value. The system resilience, $r^i(t,s)$, is calculated as follows:

$$r^i(t,s) = 1 - \left(\frac{\rho^i(t,s)}{P^i_0 \times (t - t_0)}\right) \tag{14.2}$$

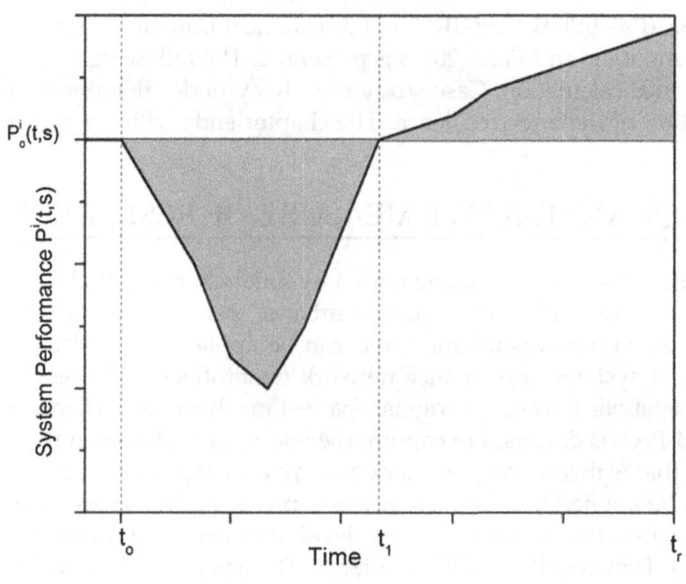

FIGURE 14.1 System performance.

As illustrated in Fig. 14.1, performance of a system, which is subject to a disruption, drops below the initial value, and time is required to recover the loss of system performance. Disturbance to a system causes a drop in system resilience from the value of 1 at t_0 to some value $r^i(t_1,s)$ at time t_1, see Fig. 14.2. Recovery usually requires longer time than the duration of

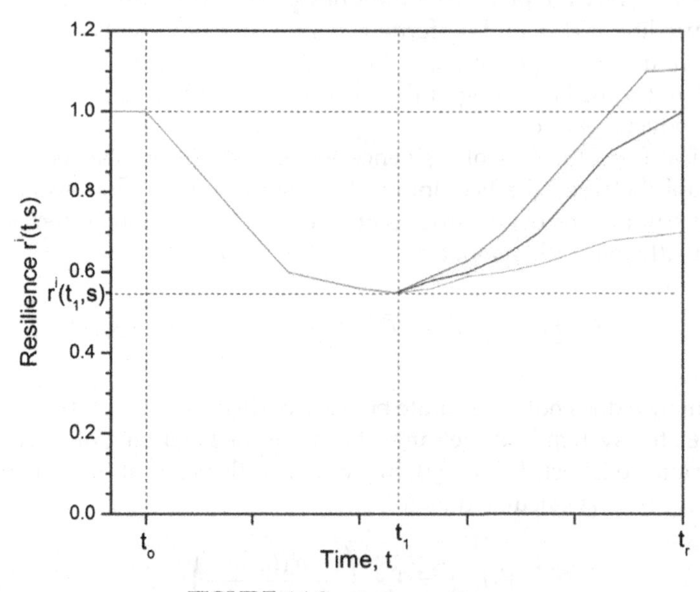

FIGURE 14.2 System resilience.

disturbance. Ideally, resilience value should return to a value of 1 at the end of the recovery period, t_r (dashed line in Fig. 14.2); the faster the recovery, the better.

The integral STDRes, overall impacts (i), is calculated using the following equation:

$$R(t,s) = \left\{ \prod_{i=1}^{M} r^i(t,s) \right\}^{\frac{1}{M}} \quad (14.3)$$

where M is the total number of impacts.

The calculation of STDRes for each impact (i) is done at each location in space (s) by solving the following differential equation:

$$\frac{\partial \rho^i(t)}{\partial t} = AC^i(t) - P^i(t) \quad (14.4)$$

where AC^i represents AC—capacity to handle the shocks or impacts—with respect to impact i. The AC of the system is an integrated behavior of the various components of the system that varies with time and space. It can be measured using four performance indicators: robustness, redundancy, resourcefulness, and rapidity (4R's) first introduced by Bruneau et al. (2003). These performance indicators are defined as (1) robustness ($R1$)—the ability of a system to resists external shocks without suffering any damages; (2) redundancy ($R2$)—the readiness of the system in which it resists external shocks beyond its natural capacity; (3) resourcefulness ($R3$)—the ability of the system to disseminate the resources during external shocks; and (4) rapidity ($R4$)—the ability of the system to recover (in terms of time) from damages caused by external shocks. The AC of a system is a function of both time and space. It is mathematically expressed as a function of 4R's as follows:

$$AC^i(t,s) = f(R_j(t,s) \quad j = 1,2,3,4 \quad (14.5)$$

where AC^i is the AC of the system; $f()$ is the mathematical function combining the effects of 4R's; j is the index for each R; t represents the time period; and s represents the spatial location.

The STDRes integrates resilience types, dimensions, and properties by solving for each point in space (s):

$$\frac{\partial R(t)}{\partial t} = AC^i(t) - \prod_i P^i(t) \quad (14.6)$$

The presented framework is implemented using the system dynamics (SD) simulation approach together with spatial analysis tool (Srivastav and Simonovic, 2014a; Peck et al., 2014). Assessment of impacts is driven by the selected performance measures and a set of system disturbance sources (hazards). In other words, traditional concepts of vulnerability and exposure will be used and adapted to each performance measure. In the case of natural disasters, change in climate and nonclimate drivers affects vulnerability and exposure of system components. At the same time, adaptation measures under consideration modify them

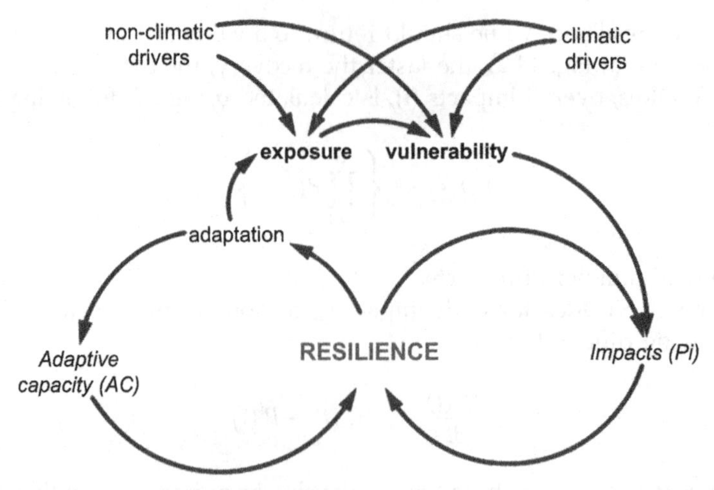

FIGURE 14.3 A schematic system dynamics model for the implementation of space–time dynamic resilience.

too. Interaction of numerous feedback mechanisms underlining the high-level model structure, as shown in Fig. 14.3, will generate system behavior. In this way, the STDRes describes the integral system behavior.

The SD simulation model of STDRes can be used to design and evaluate adaptation policies for improvement of system performance that will lead to levels of resilience higher than the predisruption levels. Adaptation policy design involves various approaches such as (1) change of model parameters and/or (2) creation of new strategies, structures, and decision rules. Regardless of the approach used, the aim of any SD simulation model experimentation is the exploration of model behavior using simulation. The purpose is to observe how the modeled system behaves normally and then how changes in policies or physical parameters alter that behavior.

It has been shown (Simonovic, 2009, 2011) that the feedback structure of a system determines its dynamic behavior over time. Most of the time, high leverage policies involve changing the dominant feedback loops by redesigning the system structure, eliminating time delays, and changing the flow and quality of information available at key decision points or fundamentally recreating the decision processes in the system. The robustness of policies and their sensitivity to uncertainties in model parameters and structure must be assessed, including their performance under a wide range of alternative scenarios. The interactions of different policies must also be considered because real systems are highly nonlinear; the impact of combination policies is usually not the sum of their impacts alone. Often, policies interfere with one another, and sometimes they reinforce one another and generate substantial synergies.

IMPLEMENTATION OF DYNAMIC RESILIENCE

The resilience framework can be implemented to a generic system such as an industry, company, or organization where the systems' resources and management strategies are used to improve resilience of the system. In the remaining text, any implementation domain (industry, company, organization) will be referred to as *company*, unless stated otherwise.

Two types of problems are addressed here: (1) a single company under the impact of natural hazard, referred as intrasystem dynamic resilience (IaSDR) problem. The possible solution to such a problem can be found through the development of multiple strategies resulting in a wide range of system performances and use of dynamic resilience for their comparison; and (2) multiple companies under the impact of natural hazard, referred as intersystem dynamic resilience (IeSDR) problem. The possible solution to this type of problem can be obtained by testing and comparing the performance of different companies in response to the disturbance using dynamic resilience.

Intrasystem Dynamic Resilience Simulation Model

Ability to handle shocks from, or impacts of, natural hazards can be used as a measure of the system performance. The IaSDR model compares a set of adaptation strategies selected by the user (referred only as strategies further in the text) for a given system disturbance. The strategies are composed of variations in AC (4R's, i.e., robustness, redundancy, resourcefulness, and rapidity). The company can adopt any strategy and find out the corresponding performance by using the resilience value calculated from performance loss and adaptation capacity. Each strategy would result in one graph showing dynamic resilience value for each location in space. All values together will present a dynamic resilience map corresponding to the change of system resilience as a consequence of a particular strategy. The output of the model is the resilience of the company, represented as a spatial map, and can be used to compare different strategies.

The following steps are involved in the implementation of generic dynamic spatial resilience framework with the IaSDR model: (1) identify all impacts associated with the system disturbance; (2) generate a number of alternate strategies (4R's obtained from the user); (3) build an SD model to simulate system performance; (4)for each of the strategies in step (2), simulate the SD model at each location in space; (5) for each strategy, calculate the resilience of the company using Eqs. (14.1–14.4); and (6) rank/select the adaptation strategy for the company based on the dynamic resilience of the system. The possible selection criteria are discussed in Decision Criteria section.

Intersystem Dynamic Resilience Simulation Model

In the real world, many companies function together and provide services in the same space. These companies will respond differently when subject to the same natural hazard shock (such as flood, storm, hurricane, etc.). The performance of the various companies can be compared using the proposed concept of dynamic spatial resilience, IeSDR model. The IeSDR model structure is similar to IaSDR except that in IeSDR each model component represents a company, whereas in IaSDR each model component represents the strategy adopted by a single company. In the IeSDR problem, depending on the company location in space and strategy adopted, the system would exhibit varying levels of dynamic resilience.

The following steps are involved in the implementation of generic dynamic spatial resilience framework with the IeSDR model: (1) identify all impacts associated with the system disturbance for each of the companies; (2) generate a strategy for each company (done by user in form of 4*R*'s); (3) for each company, build an SD model to simulate system performance; (4) simulate the system dynamic model in space for each of the companies using the strategies defined in step (2); (5) calculate the resilience for each of the companies using Eqs. (14.1–14.4); (6) rank/select the best performing company based on the dynamic resilience of the system. The presentation of possible selection criteria follows.

Decision Criteria

For the application of resilience measure in the selection of best strategy (with IaSDR) or best company (with IeSDR), a set of decision criteria is required. The resilience measure offers multiple choices that have considerable advantage in comparison to a traditional overall risk measure. The use of dynamic spatial resilience offers a plenty of insight into the system performance in response to a system disturbance. All the information can be used at the same time to make the choice of the "best solution," or various criteria for the selection of the "best solution" can be derived from the available information based on the preferences of the decision-maker. Here, four criteria have been derived to illustrate to decision-making context based on the dynamic spatial resilience as illustrated in Fig. 14.4. They are (1) the maximum resilience value (MRV)—the level of system performance achieved when the physical characteristics of the disturbed system return to predisturbance state (end of simulation period). According to this criterion the higher value of MRV is preferred; (2) the fastest recovery time of system performance (TFRV)—the time required by the system

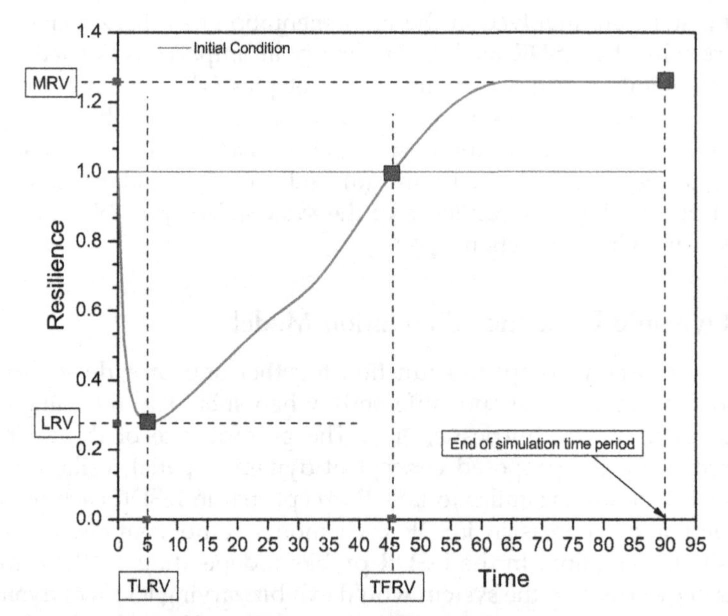

FIGURE 14.4 Decision-making selection criteria based on the dynamic spatial resilience. *LRV*, lowest resilience value; *MRV*, maximum resilience value; *TLRV*, time to lowest resilience value.

under the impact of a disaster to reach the resilience value of 1. According to this criterion the shortest time TFRV is preferred; (3) the lowest resilience value (LRV)—the maximum loss of system performance due to the disturbance over the simulation period. According to this criterion the higher value of LRV is preferred indicating the smaller loss of system performance; and (4) time to lowest resilience value (TLRV)—the time at which the system reaches the LRV.

The TLRV criterion can be used in different ways in the decision-making process. For example, one option is to select the lowest TLRV, which shocks the system fast and hard but offer longer time for recovery (mobilization of resources, assistance to population, etc.). The other option may be to select the highest TLRV, which shocks the system in a much more controlled way and leaves much shorter time for recovery.

CASE STUDY—A RAILWAY UNDER FLOODING

In this case study, the performance of the railway companies is assessed when subjected to flood hazard (Srivastav and Simonovic, 2014b). It is considered that the railway tracks pass over a river floodplain and the inundation of tracks due to floods disrupts the services offered by the railway company. The recovery of the railway after the flooding depends on the strategies adopted (for example, diverting train traffic to alternative routes, elevating tracks, etc.) for dealing with the disturbance. A good strategy would recover the railway company performance at faster rate.

In Fig. 14.5 a simple flooding scenario is shown where the flood levels start increasing at time $t = 1$ and reach a maximum at time $t = 50$. The flood recedes to normal levels at time $t = 90$. For the illustration of the use of resilience framework, two example cases are presented, one corresponding to the generic problem IaSDR and other corresponding to IeSDR. The first example illustrates selection of the best adaptation strategy by one railway company operating in the region, and the second example offers comparison of two railway companies operating under the same disturbance (flood).

The *AC* represents one of the main inputs that allow the use of spatial dynamic resilience simulation for comparison of various adaptation options. In this case study we propose the

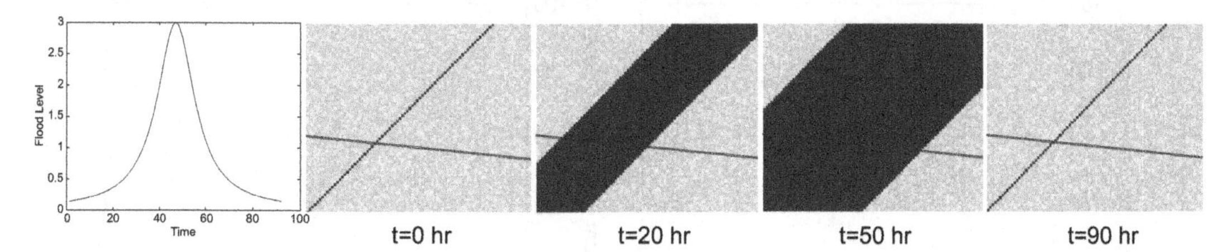

FIGURE 14.5 Flood inundation map. The *light blue parallel lines* (*gray in print versions*) indicate location of railway tracks, and *dark blue shaded area* (*black in print versions*) indicates extent of flood inundation. *After Srivastav, R.K., Simonovic S.P., 2014b. Simulation of Dynamic Resilience: A Railway Case Study. Water Resources Research Report no. 090. Facility for Intelligent Decision Support, Department of Civil and Environmental Engineering, London, Ontario, Canada. 91 pages. ISBN: (print) 978-0-7714-3089-3; (online) 978-0-7714-3090-9. http://www.eng.uwo.ca/research/iclr/fids/publications/products/90.pdf.*

use of four key variables: robustness ($R1$), rapidity ($R2$), resourcefulness ($R3$), and redundancy ($R4$) as the way of describing the AC. Theoretical framework presented here can easily accept any other form of quantitative description of AC. The $4R$'s should fully describe the adaptation strategy implemented with the system that defines its performance under the disturbance (flood). The mathematical form of adopted $4R$'s (they can be very simple, from number of hospital beds to very complex economic functions of business disruption because of disturbance, for example) can be very different, and it will directly affect the system performance under changing conditions.

Simplified (arbitrarily selected) time-dependent functions for the $4R$'s used in this case study are shown in Fig. 14.6. In this case the $4R$'s represent (1) the ability to maintain services and railway tracks and equipment—robustness function; (2) multiple response units and alternative transport arrangements—redundancy function; (3) arrangements for reimbursement or insurance—resourcefulness function; and (4) procurement of new equipment to return to normal operations—rapidity function. The functions can take any form; however, for illustration, these are assumed linear as shown in Fig. 14.6. The change of these functions with time is the consequence of interdependence of disturbance impacts.

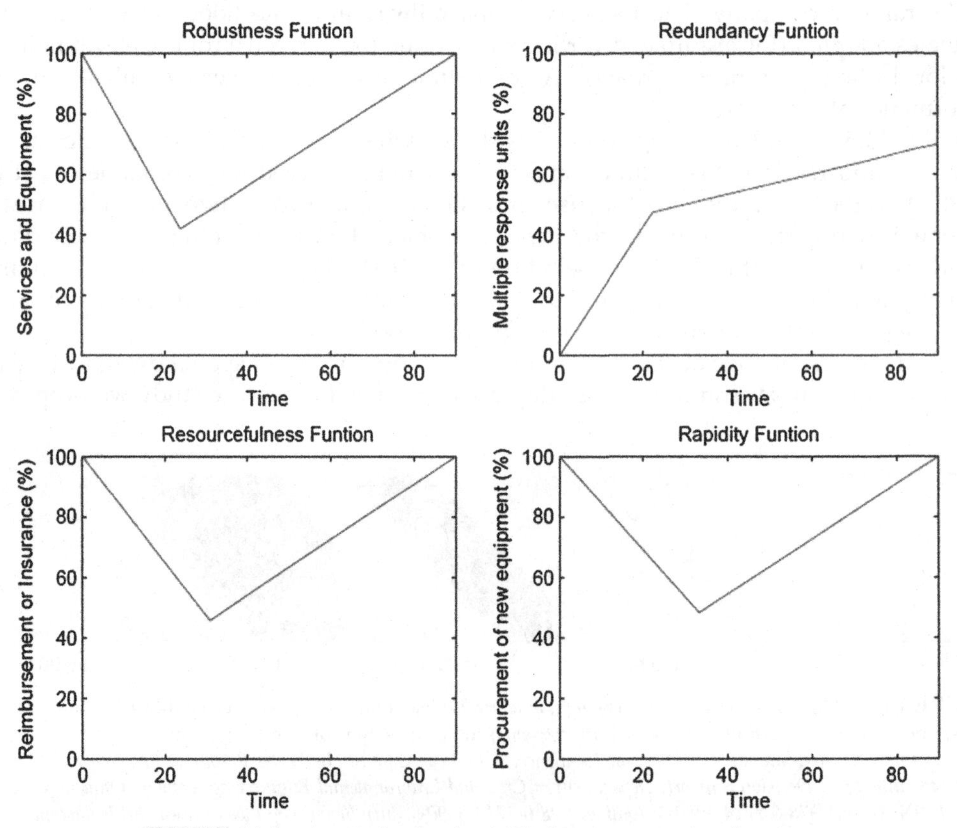

FIGURE 14.6 A typical example of $4R$'s functions at a given location.

Comparison of Various Adaptation Strategies for One Railway Company (Intrasystem Dynamic Resilience Problem)

In this case study, the spatial resolution of the study area is assumed 90 × 90 spatial units for illustrative purposes only. In the event of flooding, the railway tracks (1) could get inundated and/or (2) damaged. Under these conditions, the functionality of the railway is affected, and this in turn affects the performance of the railway company. The dynamic resilience framework provides an effective tool for analyzing the performance of the railway system under the disturbance—flood. The strategies developed to represent AC of the system can be widely simulated and can provide the decision-makers with valuable information to be used in emergency management and mitigation planning. For illustrative purposes, 100 alternate strategies are hypothetically generated for a railway company subjected to flood hazard. Fig. 14.6 shows one of the alternative solutions for 4R's. As discussed earlier, the 4R's can take any functional form depending on the company's resources, alternate arrangements, and procurement of new equipment. In this example, a number of such alternate performance indicators are hypothetically generated to simulate the performance of the railway company.

From all simulation results, one map with temporal and spatial variation of flood inundation and system resilience is shown in Fig. 14.7 for the simulation time period $t = 50$ h. The figure consists of four panels. The top panel provides the temporal change of dynamic resilience value—multiple lines correspond to multiple adaptation option alternatives (values of

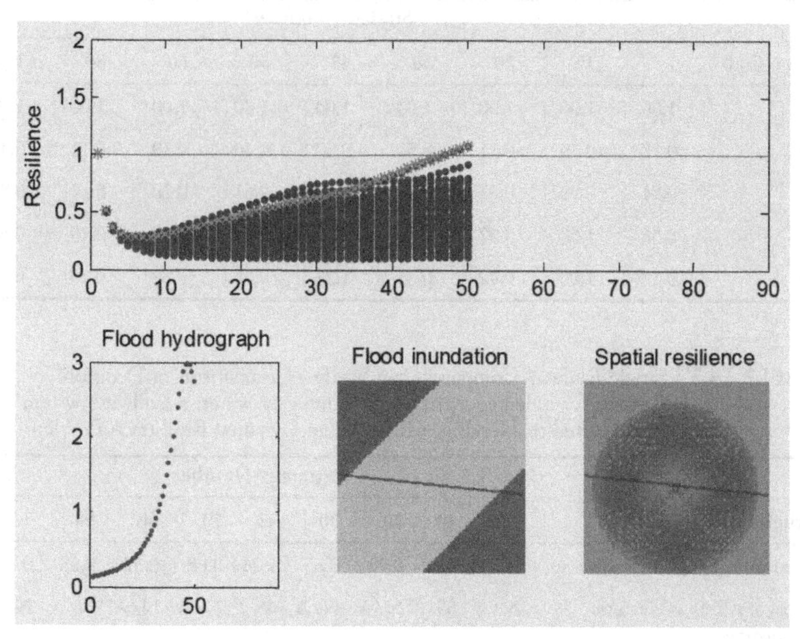

FIGURE 14.7 Intrasystem dynamic resilience model resilience map at time period $t = 50$ h. The spatial resilience is shown for strategy 10 (shown in *red asterisk (gray in print versions)*). *After Srivastav, R.K., Simonovic S.P., 2014b. Simulation of Dynamic Resilience: A Railway Case Study. Water Resources Research Report no. 090. Facility for Intelligent Decision Support, Department of Civil and Environmental Engineering, London, Ontario, Canada. 91 pages. ISBN: (print) 978-0-7714-3089-3; (online) 978-0-7714-3090-9. http://www.eng.uwo.ca/research/iclr/fids/publications/products/90.pdf.*

4*R*'s). This panel captures the overall performance of the railway system. The second panel (bottom left) in the figure shows a hypothetical flood hydrograph used in this case study. The third panel (bottom middle) consists of a spatial inundation map with 90 × 90 spatial units resolution and a railway track crossing the river. The fourth panel (bottom right) illustrates the spatial resilience due to flooding of the tracks. In IaSDR this map shows the change of resilience at one point in space (intersection of railway tracks and the river). We have selected, for illustrative purposes only, to show the change of resilience value using circles of different diameter. However, this value is the resilience value at one point (intersection point) only.

The dynamic resilience for a railway system is evaluated for 100 alternative adaptation strategies. Once the area is subjected to a flood, the resilience value changes according to the behavior and interactions between the 4*R*'s and system performance. The flood inundation and the spatial resilience maps are showing the values corresponding to one selected strategy (shown in red in panel 1, Fig. 14.7). The overall system resilience for selected strategies and time intervals is compared in Table 14.2. After the complete simulation, it is observed that only two strategies, no. 10 and 43, were able to recover the system to the initial state, i.e., the resilience index has reached and/or crossed the value of one. Table 14.3

TABLE 14.2 Single Railway Company Case Study—Comparison of Resilience Value Under Flooding for Selected Adaptation Strategies—Intrasystem Dynamic Resilience Problem

	Strategy Number								
Time (hrs)	**1**	**10**	**20**	**30**	**43**	**50**	**60**	**80**	**100**
0	1.00	1.00	1.00	1.00	1.00	1.00	1.00	1.00	1.00
25	0.47	0.49	0.21	0.51	0.57	0.40	0.19	0.11	0.46
50	0.54	0.93	0.31	0.61	1.10	0.61	0.20	0.12	0.80
75	0.56	1.00	0.37	0.64	1.21	0.67	0.26	0.19	0.86
90	0.57	1.00	0.42	0.65	1.21	0.68	0.32	0.25	0.86

TABLE 14.3 Single Railway Company Case Study—Comparison of Decision Criteria for Selected Adaptation Strategies When a Railway System Is Subjected to Flooding—Intrasystem Dynamic Resilience Problem

	Strategy Number								
Criterion	**1**	**10**	**20**	**30**	**43**	**50**	**60**	**80**	**100**
Maximum resilience value	0.57	1.00	0.42	0.65	1.21	0.68	0.32	0.25	0.86
Recovery time of system performance	N/A	55	N/A	N/A	48	N/A	N/A	N/A	N/A
Lowest resilience value	0.25	0.26	0.17	0.26	0.28	0.24	0.17	0.11	0.26
Time to lowest resilience value	6	5	6	6	5	6	8	16	5

Shaded values identify best strategies for different decision criteria.

illustrates the use of four decision criteria for comparison of 100 strategies adopted in this case study. From Table 14.3 it is observed that: (1) MRV: Strategy no. 43 outperforms all other strategies because it has the highest MRV value. The next best strategy is no. 10. All other strategies do not recover the system performance to preflood level and have resilience value lower than one; (2) TFRV: Strategy no. 43 outperforms the other strategies because it has the smallest TFRV value. The next best strategy is strategy no. 10 again; (3) LRV: Strategy no. 43 outperforms the other strategies because it is results in the smallest railway performance loss due to flooding. Strategies no. 10, 30, and 26 have the next best performance; and (4) TLRV: In case of railways, the companies require more time to recover the system (for example, to repair the flood-damaged rail tracks), and hence the shortest time to minimum resilience value is adopted as preferable. Strategies no. 10, 43, and 100 outperform the other strategies.

In this example, strategy no. 43 is recommended, because it is able to show better performance of the railway system when compared with the other strategies.

Comparison of Two Railway Companies for One Adaptation Strategy (Intersystem Dynamic Resilience Problem)

This hypothetical example compares the performance of two railway companies (P and Q) operating in the same region under the flooding conditions. In this example, the railway tracks of the two different companies cross a flood prone river as illustrated in Fig. 14.8. Although the companies are subjected to similar flooding conditions, the adaptation strategies and response to disturbance can be very different between the companies. Therefore, they may have a very different level of resilience to flooding.

The railway companies will recover after the flood differently because of difference in their response (adaptation) strategies. For example, during a flood event, company P has an option to detour the trains using alternate routes, and company Q has no such alternates. In this case, company P would be more resilient to flood hazard than the company Q. The dynamic resilience framework is proposed here to be used for comparing the performances of two different companies.

From all simulation results, one map with temporal and spatial variation of flood inundation and resilience value is shown in Fig. 14.8 for time period $t = 50$ h. Again, Fig. 14.8 consists of four panels. The top panel provides the temporal change of dynamic resilience value—the two lines correspond to the overall performance of the companies P and Q, respectively. The second panel (bottom left) in the figures shows a hypothetical flood hydrograph used in this illustrative case study. The third panel (bottom middle) consists of a spatial inundation map with 90×90 spatial units resolution and two railway tracks of company P and Q crossing the river. The fourth panel (bottom right) illustrates the spatial resilience due to flooding of the tracks. For IeSDR problem, this map shows the change of resilience at two points in space (intersection of two railway tracks and the river). We have selected, for illustrative purposes only, to show the change of resilience value using circles of different diameter. However, the value shown in Fig. 14.8 is the resilience value at intersection point for company P tracks and company Q tracks, respectively. The dynamic resilience in space (locations of intersections of tracks and the river) and time is evaluated for both companies. In the event of a flood, the resilience map shows the change of resilience value for each

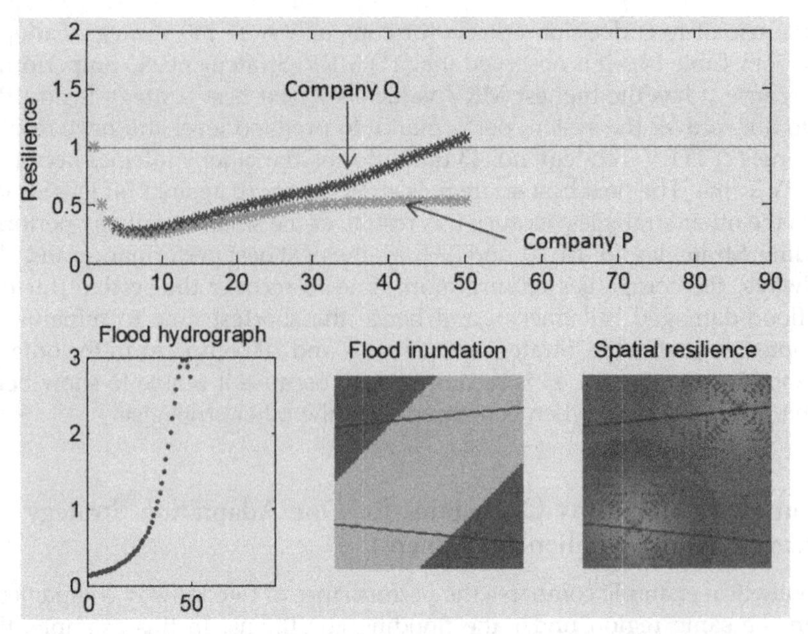

FIGURE 14.8 Intersystem dynamic resilience problem resilience map at time period $t = 50$ h. Company *P* is shown in *red color (gray in print versions)*, and company *Q* is shown in *blue color (black in print versions)*. *After Srivastav, R.K., Simonovic S.P., 2014b. Simulation of Dynamic Resilience: A Railway Case Study. Water Resources Research Report no. 090. Facility for Intelligent Decision Support, Department of Civil and Environmental Engineering, London, Ontario, Canada. 91 pages. ISBN: (print) 978-0-7714-3089-3; (online) 978-0-7714-3090-9. http://www.eng.uwo.ca/research/iclr/fids/publications/products/90.pdf.*

company at the location of intersection of company's tracks and the river. Similar to the IaSDR case, for illustrative purposes only, the change of resilience value using circles of different diameter is used. However, this value is the resilience value at one point (intersection between the railway track and the river) only.

The overall system resilience for the two companies and selected time intervals is compared in Table 14.4. It is observed from complete set of simulations that only company *P* is able to recover its performance to predisaster level, i.e., the resilience index has reached and/or exceeds the value of one. Table 14.5 shows the values of four decision criteria for comparison of two companies, *P* and *Q*. From Table 14.5 it is observed that (1) MRV: Company *Q* outperforms the other company because it has the higher MRV value. The strategy adopted by company *P* does not recover the system performance to preflood level and has resilience value lower than one; (2) TFRV: Company *Q* outperforms company *P* because it has the smaller TFRV value; (3) LRV: Company *Q* outperforms the company *P* because it results in the smaller railway performance loss because of flooding; and (4) TLRV: In case of railways, the companies require more time to recover the system (for example, to repair the flood-damaged rail tracks), and hence the shortest time to minimum resilience value is adopted as preferable for this case study. Both the companies show similar performance level in terms of TLRV.

In this example, the performance of company *Q* dominates the performance of company *P*.

TABLE 14.4 Two Railway Companies Case Study—Comparison of Flood Resilience Value—Intersystem Dynamic Resilience Problem

	Company	
Time	P	Q
0	1.00	1.00
25	0.47	0.57
50	0.54	1.10
75	0.56	1.21
90	0.57	1.21

TABLE 14.5 Two Railway Companies Case Study—Comparison of Decision Criteria—Intersystem Dynamic Resilience Problem

	Company	
Criterion	P	Q
Maximum resilience value	0.57	1.21
TFRV	N/A	46
Lowest resilience value	0.25	0.27
Time to lowest resilience value	5	5

Shaded values identify best strategies for different decision criteria.

CONCLUSIONS

This chapter presents the application of generic resilience framework for implementation of dynamic resilience in modeling complex system behavior as a consequence of system disturbance. An original framework for quantification of resilience through spatial SD simulation, STDRes, is presented. The quantitative resilience measure combines system performance and system *AC*. It can be considered in system performance economic, social, organizational, health, physical, and other impacts caused by system disturbance—for example, some catastrophic event. The framework is designed to provide for (1) better understanding of factors contributing to system resilience and (2) comparison of adaptation options using resilience as a decision-making criterion. The STDRes defines resilience as a function of time and location in space. This framework is being implemented through the SD simulation modeling in an integrated computational environment (SD simulation software is integrated with GIS software).

The implementation of the dynamic resilience framework is illustrated for two problems: (1) IaSDR problem that compares different response/adaptation strategies for a given system subject to disturbance and (2) IeSDR problem that compares the performance of two systems based on their response/adaptation strategies to similar disturbance conditions. Four decision criteria are identified and used for the assessment of system performance based on the implementation of dynamic resilience framework. The decision criteria are based on (1) MRV; (2) fastest recovery time; (3) LRV; and (4) TLRV.

The utility of the proposed framework is demonstrated using two hypothetical case studies of the railway companies under flooding conditions. In the first example, the framework is implemented for the comparison of a single railway company performance under various response/adaptation strategies to flooding conditions. In the second example, the performance of two railway companies is compared in response to the same flood disaster. The modeling framework presented for the railways can be easily extended to the analyses of multiple companies under various natural and nonnatural hazards.

Acknowledgments

Presented work was supported by MITACS, Montpellier Re, and Property and Casualty Insurance Compensation Corporation through the project "Simple Proxies for Risk Analysis and Natural Hazard Estimation." Computer programming assistance of Dr. R. Srivastav is acknowledged.

References

Bruneau, M., Chang, S.E., Eguchi, R.T., Lee, G.C., O'Rourke, T.D., Reinhorn, A.M., et al., 2003. A framework to quantitatively assess and enhance the seismic resilience of communities. Earthquake Spectra 19 (4), 733–752.

Chang, S.E., Shinozuka, M., 2004. Measuring improvements in disaster resilience of communities. Earthquake Spectra 20 (3), 739–755.

Cutter, S.L., Barnes, l., Berry, M., Burton, C., Evans, E., Tate, E., 2008. A place-based model for understanding community resilience to natural disasters. Global Environmental Change 18, 598–606.

Gunderson, L.H., Holling, C.S. (Eds.), 2001. Panarchy: Understanding Transformation in Human and Natural Systems. Island Press, Washington.

Park, J., Seager, T.P., Rao, P.S.C., Convertino, M., Linkov, I., 2013. Integrating risk and resilience approaches to catastrophe management in engineering systems. Risk Analysis 33, 356–367.

Peck, A., Neuwirth, C., Simonovic, S.P., 2014. Coupling System Dynamics with Geographic Information Systems: CCaR Project Report. Water Resources Research Report no. 086. Facility for Intelligent Decision Support, Department of Civil and Environmental Engineering, London, Ontario, Canada, 60 pages. ISBN: (print) 978-0-7714-3069-5; (online) 978-0-7714-3070-1. http://www.eng.uwo.ca/research/iclr/fids/publications/products/86.pdf.

Rose, A., Liao, S., 2005. Modeling regional economic resilience to disasters: a computable general equilibrium analysis of water service disruptions. Journal of Regional Science 45 (1), 75–112.

Simonovic, S.P., 2009. Managing Water Resources: Methods and Tools for a Systems Approach. UNESCO, Paris and Earthscan James & James, London, ISBN 978-1-84407-554-6, pp. 576.

Simonovic, S.P., 2011. Systems Approach to Management of Disasters: Methods and Applications. John Wiley & Sons Inc., New York, ISBN 978-0-470-52809-9, pp. 348.

Simonovic, S.P., 2014. From flood risk management to quantitative flood disaster resilience: a paradigm shift. In: Electronic Proceeding of the 6th International Conference on Flood Management, pp. 8, September 16–18, Sao Paulo, Brazil. http://www.abrh.org.br/icfm6/proceedings/papers/PAP014995.pdf.

Simonovic, S.P., Peck, A., 2013. Dynamic resilience to climate change caused natural disasters in coastal megacities - quantification framework. British Journal of Environment & Climate Change 3 (3), 378–401.

Srivastav, R.K., Simonovic, S.P., 2014a. Generic Framework for Computation of Spatial Dynamic Resilience. Water Resources Research Report no. 085. Facility for Intelligent Decision Support, Department of Civil and Environmental Engineering, London, Ontario, Canada, 82 pages. ISBN: 978-0-7714-3067-1; (online) 978-0-7714-3068-8. http://www.eng.uwo.ca/research/iclr/fids/publications/products/85.pdf.

Srivastav, R.K., Simonovic, S.P., 2014b. Simulation of Dynamic Resilience: A Railway Case Study. Water Resources Research Report no. 090. Facility for Intelligent Decision Support, Department of Civil and Environmental Engineering, London, Ontario, Canada, 91 pages. ISBN: (print) 978-0-7714-3089-3; (online) 978-0-7714-3090-9. http://www.eng.uwo.ca/research/iclr/fids/publications/products/90.pdf.

United Nations International Strategy for Disaster Reduction, 2009. UNISDR Terminology on Disaster Risk Reduction. United Nations, Geneva. http://www.unisdr.org/we/inform/terminology.

CHAPTER

15

Beyond "Model Risk": A Practice Perspective on Modeling in Insurance

Andreas Tsanakas[1], *Laure Cabantous*[2]

[1]Faculty of Actuarial Science and Insurance, Cass Business School, City, University of London, London, UK; [2]Faculty of Management, Cass Business School, City, University of London, London, UK

We spend so much time trying to get the numbers correct that we tend to forget that one of the largest risks is actually model risk. It is the risk of what we do not know very much about. **James Toller, Head of Capital, Beazley; quoted in InsuranceERM (2014)**

INTRODUCTION

The operational importance and scope of quantitative modeling in insurance has been increasing in recent years. A range of models are now deeply embedded in insurance operations: from pricing and catastrophe models, all the way to the internal models employed in economic capital calculations.

At the same time, insurance practitioners and regulators show a growing concern for what is commonly called *model risk*, that is, the potentially adverse consequences arising from decisions based on incorrect model outputs (Federal Reserve, 2011). Governance processes meant to mitigate model risk are a key focus of the Solvency II regulatory regime (Cadoni, 2014, chap. 15). Deloitte's (2013) global risk management survey demonstrates not only the perceived need by insurance executives to *manage* model risk but also a lack of confidence in their ability to do so. Meanwhile, advice on how to manage model risk abounds in the practitioner literature, from Working Party reports (Aggarwal et al., 2016) to consultancy publications (e.g., EY, 2014), typically drawing heavily from the Federal Reserve's (2011) *Supervisory Guidance on Model Risk Management*.

Risk Modeling for Hazards and Disasters
http://dx.doi.org/10.1016/B978-0-12-804071-3.00015-X

Our aim in this paper is to critically assess the dominant conceptualization of model risk, as exemplified by the Federal Reserve's (2011) guidance, which views it as the risk arising from the propagation of model flaws into poor decisions, mediated by a rigid decision process (in the sequel, by "model risk," we will always refer to this particular interpretation). Using Aggarwal et al.'s (2016) cultures of model use (CMU) framework, we argue that this notion of model risk suffers from significant limitations: it reflects a partial view of rationality and fails to acknowledge the diversity of uses of (and interactions through) models taking place in insurance organizations. Our ongoing empirical research on internal capital models in the London insurance market corroborates this critique and challenges preconceived ideas about how models are used in practice. Overall, we argue that appreciating the ways in which quantitative models are used in practice is crucial to the understanding of what risks the use of such models may entail and how a governance response is to be structured.

CULTURES OF MODEL USE

As part of a working party of the UK Institute and Faculty of Actuaries, the first author of this article has developed with colleagues (Aggarwal et al., 2016; Tsanakas et al., 2016) a CMU framework that refines our understanding of how quantitative models are perceived and used (or indeed not used) in insurance organizations.

The CMU framework, which builds on an anthropological theory of risk called the *theory of plural rationalities* (Thompson et al., 1990; Verweij and Thompson, 2006), identifies four types of *rationality*—Individualism, Hierarchy, Egalitarianism, and Fatalism. Each rationality is associated with a set of social and economic relations, and a cast of supporting beliefs through which the world is perceived, explained, and responded to. In the specific context of model governance, each rationality generates a culture of model use, as is outlined in Table 15.1.

TABLE 15.1 Cultures of Model Use (Aggarwal et al., 2016)

Rationality	Culture of Model Use
Individualism	The operational usefulness of quantitative models is emphasized. A sophisticated quantitative model can give its user a competitive advantage. The model should be embedded in the organization's processes, such that decisions are, to a large extent, driven by model outputs.
Hierarchy	Although the value of modeling is acknowledged, concerns about the fitness for purpose of models are emphasized. The use of models should be tightly controlled, to limit uses for which relevant model outputs are not reliable.
Egalitarianism	The risk environment is characterized by uncertainty, structural changes, and interconnectedness, which make the quantification of risk by statistical methods an exercise doomed to fail. Model use should be heavily restricted, with more value placed on scenario-based envisioning of adverse events.
Fatalism	The emphasis is on the commercial realities and constraints under which management operates, as well as practitioners' tacit knowledge and intuition. Models are not well placed to reflect these concerns and are primarily used to justify and corroborate already-held beliefs.

These cultures are not seen as profiles of individual stakeholders but as building blocks in the arguments deployed within model governance[1].

While these four cultures are mutually conflicting, each offers valuable insights into appropriate uses and limitations of quantitative models. Furthermore, Aggarwal et al. (2016) argue that the dominance of any of the cultures in an organization can lead to a failure of model governance. As a result, it is suggested that governance should explicitly reflect insights from all four CMU and encourage multidirectional challenges between them. As we demonstrate in the next section, the CMU framework points to a broader understanding of the business problems arising in the context of using quantitative models, beyond the narrow scope of model risk.

BEYOND MODEL RISK: INSIGHTS FROM THE CULTURES OF MODEL USE FRAMEWORK

The CMU framework is a useful resource for reflecting on the notion of model risk and its limitations. In Fig. 15.1, we summarize the ways in which the culture associated with Hierarchy challenges is challenged by other cultures.

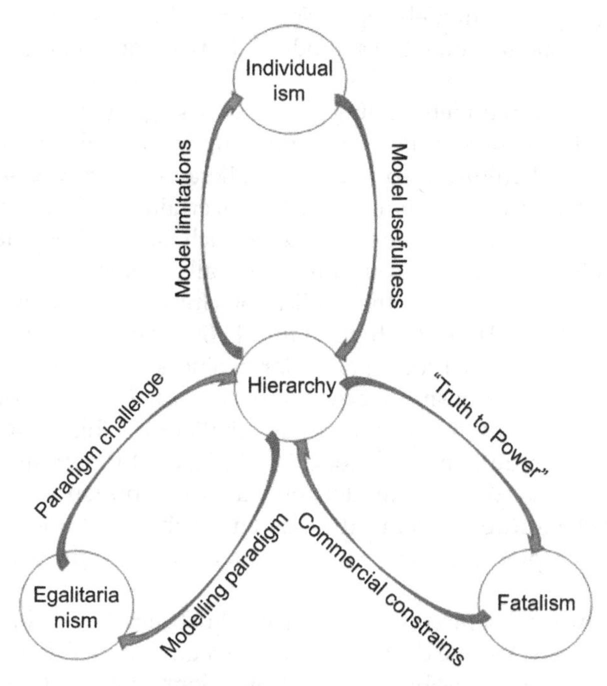

FIGURE 15.1 Two-directional challenges between the culture of model use rooted in Hierarchy and the other three cultures.

[1]While the labels given to the rationalities are not easily interpretable in the context of model governance, we use them in this article, to maintain an explicit link with the theoretical framework of Thompson et al. (1990).

The relationship between Individualism and Hierarchy is characterized by a trade-off, corresponding to the respective concerns of the associated CMU: between expanding the model's use in business decisions (Individualism) and limiting its use to those applications for which the model is deemed fit for purpose (Hierarchy). Notably, the Federal Reserve's (2011) guidance places a heavy emphasis on the fitness for purpose of quantitative models, argues for the use of prudent modeling choices in the face of uncertainty, and cautions against using models outside their design scope. Although acknowledging that excessive conservatism can become an impediment to model development and use, the guidance clearly states that limits should be placed on model use, if sensitivity to model outputs is high when inputs, such as statistical parameters, are slightly varied.

Hence, the notion of model risk, as commonly understood, is rooted in one particular interaction: a critique of an Individualist model–use culture by a Hierarchical one. Fig. 15.1 shows that this is only one critique out of many that can be formulated, through the mutual challenges that model-use cultures pose to each other.

First, a focus on model risk does not sufficiently reflect the inverse challenge from Individualism to Hierarchy. In practice, critical model outputs, such as the regulatory capital in the context of capital modeling, can be very sensitive to assumptions that cannot be empirically validated. However, severely limiting the use of models is neither a pragmatic response nor a desirable response, given such models' operational criticality. As we discuss in the next section, insurance practitioners value their models despite (and being cognizant) of their limitations.

Furthermore, the common understanding of model risk ignores the other two rationalities of Egalitarianism and Fatalism and their corresponding model cultures. From the viewpoint of Egalitarianism, a model culture associated with Hierarchy does not sufficiently allow for the challenges of modeling in conditions of deep uncertainty, within which insurers often operate. The scarcity of empirical evidence makes each of a large collection of candidate models, with potentially very different outputs, practically unfalsifiable. Hence, it is generally not possible to specify with confidence the applications for which models will perform consistently with design objectives. Thus, it is hard to fulfill Hierarchy's key aim, that is, to establish the fitness for purpose of a model for a particular business use.

From the perspective of Fatalism, decisions are taken irrespective of the model, which is used only to corroborate them. Hierarchy challenges this culture by stipulating criteria for the technical acceptability of models and placing limitations on the extent which management expectations on the model output can influence model specification. Conversely, a Fatalist culture prevents the design, calibration, and use of models that, although technically acceptable, generate outputs that could put the company at a commercial disadvantage.

The interaction between Hierarchy and Fatalism illustrates how a broader conception of model risk is needed to address the difficulties of model governance. If a wide range of model specifications are technically plausible, some specifications that yield model outputs consistent with management expectations will be found. Governance focused on model risk may not accept as legitimate those concerns of management that are neither reflected in the model nor in any normative decision principles encoded in it. However, by ignoring or indeed suppressing such concerns, they do not go away. Instead, incentives can be created for modelers to provide outputs that are convenient to management. This undermines both management's accountability and modelers' role in asking uncomfortable questions about the model's technical validity and about experts' intuitive judgments.

EVIDENCE FROM THE PRACTICE OF LONDON MARKET INSURERS

The CMU framework calls into question the generality of the assumption that model outputs are used rigidly to produce business decisions, which only holds for an Individualist culture of model use. To refine our understanding, we studied the uses, in the London insurance market, of a specific type of model susceptible to model risk—internal capital models.

Internal models are complex simulation models, reflecting the volatility of different lines of business and asset returns, the correlations between them, and the impact of business decisions such as reinsurance strategy. Other models such as catastrophe models typically provide input into them. In the context of preparations for Solvency II, capital models have been put to a number of uses, such as capital setting, reinsurance optimization, risk appetite monitoring, business planning, performance measurement, and portfolio optimization[2].

Our preliminary findings, based on research interviews with 31 practitioners in the London insurance market, including chief risk officers (CROs), capital modelers, nonexecutive directors, regulators, and consultants, are as follows. *First*, we do not find evidence that internal models are applied rigidly to drive decisions; thus our findings contradict the implicit assumption underlying the common definition of model risk. As more than one CRO stressed to us, "the model is one input out of many to the decision process." A decision suggested by the model output is not always followed, for a variety of possible reasons. For example, there may be commercial constraints not reflected in the modeling, which make a suggested course of action unpalatable to management (e.g., a big change in the amount of exposure ceded to reinsurers, given the need to maintain commercial relationships). Or there may be concerns that the model does not represent in sufficient detail the part of the business impacted by the particular decision (e.g., simplifying assumptions about how the reinsurance programme works). Furthermore, it is the case that different modeling assumptions (typically regarding correlations between lines of business), which are equally plausible, given the empirical evidence, could give rise to very different model outputs, especially in the quantification of the probability of extreme events required under Solvency II.

Second, our research shows that insurers appreciate their capital models not primarily because of their perceived accuracy as representations of the business and the risk environment. Our interviewees value models because they find them useful in structuring strategy discussions, explicitly formulating risk-reward trade-offs, challenging expert knowledge through technical evidence, and consolidating information across complex portfolios and organizations. Our interviews thus suggest that, in practice, insurers do not blindly "apply" models. Instead,

[2]In fact, it is one of the requirements of Solvency II that internal models be put to broader business uses than just calculating regulatory capital requirements (Cadoni, 2014; chap. 6).

they use models as tools, which through their specific properties, in confluence with the skills and characteristics of their users, enable certain interactions and activities to take place. We find that insurers are sophisticated model users, who experiment with their models and use them to reflect on strategic options and communicate across organizational boundaries[3]. The commitment to such use can, however, deflect attention from models' limitations, which may in turn engender excessive confidence in their technical validity.

Third, the presence of substantial uncertainty around key model parameters raises the question of how modeling choices are made in practice, if empirical evidence cannot give a sufficient basis for such choices. Many model parameters are chosen on the basis of expert judgments, which are in turn influenced by the experts' own incentives and biases. For instance, our interviews show that underwriters may sometimes be inclined to overstate the volatility of their business (to justify purchasing more reinsurance) or understate it (to make their returns on capital more attractive and thus enable them to write more business). Regulators also provide feedback to (and ultimately approval of) the modeling process and are seen by our (notably frustrated) respondents to often push toward model assumptions that are prudent and consistent across the market, reflecting their concerns for policyholder protection, and creating a level playing field. Thus, uncertainty, while forming an impediment to accurate risk measurement, in our context also performs a crucial organizational function: allowing such a negotiation, reflecting divergent concerns, and interests, to happen in the first place. The practice of model use does not resolve uncertainty, but turns it into a resource.

CONCLUSION AND OUTLOOK

In this article, we critically assessed the notion of model risk, seen as arising from a direct propagation of model flaws to incorrect decisions, through a rigid "application" of models. Although such a definition of model risk is meaningful, the CMU framework shows that it is incomplete, because it ignores a number of important interactions and challenges surrounding the use of modeling by insurers. Our empirical research on internal models in the London insurance market corroborates this insight by showing that models are used by practitioners with a sophisticated understanding of their limitations and are enmeshed in a variety of discourses, going beyond questions of technical validity. Thus, although the notion of model risk is certainly relevant to insurance risk management, model governance can only be successful if it takes into account the richer picture observed in the actual *practice* of model use.

A further issue, not addressed in detail here, is related to the potential dangers engendered by the widespread use of increasingly *similar* quantitative models, in terms of structure and parameterization, due to homogenizing regulatory feedback and reliance on a small number of external proprietary models (catastrophe models and economic scenario generators). The specter of such a *systemic model risk* clearly worries our respondents in

[3]Insofar, our findings are consistent with those of Jarzabkowski et al. (2015), who, in studying the global reinsurance market, found that the outputs of different types of models are creatively interweaved with underwriters' judgments and contextual knowledge to produce a price quote.

the London market; possible flaws in those models could mean that the market may collectively underestimate (or even remain unaware of) particular risks, as was also noted by the Systemic Risk of Modelling Working Party (2015). Furthermore, our respondents stated that the implementation of Solvency II has shifted the attention of Boards from crucial business areas (e.g., underwriting standards) to model governance and has also, in some cases, impacted on the composition of Boards, with experienced directors retiring, when they found the increased emphasis on model governance inconsistent with their own interests and skills. Whether such developments will have systemic implications for the insurance market remains to be seen.

AUTHOR BIOGRAPHIES

Andreas Tsanakas is reader in Actuarial Science at Cass Business School, City, University of London. His research areas are in capital modeling and quantitative risk management. Dr. Tsanakas is interested both in the quantification of uncertainty and its limits. A recent area of emphasis relates to the ways in which models are (not) used in the practice of decision-making and on how governance can address model risk. He has coauthored the prize-winning paper *Model Risk: Daring to Open the Black Box.*

Laure Cabantous is professor of Strategy and Organization at Cass Business School, City, University of London. In her research, she studies decision-making processes both at organizational and the individual levels. She is interested especially in decision-making under risk and uncertainty and the use of models in supporting these decisions. She has a long-lasting interest in insurance and reinsurance and the insurance industries and has coauthored the industry report *Beyond borders: Charting the changing global reinsurance landscape.*

References

Aggarwal, A., Beck, M.B., Cann, M., Ford, T., Georgescu, D., Morjaria, N., Smith, A., Taylor, Y., Tsanakas, A., Witts, L., Ye, I., 2016. Model risk: daring to open up the black box. British Actuarial Journal 21 (2), 229–296 (forthcoming).

Cadoni, P. (Ed.), 2014. Internal Models and Solvency II. Risk Books.

Deloitte, 2013. Global Risk Management Survey, Eighth Edition: Setting a Higher Bar. Available from: http://www2.deloitte.com/content/dam/Deloitte/global/Documents/Financial-Services/dttl-fsi-us-fsi-aers-global-risk-management-survey-8thed-072913.pdf.

EY, 2014. Model Risk Management for Insurers: Lessons Learned. Available from: http://www.ey.com/Publication/vwLUAssets/ey-building-a-model-risk-management-capability/$FILE/ey-building-a-model-risk-management-capability.pdf.

Federal Reserve, 2011. SR 11-7: Guidance on Model Risk Management. Available from: http://www.federalreserve.gov/bankinforeg/srletters/sr1107a1.pdf.

InsuranceERM, June 5, 2014. Are Models More Complex than They Need to Be? InsuranceERM. Available from: https://www.insuranceerm.com/analysis/are-models-more-complex-than-they-need-to-be.html.

Jarzabkowski, P., Bednarek, R., Spee, P., 2015. Making a Market for Acts of God: The Practice of Risk Trading in the Global Reinsurance Industry. Oxford University Press.

Systemic Risk of Modelling Working Party, 2015. Did Your Model Tell You All Models Are Wrong? Available from: http://www.amlin.com/~/media/Files/A/Amlin-Plc/financial/systemic-risk-white-papers-2015.pdf.

Thompson, M., Ellis, R., Wildavsky, A., 1990. Cultural Theory. Westview Press.

Tsanakas, A., Beck, M.B., Thompson, M., 2016. Taming uncertainty: the limits to quantification. ASTIN Bulletin 46 (1), 1–7.

Verweij, M., Thompson, M., 2006. Clumsy Solutions for a Complex World: Governance, Politics and Plural Perceptions. Palgrave Macmillan.

Index

'*Note*: Page numbers followed by "f" indicate figures, "t" indicate tables.'

F

L

M

N

Q

R

W

Y